Eat Light & Feel Bright

MICROALGAE SOLUTIONS FOR INDIVIDUAL AND PLANETARY HEALTH

ULTIMATE MICRONUTRITION FOR ANTIAGING, BRAIN ENHANCEMENT, CELLULAR REPAIR, INFLAMMATION REDUCTION AND IMMUNITY BOOSTING

JEFFREY BRUNO, PhD

Printed in the United States of America

ISBN: 978-0-9913925-0-6 (Paperback)

First Edition

Cover Photo Credit by Doc Searls (2009)
http://www.flickr.com/photos/docsearls/4096329287/

The cover photo displays the rich pigments of different microalgae. A number of salt evaporation ponds exist on the shores of San Francisco Bay. As the salinity of the water and oxygen levels change competing species of microalgae and other organisms display their colorful pigments. First come green algae, which darkens as orange brine shrimp feast on the microscopic planktonic algae. Finally red predominates as Dunaliella salina, a microalgae containing high amounts of beta-carotene predominates. A variety of microalgae pigment colors include red, green, orange, yellow, brown and blue. Finally, when the water is evaporated, the white of salt alone remains, visible around the edges of the ponds. The salt is harvested with machines, and then the remarkable process repeats.

2009-10_31_BOS-SFO_398

"As above, so below"

Linda and Tylia, you light up my world and inspire me to recognize how precious life can be. In memory of my father, Anthony J. Bruno, a self-educated and courageous man, who shared his conviction about evolution to me as a youth, I dedicate this book. And to my mother who urged me to write this book by asking, "If it's so good, why isn't everyone using it?" Special thanks go to Professor Cheryl Joseph, Val Maciejowska and many other friends and colleagues for editorial guidance. Chemistry professor Karl Abrams contributed most of the glossary and inspired me to gain a deeper biochemical understanding of microalgae.

Special thanks to my far-reaching microalgae community, too numerous to name all the individuals who have both inspired and encouraged me. Yet particular acknowledgement goes out to Scott Ohlgren, Mark Segars, Barbara Swanson and the late Showshawme who created the phrase "Eat Light & Feel Bright." Last but not least, to Dr. John Ray who first introduced me to microalgae and to the Kollman family who risked everything to harvest wild *AFA* microalgae from Klamath Falls, Oregon — you revealed the possibilities of a better future.

This book is offered with deep reverence to our blue-green world, over four billion years in the making. My gratitude goes forth to one of Earth's most elemental organisms, the diverse species of microalgae, which imparts a foundation for all of us to thrive. Can humanity come to appreciate our precious and utterly interdependent existence, which connects us from the tiniest of life forms to our vast planetary system?

Contents

List of Figures

Foreword

Over 12 years have passed since I wrote *Edible Microalgae: A Review of the Health Research*.[1] Since then I received numerous requests to both reprint as well as to update the information. Since the early 1980s, when I began eating edible microalgae, a nearly ten-fold increase has occurred in annual published research on microalgae. When I published my first book on Edible Microalgae, the available studies numbered in the hundreds, now there are thousands of research articles available on microalgae. What began as an updated volume has evolved a new book, with new chapters, new concepts, and the latest research findings.

The 21st century is witnessing a worldwide acceleration in published scientific research and access to research discoveries by Internet savvy, interdependent thinkers. Like the emergence of "gentleman scientists" in the 19th century, who pursued scientific study without direct backing by public institutions like universities, government or business, a new era of "citizen science" is occurring. Due to Internet access of much of the world's published scientific research, an educated citizen can read about research findings years before most graduate or medical students learn about them through class lectures. Yet individual research findings only make sense in a larger context. Nowadays science and medicine are so highly specialized that it is rare for medical researchers to refer to plant physiology, nutrition or evolutionary theory to help explain their findings. An extensive glossary, provided by Professor Karl Abrams, near the end of the book (beginning on page 362) will help most readers unfamiliar with the technical terminology. An appreciation of microalgae's world-changing health potentials requires a multidisciplinary approach. Looking at human health through a multifaceted prism of evolution, solar influences and microalgae help us tap into remarkable possibilities.

Microalgae discoveries are at the forefront of evolutionary origins in the emergence of life, as well as contributing to mass species extinctions, during the "Great Oxidation event." Over three billion years ago, our Earth lacked significant amounts of atmospheric oxygen. Life was mainly anaerobic, surviving without oxygen. Strikingly, when cyanobacteria (blue green algae) started to transform the planets' atmosphere with increasingly levels of oxygen, most anaerobic bacteria just died off. Those few organisms which were able to adapt from an oxygen deficit atmosphere to an oxygen rich atmosphere are the primordial ancestors of most of today's life forms. Not surprisingly, microscopic organisms, such as cyanobacteria and other microalgae, are being considered as global game changers. Increasingly, microalgae are being valued for a wide range of health, industrial, and commercial uses; as well as use as important biological markers for the health of our planet.

Phytoplankton, a marine species of microalgae, make unique contributions to our biosphere, largely through renewal of the atmospheric oxygen, trapping of huge amounts of CO_2 and their synthesis of organic compounds through photosynthesis. "This function makes phytoplankton crucial primary producers on which every food chain relies. While accounting for less than 1% of Earth's biomass, phytoplankton is responsible for more than 50% annual net biomass production. Marine organisms, including phytoplankton, constitute a diversified source of compounds useful for health, industry, cosmetics, and new biofuels sources. It is estimated that approximately 20,000 active substances originating from marine organisms are used in the industry and can be used as an alternative source of chemically synthesized medicine."[2]

Tremendous advancements in the field of algal genomics have occurred over the past decade as well. Complete genome sequences have been obtained for certain cyanobacteria, red alga, and diatoms, while large scale DNA sequencing projects are being carried out for many more algal species.[3] Advances in phycology (the study of algae) and in biotechnology are exploring cyanobacteria chloroplasts in the hopes of designing more efficient solar panels and to further improve crop productivity.[4]

DNA research has confirmed that cyanobacteria (blue green alga), the first organism known to have harvested the sun's rays to produce fuel, passed on these ancient secrets to all plants on earth. Cyanobacteria and photosynthetic bacteria are increasingly studied to better understand photosynthesis; the metabolism of nitrogen, carbon, and hydrogen; cells ability to overcome environmental stresses; and even to understand the molecular basis of evolution.[5]

Perhaps one of the most important research areas is the development of algae as bio fuel. Exon Mobile, Virgin Fuels, and other multinational corporations, have announced a multi-billion dollar investment to develop algae-based fuel technologies. The United States Air Force and Navy, along with several commercial airlines are investing in the development of algae based jet and marine fuels. Microalgae derived fuels such as bio diesel, bio petroleum, and bio hydrogen are likely to play an increasingly vital role to shift away from a fossil fuel era to a greener bio energy era.

The transformation from a fossil fuel based economy to a microalgae and solar powered economy will alter the very fabric of society. Most people do not realize the far-reaching extent that fossil fuel is connected to nearly every consumer product, from I-phones to aspirin. Yet, a quantum shift is possible to a new era of bio plastics, bio pharmaceuticals, bio energy, bio fertilizers and a range of bio chemicals derived from microscopic solar powered organisms.

Algae bio fuels and restorative biotechnologies offer a number of innovative answers. Global climate data collection and forecasting continue to strengthen the evidence that human-related activities are changing the temperature of our planet. Drought in the Great Plains, flooding along the Mississippi River and deep freezes in the South pushed insurance payments (losses) by companies including Wells Fargo and Chase to a record $9.1 billion in 2011, according to National Crop Insurance Services.[6]

Altogether, rising temperatures and expanding human populations, depletion of freshwater aquifers, shifting agricultural temperate zones, destruction of prime agricultural and marine fishing stocks, plus increasing world-wide demands for food and energy suggests humanity requires powerful ecofriendly solutions to our 21st century challenges. The consequences of global climate change, which could severely impact hundreds of millions of people, might be mitigated by embracing a microalgae biotechnology revolution. Ironically, we may need to reclaim ancient biological wisdom to better adapt to our modern technologically-driven times.

All of the most promising health and environmental notions first catalogued in my 2000 edition of *Edible Microalgae* continue to be validated and bolstered by ongoing scientific studies. Omega 3 algae oils in the form of DHA, and microalgae components are increasingly being used in a wide variety of food products. Microalgae is found in greater numbers of green drinks, multivitamin pills, smoothie products and even enhanced DHA eggs in grocery stores. New uses of algae have emerged in adult stem cell enhancing products, based on innovative experimental research. The potential of microalgae to help address neurobehavioral and metabolic disorders, such as diabetes, shows considerable promise.

Even an entirely new field of brain research, optogenetics, based on the use of light sensitive molecules found in algae and microbes, are allowing scientists to activate or to shut down neural pathways, to alter behavior and to illuminate brain pathways found in psychiatric conditions. Photons and photon absorbing molecules are likely to play increasingly important roles in medicine. Additionally, algal light sensitive molecules, such as phycobiliproteins are already used for fluorescent labeling, and particularly for flow cytometry.[7] Flow cytometry is a technology that measures the characteristics of microscopic particles such as cells. Some of the clinical uses are diagnosis of blood and lung cancers, use in immune studies, or in monitoring stem cell content for transplants.

While microalgae components are increasingly used on the front-lines of modern medicine, microalgae have been used for centuries by humans (e.g., food and fertilizers). Yet, large-scale commercial production began only recently, a few decades ago. Microalgae are harvested from open-culture systems such as lakes or from highly controlled closed-culture systems such as ponds or chambers. The main edible microalgae are the green algae (chlorophyta) and the blue green algae (cyanobacteria).

Significant and growing interests in the commercial applications of algae chemical components for biofuel production, agro- and horticulture, environmental protection and restoration, pharmaceuticals, nutraceuticals, food products, and cosmetics, will continue to expand. Microalgae contain substances of high biological value, antioxidant pigments, lipids and fatty acids, amino acids and proteins, polysaccharides and phenolics. Cyanobacteria such as *Spirulina*, *Aphanizomenon flos-aquae*, *Nostoc* and *Oscillatoria* are capable of producing a wide variety of secondary metabolites.[8] Cyanobacteria have been identified as one of the most promising group of organisms from which novel and biochemically active natural products can be made.

Thousands of undiscovered and potentially useful molecules still await discovery, as much biochemical diversity exists between and within algal species. Yet, most individuals remain woefully ignorant about the remarkable health benefits of edible microalgae. Why? The majority of Americans increasingly, rely on pharmaceutical solutions to address chronic health ailments, both physical and mental. Yet the right combination of optimal nutrients, attitude, and exercise will improve many health problems. Certainly physicians are now better informed about the benefits of nutrition. However, the medical profession has yet to embrace the ancient concept of "food as medicine" as a front-line treatment for most chronic health conditions.

Edible microalgae in particular act as a powerful immune-enhancer and adaptive bio-modulator to stimulate healthy human and animal physiology. Edible microalgae exhibit multiple therapeutic functions such as antioxidant, antibacterial, antiviral, anticancer, anti-inflammatory, antiallergic and antidiabetic effects. Edible microalga offers a plethora of useful functions from brain to cardiovascular benefits, and even supports the growth of beneficial intestinal micro flora.

This book is written especially for those new innovators, who envision or are actively working towards building a sustainable civilization. Can we create a society that does not greedily deprive future generations of Earth's more limited resources? Can we build a healthier society that harnesses solar and biological energies for an ecological-based system of commerce and production? Can we become part of a global humanity that values and protects the broadest range of life, including our own diverse human cultures?

Surprisingly, microalgae offer many potent solutions to an enormous world of challenges.

— Jeffrey J. Bruno on Thanksgiving Day, 2013,
at my family gathering in Reno Nevada.

Figure 1. A Tree of Life

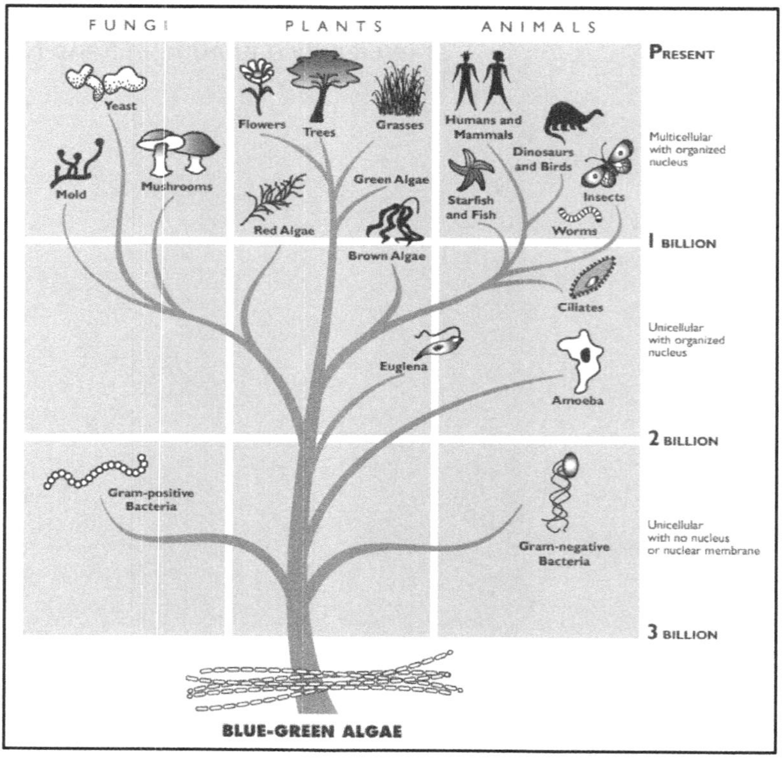

The Tree of Life illustrates the relationship among various organisms, both living and extinct. All species are believed to be connected by the passage of genes along and even across the branches of the phylogenetic Tree of Life. While Blue-Green Algae, cyanobacteria, were not the first organisms; they represent some of the earliest and longest surviving species of life. The roots at the bottom of the Tree of Life are believed to include the Last Universal Common Ancestor (LUCA) from which cyanobacteria and all other species originated.

What Happens In A Few Billion . . .

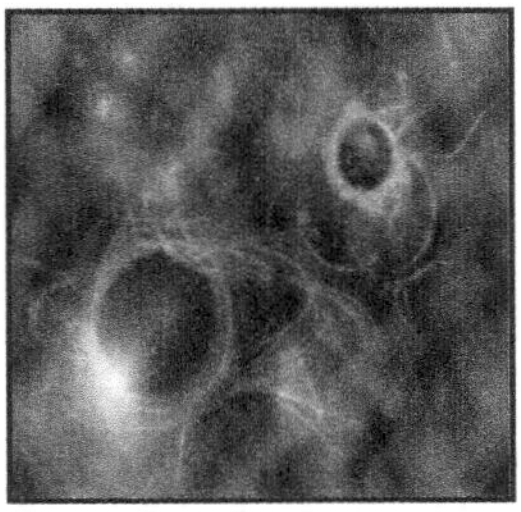

- 13.7 times one billion years ago is the estimated age of the *Universe*.
- 4.5 times one billion years ago is the estimated age of our planet, *Earth*.
- 3.5 times one billion years ago, *cyanobacteria* first appeared on Earth and helped to create our planet's oxygen-rich atmosphere.
- One billion years ago, the *first multicellular organisms* appeared on Earth.
- One billion months ago, *dinosaurs* walked the earth during the late Cretaceous Period.
- One billion days ago, *Australopithecus*, an ape-like creature considered a relative of humans, roamed the African savannas.
- One billion hours ago, ancestors of modern human were living in the *Stone Age* (the Middle Paleolithic).
- One billion minutes ago, the *Roman Empire* was flourishing.
- One billion seconds ago, *the parents of middle school children* were themselves in elementary school.

Introduction

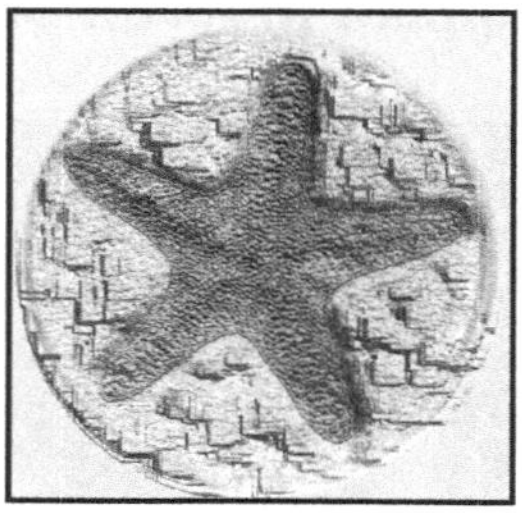

"The history of animals that we've recorded from fossils is really only the last 15 percent or so of the recorded history of life on this planet. The deeper history of life and the greater diversity of life on this planet is microorganisms—bacteria, protozoans, algae. One way to put it is that animals might be evolution's icing, but bacteria are really the cake."

— Knoll A, 2004

"We are the product of 4.5 billion years of fortuitous, slow biological evolution. There is no reason to think that the evolutionary process has stopped. Man is a transitional animal. He is not the climax of creation."

— Sagan, C

Way Deep Evolutionary Time

Earth's present condition is the outcome of nearly 4.5-billion years of evolution! Unimaginable, global-scale geological and extraordinary biological events have transformed our planet. One of the earliest and greatest of these events was a substantial increase (by orders of magnitude) in the abundance of atmospheric oxygen, sometimes referred to as the "Great Oxidation Event." Since Earth was a relatively small planetary mass, nearly all hydrogen might have been lost forever in cosmic space, without the binding of hydrogen to oxygen. Oxygen was essential to keep hold of hydrogen and to convert a molten rocky planet into a watery blue-green world.

According to John Waterbury (2004), an American scientist at Woods Hole Oceanographic Institution, "For a billion years, growing and multiplying in the sea, they slowly raised the oxygen level in the atmosphere to 20 percent, the level that supports oxygen breathing life. Cyanobacteria alone, directly or indirectly, are responsible for all of the oxygen in our air."[1]

Cyanobacteria, blue-green algae, played an essential role in the "Great Oxidation Event," which transformed our planet's atmosphere, and set the stage for the shape of most life that survived this slow world altering event (a shift from anaerobic to aerobic life). Furthermore, cyanobacteria during a time period of relative isolation, a far-reaching one billion years, developed and then passed along quintessential building blocks (e.g., amino acids, essential fatty acids, chelated minerals, and powerful antioxidants), which continue to serve as foundational organic biochemical and nutritional molecules for most species of life, today. Not only the root for our planetary food cycle; two specific microalgae, *Synechococcus* and *Prochlorococcus*, happen to be "the two most abundant organisms on Planet Earth," and "may represent fully half the total photosynthetic production" in the world's oceans.[2]

To grasp the mystery of why blue-green algae contain so many biological marvels found in higher organisms (including humans) will provide an understanding of how deep evolution has shaped all living creatures. For example, most life forms have circadian rhythms, linked to light and dark cycles; these rhythms have been evident in cyanobacteria from the very dawn of life. Most humans mistakenly believe that "evolved" means superior. Yet, hidden within the cells of higher mammals, reside an ancestral biological architecture, extending back to cyanobacteria, and beyond. Even how blue-green algae releases carbon monoxide (by means of chlorophyll) and produces phycocyanin (blue pigment); is retold by how blood (heme) releases carbon monoxide and produces bilirubin.

The molecular structure of chlorophyll is quite similar to that of heme. The two bioactive molecules have only different metal ions in the middle. Heme has iron whereas chlorophyll has magnesium. Still visible within all plant chlorophyll, as verified by evolutionary biologists, is the prehistoric architecture from cyanobacteria. The foundational oxidation-reduction (REDOX) reactions that help fuel our bodies today are comparable to ones that occurred first within cyanobacteria, billions of years ago. Even the amino acid building blocks of cyanobacteria provide the essential alphabet of proteins found in living creatures today.

For readers who feel that the science of evolution goes against religion, it is noteworthy to read a statement by Pope John Paul II, former pontiff of the Holy Roman Catholic faith.

On Evolution – In 1996, Pope John Paul II declared, "In his encyclical Humani Generis (1950), my predecessor Pius XII had already stated that there was no opposition between evolution and the doctrine of the faith about man and his vocation . . . It is indeed remarkable that this theory has been progressively accepted by researchers, following a series of discoveries in various fields of knowledge. The convergence, neither sought nor fabricated, of the results of work that was conducted independently is in itself a significant argument in favor of this theory." – Pope John Paul II – Message to the Pontifical Academy of Sciences (Oct. 22, 1996).[3]

Pope John Paul II must have anticipated that the scientific basis of evolution was here to stay and Church doctrine would eventually have to accept that fact. A parallel historic situation occurred when the Church at long last adapted to the scientific discovery that the Earth was not the center of the Universe, as one-time believed. Nor do people hold firm to the medieval idea that heaven is up skyward in the clouds and that hell is below ground in volcanic caverns. Human understanding of our place in the cosmos has changed over time.

Figure 2. Schematic of AFA Cell

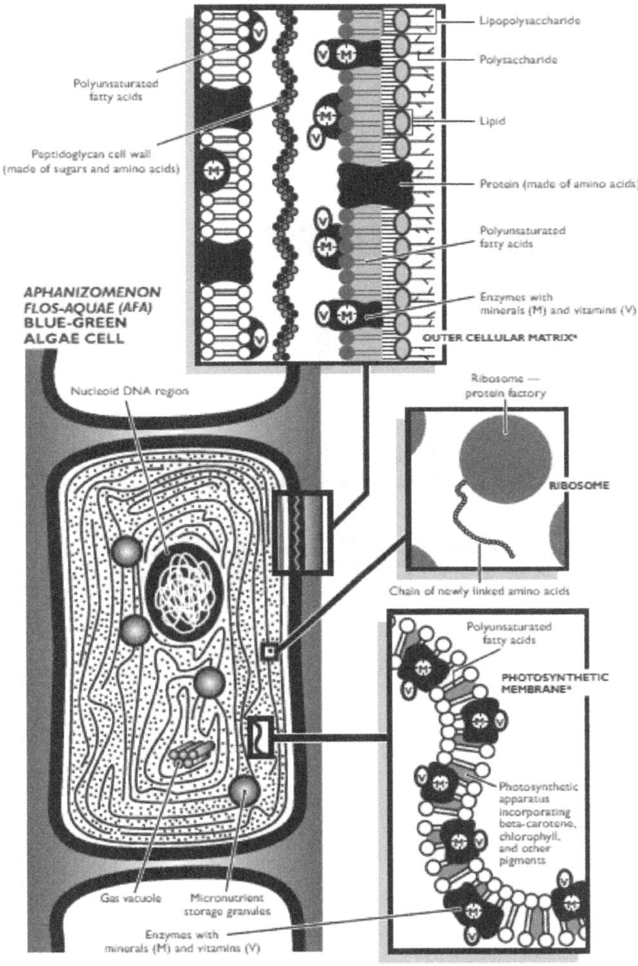

Schematic diagram of *AFA* microalgae cell structures

PART

I

Algae to the Rescue: Earth's First Food

CHAPTER

1

Drawing from the Past to Create Our Future

"Imagine our future living in cities where buildings are covered with photosynthetic membranes and vertical gardens, collecting the sun's energy and producing food and bio products for urban citizens. Imagine greening desert coastlines and producing food for millions of people. Imagine algae systems that recover and recycle polluting wastes into high value animal food, fuel and bio fertilizers."[1]

— Henrikson R, Edwards M, 2012

"Recently, great attention and extensive studies have been devoted to evaluate blue-green alga's therapeutic benefits on an array of diseased conditions including hypercholesterolemia, hyperglycerolemia, cardiovascular diseases, inflammatory diseases, cancer, and viral infections."[2]

— Deng R, Chow T, 2010

To come to know microalgae, especially cyanobacteria, is to reach deep into the mysteries of life. Devoting years of study on something that can only really be seen under a microscope, may seem incomprehensible to some. Yet, practically all that is necessary to grasp about nutrition can be found within the study of microalgae.

Most people simply do not recognize, much less appreciate, the amazing gifts of microalgae. For me, microalgae capture the ecstasy expressed by William Blake, when he said:

To see a World in a Grain of Sand
And a Heaven in a Wild Flower
Hold Infinity in the palm of your hand
And Eternity in an hour

Figure 3. Actinoptychus splendens

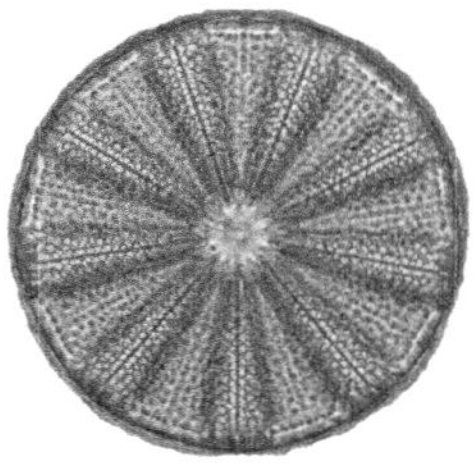

Microalgae Diatom

Most folks take a while to get beyond their image of pond scum: Isn't that what grows inside the glass walls of fish tanks? Common reactions that people have upon learning of the many remarkable health benefits of edible microalgae generally fall into one of three categories: 1) "If it's that good, why hasn't my doctor told me about microalgae?" 2) "So why isn't everyone eating microalgae, then?" or 3) "Why haven't drug companies created medications from microalgae?" This book is written to especially help with the first two categories of "why" objections.

For those interested in drugs derived from algae, I recommended reading Dr. Srinivasakumar and Professor Rajashekhar's 2010 book, *Drugs from Marine Microalgae: Investigating Medicinal Properties of a Few Marine Microalgae*. According to these medical scientists, "Numerous therapeutic compounds have been isolated from microalgae and have been tested for different types of bioactivity with positive effects . . . However [only] a small percentage of the marine microalgae population has been investigated thoroughly for their pharmaceutical potential."[3]

I have consumed, researched, and advocated the use of edible microalgae as a health resource for nearly quarter of a century. During these decades of exploration, edible microalgae has benefited many hundreds of thousands of people, with a variety of acute or chronic health challenges, and these benefits have been well proven to my own personal satisfaction. Yet, the fundamental question remains: What makes edible microalgae so effective in helping to alleviate or reduce symptoms across so many different health conditions?

This is no small point. The central problem in approaching all of the health research compiled within this book is that few people know anything that rivals the far-reaching benefits of microalgae. My mother joked that microalgae sounds like "aspirin," which is not yet a "drug", but reportedly helps reduce the risk of some cancers, stroke, heart attacks, various pains and of course headaches. Microalgae may be able to do all that and a lot more. Both resources can lessen inflammation, decrease COX-2 [an enzyme involved with inflammation and pain], and prevent blood platelets from sticking together. That's about where the similarities end; except these healing wonders were both discovered in nature (aspirin originates from willow bark).

What is more, microalgae can improve our immune system; protect against radiation exposure; detoxify heavy metals and harmful chemicals; neutralize various viruses, bacteria and fungus; and help to stimulate wound healing. In addition, microalgae decreases certain allergic reactions; acts as a highly nutritious food source; helps lower greenhouse CO^2 gas; can be used as fertilizer and animal feed; aids in water purification and toxic clean-up; and just might power your car in the future. It is easy to understand why the biggest challenge for most people is to even image that any one thing could possibly do all that! It may be helpful to envision the variety of species of microalgae as a vast pharmacopeia comparable to the massive array of plants found in the rainforests of the world.

Several health experts and authors have provided their own accounts of why edible microalgae are so effective. These various authorities have used their own experiential perspective or theoretical lens to account for the empirical evidence. Their assorted points of view have included dietary nutritional perspectives (Elkins, R., *Blue-Green Algae,* 1995; McKeith, G. *Miracle Superfood,* 1998); oriental energetic perspectives (Gagné, S., *The Energetics of Food,* 2008); raw food perspective (Wolfe, D., *Superfoods,* 2009) and whole food perspectives (Sandoval, D., *The Green Foods Bible,* 2007).

While useful, the most powerful framework for understanding the power of edible microalgae for me has been an evolutionary health perspective. What is essential about microalgae, especially the primordial cyanobacteria (blue-green algae) is the range of foundational roles they played in the evolution of multicellular life while terraforming our planet's atmosphere and dramatically shaping our entire ecosystem. Without this remarkable evolutionary perspective, we might think of edible microalgae as just protein, mineral, or lipid sources.

Estimates vary widely about how many species of microalgae exist. Yet according to Dr. Carlos Fernandez, a crop physiologist at Corpus Christi, Texas, "There are an estimated 200,000 to 800,000 species of microalgae."[4] Microscopic algae live nearly everywhere, in the water, ice, soil, clouds, and even inside of rocks. Wherever there is life, microalgae are likely to be there or nearby. Yet, medical researchers have screened only about one thousand species for potential useful applications. Moreover, the vast majority of health-related research has focused primarily upon a dozen or so species of microalgae.

Microalgae can provide a vast range of active biomolecules. Microalgae have colonized nearly every type of ecological niche – from polar ice caps to bubbling hot springs – and they make up one of the most dominant group of living organisms in overall species diversity, especially in marine waters. Microalga species employ a wide variety of metabolism strategies; which creates compounds of far greater chemical diversity than those found in land plants.

While hundreds of thousands of different microalgae species exist on Earth, less than a dozen microalgae species are domestically used by humans. Microalgae used for nutritional and oil purposes include: *Aphanizomenon flos-aquae* (blue-green), *Spirulina* (blue-green), *Chlorella* (green), *Nostoc* (blue-green), *Haematococcus* (green),

Dunaliella (red), and *Schizochytrium* (marine). A brief description of each species follows:

Aphanizomenon flos-aquae (AFA) are nitrogen-fixing blue green algae. An edible microalga *AFA* is currently only harvested wild from Klamath Lake in Oregon, U.S.A. Because of its unique pigment complex *AFA* has been described as beneficial for neurological, mental, and emotional wellness. *AFA* is sold as a nutritional supplement and added to nutraceuticals formulas to promote athletic performance, reduce inflammation, enhance stem-cell production, improve immune function, and generally to support overall health.

Chlorella was one of the first microalgae to be grown commercially in Japan and sold as a food supplement in the 1970s. *Chlorella* is a genus of single-cell green algae, belonging to the phylum Chlorophyta. The name *Chlorella* is taken from the Greek chloros, meaning *green*, and the Latin diminutive suffix ella, meaning *small*. In 1961, Melvin Calvin received the Nobel Prize in Chemistry for his research on the pathways of carbon dioxide assimilation in plants using *Chlorella*. Scientists in the 1960s found *Chlorella* was impossible for humans and other animals to digest in its natural state due to the tough cell walls encapsulating the nutrients, which restricted its use in food production. However, Japanese scientists developed a method to break-down the cell-wall and *Chlorella* is a popular health supplement.

Dunaliella is a green microalga that grows in water even saltier than the oceans. A remarkably hardy organism; a new species of *Dunaliella* was discovered in the Atacama Desert in 2010. This region is so dry usually the only water comes from fog that gathers from a distant coast at night. Some of this moisture condenses on spider webs as dew. This dew, scientists believe, is enough to keep the *Dunaliella* microalgae alive. Because marine *Dunaliella* is too salty to consume as a whole food, its beta carotene is extracted as an oil or powder and sold as a powerful antioxidant and color for human and animal feed purposes.

Haematococcus is a microalgae cultivated to extract astaxanthin, a reddish carotenoid pigment. A widespread carotenoid in marine creatures, astaxanthin provides crustaceans, shellfish, and salmon with their pink-red coloration. Importantly, astaxanthin is a powerful antioxidant for humans and is used in aquaculture and for cosmetic colors. Studies have shown astaxanthin to be over 500 times stronger than vitamin E and more potent than other carotenoids such as lutein, lycopene and β-carotene. Astaxanthin, unlike most carotenes is not converted to vitamin A (retinol) in the human body. Astaxanthin reportedly has shown beneficial effects in a variety of health-related conditions.

Nostoc is a genus of cyanobacteria that can be found in soil, on moist rocks, and in both fresh and saltwater habitats. Nostoc species were recorded and used as food 1600 years ago in China. A species of *Nostoc, N. commune* is also consumed in the Andes. *Nostoc commune* is eaten as a salad in the Philippines and eaten as food in Indonesia, Japan and Taiwan. A variety of *Nostoc commune*, known as "facai" in China, is part of traditional food served at the Lunar New Year. Having a remarkable thick-walled cell (akinete) some species of *Nostoc* have the ability to survive with no water for long periods of time. After 70 years of dry storage, the akinete of one species was able to germinate into a filament when moistened.

Schizochytrium is a unicellular marine microalga. These organisms were once thought of as a kind of fungus, but are now assigned to an algal group that contains a diverse range of micro-algae and macro-algae. Several different forms occur in the life cycle. *Schizochytrium* is grown in vats by fermentation to produce significant amounts of docosahexaenoic acid (DHA). DHA is an essential fatty acid that especially supports brain and eye function and is often added to baby formula and oil for biofeeds, biomass and use as biofuels.

Spirulina is a blue-green algae and the most widely cultivated microalgae since the 1970s. Yet human consumption of Spirulina was observed as early as the 14th century Aztec civilization. World-wide use of *Spirulina* has grown and thousands of tons are sold each year. China is the world's largest producer of *Spirulina*, providing nearly 50% of the world demand. Most of the health research on microalgae has been done on species of *Spirulina (maxima, fusiformis, and platensis)*. Yet, as *Spirulina* is a blue-green cyanobacterium like *AFA*, for the most part these research findings will likely apply to *AFA* as well.

Other species of microalgae, like ***Botryococcus braunii*** and ***Nannochloropsis*** are being grown for algae biofuel production.

Medical researchers are just starting to appreciate the important roles that beneficial intestinal bacteria play in supporting health. Likewise, microalgae health solutions can support our lives in even more far-reaching ways. For example, cyanobacteria secondary metabolites have been found to be a fruitful source of prime compounds in drug discovery efforts.

Micro Algal Nutrients Work in Numerous Ways Because Nature Retains Effective Solutions

One of the reasons microalgae nutrients work in so many areas is that nature is conservative in its designs. Solutions that work are retained. For example, chlorophyll an "invention" that allows organisms to capture sunlight and produce sugars, first appeared in blue-green microalgae billions of years ago and is now used as a survival strategy by all higher plants. Animals, in turn, depend upon chlorophyll-containing plants, directly or indirectly, as a food source.

All cyanobacteria are photosynthetic and can therefore manufacture their own food. Cyanobacteria have been dubbed "blue-green algae," as cyano means blue. Yet not all cyanobacteria are actually blue-green in color. A remarkable diversity of cyanobacteria photosynthetic pigments impart a rainbow of possible algal colors, including shades of red, pink, yellow, orange, green, blue-green, deep blue, and violet.

Primordial evolutionary threads are repeated countless of times throughout nature. Ancient organic molecules, such as amino acids, which were found in primordial microalgae at the dawn of life, now act as basic building blocks for all of earth's creatures. Genes that encode heme oxygenase, an enzyme crucial for cellular respiration, are conserved by nearly all living systems from cyanobacteria to humans.[5] Several families of potent antioxidants (e.g., beta carotene or glutathione) that originated in primitive microalgae are conserved and widely used across nature.

An excellent dietary source for vitamin D is fish. Yet, fish are unable to synthesize vitamin D. Fish obtain their vitamin D from phytoplankton, microalgae. In the ocean it appears that the "conversion of provitamin D_2 to vitamin D_3 can take place only under the influence of ultraviolet-B radiation in algae."[6] A provitamin is a substance that can be converted into a vitamin; like ergosterol that gives rise to vitamin D_2. Over the past 500 million years or so, photosynthesis conversion of vitamin D by phytoplankton has occurred in the oceans. Vitamin D can only be synthesized through a photochemical process.

Therefore, land vertebrates typically obtained vitamin D either from eating marine foods, containing microalgae vitamin D, or through exposure to sunlight to convert vitamin D within their skin. Plants with the exception of certain mushrooms do not provide adequate amounts of vitamin D. Scientists believe that a difference of skin color pigmentation in humans (from lighter to darker) is an evolutionary adaption to optimize levels of vitamin D production in the skin epidermis related to intensity of geographic sun exposure. This phenomenon helps to account for the diversity of human skin colors and to the biological imperative of maintaining adequate levels of vitamin D.

Likewise, while most people have heard about the remarkable benefits of fish oils, especially omega-3 and DHA fats; few people know that fish do not directly produce these oils, but rather collect and concentrate them from eating microalgae rich sources. These essential fatty acids (EFAs) are critical structural components of cell membranes and play a foundational role in our brain chemistry. Microalgae are the primary source of EFAs and vitamin D in the food chain!

Microalgae has a Superior Trace Mineral Content

The trace mineral content of microalgae is extraordinary and has been credited as "the root of most of their healing properties." The diversity of trace minerals found in aquatic microalgae is far greater than that of typical land-based plants. Also, these trace minerals are "bioavailable" in chelated and colloidal forms, which makes them especially absorbable and beneficial for human health purposes. Dr. Alexis Carrel, the winner of a Nobel Prize in Medicine (1912) stated that soil is the basis of all human life. All life is dependent upon the fertility of the soil. Dr. Carrel advocated that illness were created chiefly by destroying the harmony reigning among mineral substances present in tiny amounts in air, water and food but most importantly the soil.

A 2004 study by Donald David and his team of researchers at the University of Texas at Austin's Department of Chemistry and Biochemistry using U.S. Department of Agriculture data found that nearly half of nutrients measured, "six out of 13 nutrients showed apparently reliable declines between 1950 and 1999. These nutrients included protein, calcium, phosphorus, iron, riboflavin and ascorbic acid. The declines which ranged from 6 percent for protein to 38 percent for riboflavin raise significant questions about how modern agriculture practices are affecting crops."[7]

Several other studies have reported similar findings: "A Kushi Institute analysis of nutrient data from 1975 to 1997 found that average calcium levels in 12 fresh vegetables dropped 27 percent; iron levels 37 percent; vitamin A levels 21 percent, and vitamin C levels 30 percent. A similar study of British nutrient data from 1930 to 1980, published in the British Food Journal, found that in 20 vegetables the average calcium content had declined 19 percent; iron 22 percent; and potassium 14 percent. Yet another study concluded that one would have to eat eight oranges today to derive the same amount of Vitamin A as our grandparents would have gotten from one."[8]

According to Davis "efforts to breed new varieties of crops that provide greater yield, pest resistance and climate adaptability have allowed crops to grow bigger and more rapidly, but their ability to manufacture or uptake nutrients has not kept pace with their rapid growth." In addition loss of minerals in the soil can be related to large-scale and commercial agriculture methods that deplete the soil.

"Over decades, the annual plowing and disking of fields reduce the essential organic carbon matter of the soil. How much is lost? The world over, 50 to 80 percent of soil's humus, or organic matter has been lost over time. Tilling not only loosens soil but virtually earmarks it for erosion; it also disturbs the delicate balance of micro-organisms living within it. Soil is a living thing."[9]

Over time, nutrient loss and soil microorganism destruction are not adequately replaced by the use of petrochemical fertilizers. Improving the nutritional quality of our food requires healthy soils and plant crops ability to draw out vital minerals from the Earth.

If the biblical "Garden of Eden" ever existed on Earth, humans have long since become the victims of the water cycle. This is not a reference to a single "great flood." Rather, soil erosion leads to depletion of essential mineral elements that wash from the land into rivers, lakes and into the oceans. To grasp how extensive the time-scale impact of water cycles has been on mineral erosion, all you have to do is taste the ocean. The chief reason the ocean is so salty and terra-firma (land) is not, has to do with the majority of the salt on land washing into the oceans. But this happens not only to salt, but continues to occur to many trace minerals found on the surface of land. Over time our topsoil becomes depleted of essential minerals, while aquatic microalgae get a banquet of macro and micro-minerals to absorb.

Balance is important to all areas of our life and health, but is especially important when it comes to minerals and trace minerals. Almost 4,900 known mineral species have been approved by the International Mineralogical Association (IMA).[10] Yet only about 60 of these mineral species have been identified and studied in body tissues and fluids. Minerals are chemical constituents used by the body in many ways. Although they yield no energy, they have important roles to play in many activities in the body. The mineral elements are separate entities from other essential nutrients like proteins, fats, carbohydrates and vitamins. Minerals are broadly classified by the amounts required for health; macro (major), micro (trace) and ultra-trace (infinitesimal). Examples of ultra-trace minerals would include nickel, boron, vanadium or silicon, which have been found in humans, and are believed to be essential for health.

"When a trace element is deficient, a characteristic syndrome is produced which reflects the specific functions of the nutrient in the metabolism of the animal. The trace elements are essential components of enzyme systems. Simple or conditioned deficiencies of mineral elements therefore have profound effects on metabolism and tissue structure . . . The significance of the mineral elements in human, animal and plant nutrition cannot be overemphasized."[11]

Microalgae Supports Our Intestinal Ecosystem, the Microbiota.

Gut microbiota (sometimes called intestinal flora) is the name used to refer to the microbe population living in our intestine. Humans have trillions of beneficial, as well as some harmful, bacteria that reside in different parts of our body. Amazingly, scientists estimate that the number of bacteria cells found within the human body even outnumber (by at least ten times) the number of mammalian cells that make up our body! The hugest population of microbiota exist within our intestinal tract, several pounds of bacteria and microorganisms. Moreover, the gut microbitota of each individual is unique, just like our fingerprints. Our gut microbitota may even be considered a functional organ system, essential to proper digestion, immune function, production of vitamins (B and K), and creating a barrier to toxins, allergins, and other pathogenic microorgainsms.

Recently, research teams from around the globe are reporting that gut microbiota may influence mammalian brain development and adult behavior. Gene profiling in mice studies demonstrate that brain-related genes involved with learning, memory and motor control were affected by dramatic alterations in gut bacteria. According to Swedish scientist Dr. Rochellys Diaz Heitz (2011), the data suggests "a critical period early in life when gut microorganisms affect the brain and change the behavior in later life . . . Not only are signal substances like serotonin and dopamine [major neurotransmitters] subject to regulation by bacteria, synapses function also appears to be regulated by colonizing bacteria."[12-14]

Shifts in bacterial growth within algal cultures are a common phenomenon because algae create organic compounds that can both facilitate the growth and inhibit the growth of bacteria. Symbiotic relationships (both stimulating and inhibitory) between microalgae and bacteria have evolved over hundreds of millions of years. Particular microalgae oral use may support beneficial gut microbiota. Dr. Robert Kay reported AFA increased beneficial colonic microbiotia, like acidophilus or bifidum bacteria several-fold when combined with human consumption of probiotic bacteria, versus probiotic intake alone.[15] In separate research (2009), supplementation of *Aphanizomenon flos-aquae* (AFA) demonstrated modest increases on the growth of beneficial microbiota in experimental mice compared to controls.[16]

Ancient algal lineages are remarkably interactive with their greater symbiotic environments. For example, people in Japan have acquired a particular gene within a common gut bacterium that most human populations don't have. This gene supports enzyme production to better digest seaweeds. Reportedly a marine bacteria co-existing on the algae was consumed by the Japanese for so many generations, that their gut bacteria eventually acquired the gene. Bacteria frequently exchange genetic information. This particular metabolic gene found its way into the Japanese microbiome and now allows Japanese people to digest seaweeds more completely than most other people.[17] Marine seaweeds are typically far more difficult to digest than freshwater microalgae, like *AFA* or *Spirulina*.

Simply put, there is hardly any area of nutritional medicine – from amino acids to microbiota — that cannot be better understood from a further and comprehensive study of microalgae. In short, microalgae at the bottom of the food chain provide an ancient bimolecular "pharmacopoeia" upon which most of cellular life now depends.

CHAPTER

2

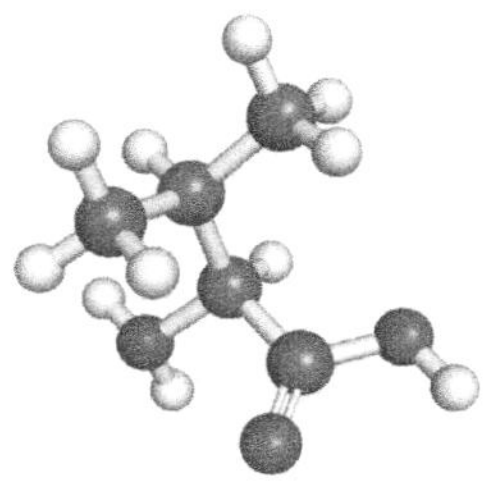

An Evolutionary Perspective on Amino Acids

We require a deeper vision of evolutionary history to understand the very origins of nutrient biosynthesis. Evidence is growing that some prebiotic "seeding" may have occurred early in Earth's history, as a result of collisions of comets and or asteroids. The discovery of amino acids from outer space, as found within the Murchison meteorite, provides powerful support for this idea. Perhaps both cosmic seeding and prebiotic compounds interacting with geological elements all helped to create the stage for amino acid evolution through biological life.[1]

All life on Earth relies on a standard set of 20 to 22 molecules called amino acids to build the proteins that carry out life's essential actions. These molecules are built primarily from carbon, hydrogen, oxygen, and nitrogen.

To date, scientists have found more than 500 amino acids in nature, but only 20 result from genetic translation. Translation occurs when the nucleotide sequence of messenger RNA (mRNA) is used as a blueprint for amino acid sequences for new proteins. Protein synthesis consumes a large part of the energy produced in the cell, and all proteins are built from these ancient amino acid building blocks.

Figure 4. Origins of Earth's First Food

Between 4 and 3.5 Billion years;
Earliest life appears.
Amino acids by chemical synthesis

↓

Around 3.5 billion years;
Last Universal Common Ancestor
(LUCA)

↓

Cyanobacteria develops metabolic
pathways for amino acid biosynthesis

↓

Around 1.8 Billion years ago multicellular
life appears (Eukarya)

↓

Specialization of metabolic pathways;
results in some **amino acid pathways
lost in many multicellular lineages.**

Nutritional scientists generally recognize twenty-two amino acids as providing the building blocks for proteins: the twenty common ones and two more, selenocysteine and pyrrolysine. Amino acids serve multiple functions. Their chief function is to act as single units necessary for protein synthesis. Amino acids are energy metabolites, essential nutrients, involved in energy transformation within cells. Amino acids function as chemical messengers in communication between cells. For example, hormones and neurotransmitters are derived from amino acids. Finally, amino acids support molecules necessary for enzyme chemical reactions.[2] Life as we know it requires numerous amino acids.

An Ancient Foundation for All Earth's Life

According to Dr. Freeland, of the NASA Astrobiology Institute at the University of Hawaii; life has used the same standard set of 20 amino acids to build proteins for more than 3 billion years. This happened despite the fact that numerous other amino acids choices might have been incorporated into RNA and DNA strands, as well. Remarkably, when Dr. Freeland and his team of NASA scientists used computer models to estimate the fundamental properties (e.g., size, charge, or hydrophilicity) of the 20 amino acid found within Earth's living organisms, they discovered that life did not appear to select these 20 amino acid building blocks randomly. "We found that chance alone would be extremely unlikely to pick a set of amino acids that outperforms life's choice."[3] In a sort of chicken versus egg, puzzle, the amino acid profiles utilized by some of life's most ancient organisms, cyanobacteria, gave rise to the foundational toolbox for nearly all of Earth's subsequent life forms.

Figure 5. Stromatolites, living fossils, Shark Bay Australia

Stromatolites (literally, 'layered rock') are rock-like formations created by cyanobacteria (blue-green algae)

Biologists have found stromatolites of living blue-green algae thriving in Shark Bay, Australia and off islands in the Bahamas. The cyanobacteria form colonies and trap sediment against their surface coatings. Over time, the trapped sediment reacts to calcium carbonate in the water to form limestone. These limestone deposits grow very slowly; a one meter stromatolite might be 2,000 years old.

Yet, similar fossilized stromatolites of cyanobacteria have been located and date back 2.8 billion years and may have existed as far back as 3.5 billion years ago.[4] Although Shark Bay's stromatolites are just 2,000 – 3,000 years old, the cyanobacteria that build them are similar to fossilized forms found on Earth billions of years ago!

Despite the various theories of how amino acids initially formed – whether from seeding by comets or asteroid fragments or chemical synthesis from ancient prebiotic molecules, a "primordial soup," during the first half billion years of Earth's existence — amino acids are a living testament to an ancient code of life. As the building blocks of proteins, amino acids are necessary for the maintenance of life.

Cyanobacteria Makes Nitrogen Essential for Life

The root of the word "amine" indicates that nitrogen is the central atom in amino acid structures. Atmospheric nitrogen (N_2) is the ultimate source of nitrogen for the biosynthesis of amino acids. While nitrogen is a crucial chemical element for living systems, because of its triple bond (N≡N), nitrogen is extremely difficult to break apart. Atmospheric nitrogen, an inert gas, must be reduced to be metabolically useful. Only blue-green algae and certain types of bacteria can accomplish this biological transformation, termed nitrogen fixation.

Figure 6. Blue Green Algae Heterocyst Cells

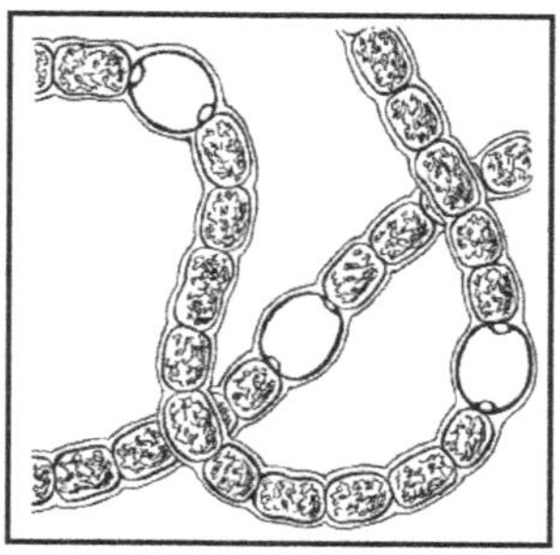

Heterocysts are specially differentiated cells, illustration as white oval cells, which capture nitrogen from the air

Most organisms are thereby dependent upon the reduced nitrogen (NH_3) produced through the power of bacteria, cyanobacteria, and lightning. Heterocyst cells within blue-green algae perform the task of fixing nitrogen from the air and making nitrogen available to the soil and thus to many plants, both aquatic and land-based. For example, blue-green alga was one of the original organic fertilizers used in rice cultivation.

Humans' Lost Ability to Produce Certain Amino Acids

Over vast epochs of time, as living organisms specialized and adapted to certain environments, they lost their ability to synthesize all of the amino acids required by their genetic code. Mammals must obtain a number of specific amino acids from their diets, and these amino acids are termed "essential." Amino acid use varies widely among species because different amino acid synthesis pathways evolved to fulfill unique metabolic needs. At long last, humans lost the ability to synthesize essential amino acids and are dependent on eating other organisms to obtain sources of amino acids. Thus, amino acids can be divided into two main categories: essential and nonessential.

Essential Amino Acids mean Eating is required

Essential amino acids are "essential" not because they are more important than other amino acids, only because our body cannot generally make them. Essential amino acids must be present in our diet to be available in our bodies. Essential amino acids for humans are phenylalanine, valine, threonine, tryptophan, isoleucine, methionine, leucine, lysine, and histidine. Additionally, cysteine, taurine, tyrosine, and arginine are considered semi-essential amino acids in children (though taurine is not technically an amino acid). In children, the metabolic pathways that synthesize these so-called "semi-essential" amino acids are not yet fully developed.[5]

Figure 7. Classification of Amino Acids

Classification of Amino Acids

Essential Amino Acids		Conditionally Essential Amino Acids	
Critical	Limiting	Non-essential	Semi-Essential
Lysine	Lysine	Alanine	Cysteine*
Methionine	Methionine	Asparatic acid	Taurine
Tryptophan		Asparagine	Arginine
Threonine	Threonine	Glutamic acid	Tyrosine*
Isoleucine		Glycine	Glutamine
Leucine			Ornithine
Valine			Proline
Histidine			Selenocysteine
Phenylalanine			Serine

Requirements for tyrosine can be met by tyrosine or phenylalanine.
Requirements for cysteine can be met by cysteine or methionine.

Certain essential amino acids are typically scarce and are known as critical amino acids. Lysine, methionine, and threonine, are the most deficient amino acids in agricultural food, so they are referred to as rate-limiting amino acids. Lysine is an amino acid commonly deficient in a wide variety of food sources (e.g., soybeans). Threonine is only synthesized by microorganisms and requires co-factors of B-6, niacin, and magnesium for optimal utilization and metabolism in the body. A deficiency of methionine, more common in vegans and vegetarians, can lead to increased risks for depression. When limiting amino acids are deficient in one's diet, the ability to utilize all other amino acids becomes significantly reduced.

Cysteine a "nonessential" amino acid is well known as a limiting amino acid for protein synthesis.[6] Cysteine is adversely affected by heat and is one of the most poorly digested amino acid within food sources.[7] It is noteworthy that the nonessential amino acid glycine can exhibit the same behavior as a limiting amino acid. For example, when a person is exposed to large amounts of environmental toxins and pollutants, the amount of glycine (a nonessential amino acid) made by the body may not be adequate. Glycine is metabolically related to serine, methionine, cysteine, and threonine. When our dietary intake of these various limiting amino acids becomes inadequate, due to food processing or other factors, our ability to synthesize protein and to use other amino acids is significantly reduced.

Medical Conditions and Food Processing can Negatively Impact our Protein Utilization

Excessive food processing and over-cooking can destroy heat-sensitive amino acids or make them resistant to digestive enzymes. When food is cooked at high temperatures or heavily processed to extend its shelf life, for convenience and profit, various amino acids can be degraded. Lysine, cysteine, histidine, aspartic acid, and, to a lesser extent, arginine are the most heat-sensitive amino acids.[8] Tryptophan is also a "sensitive amino acid," vulnerable to food processing, as are aspartic acid and glutamic acid reported to be heat sensitive.[9]

Certain medical conditions may negatively impact our protein utilization status. For example, protein digestion and metabolism involves the stomach, pancreas, and liver. Hydrochloric acid, secreted by the stomach, is necessary for the initial digestion of protein. Pancreatic enzymes participate in the breakdown of protein. And the liver controls amino acid metabolism.

The ability of our body to manufacture certain nonessential amino acids may be hampered by a deficiency of vitamin B6. Also, individuals exposed to bacterial or viral infections or those who have experienced severe physical trauma rapidly use up their protein stores and may need to increase their intake of protein. The amino acids arginine, cysteine, glycine, glutamine, histidine, proline, serine, and tyrosine are considered "conditionally essential," meaning they are not normally required in the diet. Yet for specific genetic/medical groups even these conditionally essential amino acids, must be supplied externally, as these people may not synthesize them in adequate amounts.[10]

Therefore in short, deficiencies of various amino acids can occur due to food processing, digestive problems, or prolonged illnesses even in people who regularly eat meat, as well as in vegetarians and vegans. The body is only able to make all the proteins it needs when there are sufficient quantities of all the necessary amino acids, the so-called "amino acid pool." Additionally, there are circumstances in which the body cannot make enough of the nonessential amino acids. For all of these reasons, the distinction between what is an essential amino acid and a nonessential amino acid can become blurred. It may be helpful to think about most nonessential amino acids as "conditionally essential."[11]

Under various physiological circumstances, our bodies might be unable to manufacture enough of the so-called "nonessential" amino acids, so we must obtain greater amounts through diet (or via supplementation). This concept of "conditionally essential amino acids" tells us that all amino acids are important. Fortunately, within edible blue-green algae, we have the widest range and the most bioactive, digestible, forms of the amino acids (essential and so-called nonessential) available in a single whole food.

One of the most remarkable nutritional aspects of microalgae is its high content of usable protein—ranging from 50% to 70%! This is a far higher percentage than the choicest edible parts of any higher plant or animal.

Amino acid researchers Eric Braverman, M.D. and Carl Pfeiffer, M.D., have cautioned against relying upon separate amino acid supplements. "Because many of the amino acids are absorbed and metabolized in a similar fashion, there is a great deal of competition between molecules. Sometimes an amino acid can cancel the effect of others. This adds to the overall complexity of prescribing amino acids to treat disease."[12] The use of wide-spectrum organic amino acids within an ideal ratio for human consumption, as found within certain microalgae, offers numerous advantages over isolated amino acid supplements.

CHAPTER

3

Remarkable Nutritive Qualities of Microalgae

Gram-for-gram microalgae may be the most nutrient dense food on Earth. One of the most remarkable nutritional aspects of microalgae is its high content of usable protein — ranging from 50% to 70%! This is a far higher percentage than the choicest edible parts of any higher plant or animal. Algal protein has shorter and less complex polypeptide chains — making it easier to digest than plant or animal protein. Red meat has a surprisingly low net protein utilization index of 18%, compared to *Aphanizomenon flos-aquae (AFA*'s) 75%.[1]

The net protein utilization index is a measure of how completely amino acids are assimilated by humans. In fact, some microalgae, such as *AFA*, contain all ten essential amino acids that humans require from their diets — in a profile similar to that recommended by the National Academy of Sciences.[2]

The essential amino acid profile of cyanobacteria (blue-green algae) is fairly similar to human breast milk or egg whites, considered the "ideal amino acid profile" for human consumption. Also, most legitimate microalgae growers and harvesters of edible microalgae (primarily *Aphanizomenon flos-aquae [AFA]*, *Spirulina*, or *Chlorella*), will avoid the use of high temperature drying techniques to better preserve all the essential amino acids, and minimize the loss of heat sensitive antioxidant pigments and enzymes. Multiple nutritional experts have described edible microalgae as nature's perfect food or the most nutrient-dense food on the planet.

Figure 8. Essential Amino Acid Profile of *Aphanizomenon flos-aquae* (*AFA*)

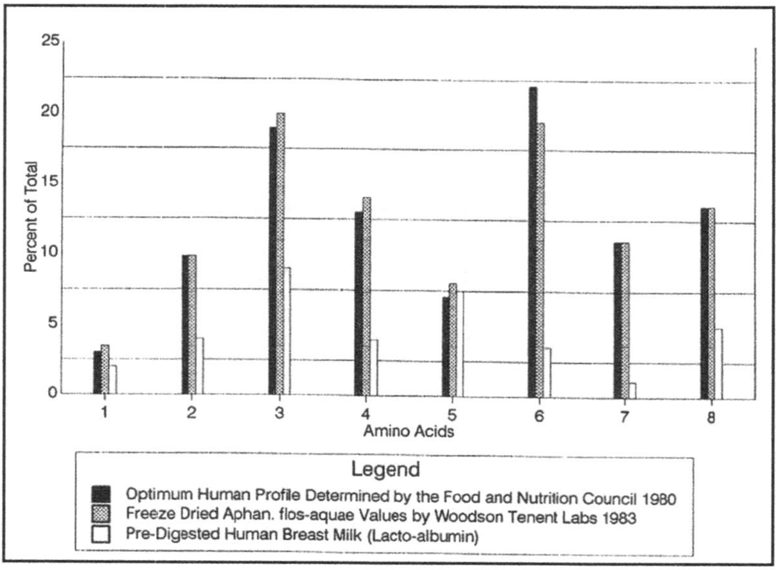

A nearly perfect ratio of essential amino acids
Number Key: 1. Histidine; 2. Isoleucine; 3. Leucine; 4. Methionine plus Cysteine; 5. Phenylalanine plus Tyrosine; 6. Threonine; 7. Tryptophan; and 8. Valine

Why Edible Blue-Green Algae is a Superfood

- One of the most remarkable nutritional aspects of *AFA* microalgae is their **high content of usable protein**—ranging from 60 to 70%. *(Also, algal protein has shorter and less complex polypeptide chains, making it easier to digest than plant or animal protein).*

- Aquatic microalgae contain the **widest range of trace minerals**, more than most land-based plants. Why? In part, because minerals on the land eventually wash into rivers, lakes and the seas. These myriad of mineral electrolytes, found in a superior colloidal and chelated forms, **support essential enzyme functions**.

- Furthermore, microalgae exhibit **superior photosynthetic pigment** efficiency, using light approximately three times more efficiently than higher plants.[3] Microalgae are among the most productive organisms on the planet. Perhaps, as they grow underwater, they contain a far wider range of light sensitive molecules than most land plants. Microalgae are evolutionary older than land plants, so their **antioxidants** appear more potent having adapted to extreme climatic conditions.

- In addition to remarkable pigments like phycocyanin, blue green algae "also contain vitamin C, E, and BG_{12}, carotenoids, phenolic compounds, and minerals that can function as antioxidants. Due to the presence of multiple antioxidant components, the administration of whole algae may result in **synergistic antioxidant effects**."[4]

Photosynthetic pigments in microalgae impart a rainbow of possible colors; yellow, red, violet, green, deep blue and blue-green pigments are known. Unique combinations of microalgae species specific photosynthetic pigments have emerged with medical and industrial applications. For example, an *AFA* microalgae light-harvesting complex was reported by Italian researchers (2008) as applicable for the treatment of neurological, neurodegenerative and mood disorders.[5] In general, microalgae light-harvesting pigments are reported to have numerous health benefits for the skin, gums, eyes, heart, lungs, liver, brain and most organ systems, practically all cells of the body.

- The primitive character of microalgae's cellular organization gives it a number of advantages over higher plants and animals as a food source. Cyanobacteria cells are about **20 to 30 times smaller than the cells within the food we usually eat.** Because of this, cyanobacteria contain 20 to 30 times the membrane surface area.[6]

The cell membrane is where many important nutrients and pigments are concentrated. Blue-green alga produces more cell membrane material without getting larger by creating a vast system of membrane inpouchings similar to the brain's infoldings (called sulci). In other words, if the cell membrane were ironed flat, it would be many times the apparent size of the cell. Because of these structural arrangements microalgae are **remarkably nutrient-dense**, compared to other foods.

Figure 9. Surface Area to Volume Ratios

Surface Area	Volume	S/V
6 cm2	1 cm3	6/1
48 cm2	8 cm3	6/1
24 cm2	8 cm3	3/1

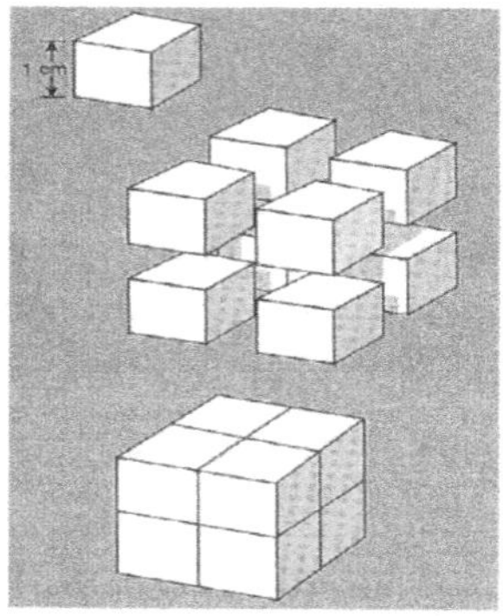

Surface-to-volume ratio (S/V ratio) of particles influence how much and how rapidly materials flow in and out of cells. Smaller cell-sized organisms, like microalgae, with higher S/V ratios have better nutrient absorption when ingested, compared to larger multicellular foods (plants or animals)

- Practically the entire organism can be nutritious, with minimal indigestible structures. By contrast, typically less than half of the dry weight of plants and animals has nutritional value. Most people don't eat the bones of animals or the roots of most plants. Primitive blue-green algae are composed almost entirely of **nutritionally useful and uniform cells.**

- Not least, "microalgae are considered to be **the primary source of unsaturated fatty acids in the food chain**."[7] Eicosapentaenoic acid (EPA) and docosahexaenoic acid (DHA) are two relatively rare and valuable fatty acids found in microalgae.[8] The reason that fish oils are so rich in polyunsaturated fatty acids (PUFAs) is that microalgae are abundant in their food chain. Unlike seafood, microalga oils are cholesterol free. The nutritional value and therapeutic merits of PUFAs have been widely documented.[9]

- Another useful **fatty acid found abundantly in blue-green algae is gamma linolenic acid (GLA).** "Gamma linolenic acid (GLA) is synthesized from linoleic acid, and from GLA the body makes a very important hormone-like substance called prostaglandin E1 (PGE1). PGE1 helps to prevent heart attacks and strokes, helps to remove excess fluid, imp roves circulation, slows down cholesterol production, improves nerve function, and regulates cell division. PGE1 is anti-inflammatory: it is vital to maintaining a healthy balance in our joints, helping to prevent inflammation and pain. Groups of arthritis sufferers have shown significant improvement after taking GLA supplements. GLA has also been found to be an important nutrient for the prevention of certain skin disorders, such as psoriasis; and it appears to alleviate premenstrual syndrome (PMS)."[10]

- Because microalgae don't eat other organisms, like animals do, they can't rely on obtaining vitamins from what they eat. Thus, microalgae tend to **produce more of their own vitamins** than most plant sources. For example, certain blue-green may provide one of the only vegetarian sources of B-12.

Microalgae contain a wide spectrum of prophylactic and therapeutic nutrients that include rich sources of B-complex vitamins, trace minerals, amino acids, gamma-linolenic acid, essential fatty acids (omega-3 and omega-6), super antioxidants such as phycobilins & beta carotene, immune-enhancing sulfated polysaccharides, phenols, and a number of unexplored bioactive compounds.

CHAPTER

4

How Cyanobacteria Captured Sunlight for Food and Forever Changed our Atmosphere

"There is diverse, yet controversial fossil evidence for the existence of photosynthesis 3500 million years ago. Among the most persuasive evidence is the stromatolites described from low grade metasedimentary rocks in Western Australia and South Africa . . . The evidence allow further interpretation that cyanobacteria were involved."[1]

— *Awramik SM, 1992*

"Unlike flight or vision, which evolved independently many times, oxygenic photosynthesis only ever evolved once. All algae, all plants, the entire green planet, use exactly the same system. All of them inherited it from the cyanobacteria, which invented it once, perhaps 3.5 billion years ago. No other cells on Earth ever learnt to split water."[2]

— *Lane N, 2002*

Photosynthesis provides color, atmosphere, water and useable energy to our world. Without the benefits of photosynthesis our planet might look like Mars or an alien planet, largely devoid of any signs of life. "Photosynthesis is the only significant solar energy storage process on

Earth and is the source of all of our food and most of our energy resources."[3]

Compelling evidence exists that photosynthesis is very ancient, evolving not long after the origin of life on Earth. Scientists have found evidence that oxygenic photosynthesis evolved within cyanobacteria, around 3 billion years ago. Evolutionary biologists have nearly universally accepted the evidence that eukaryotic photosynthesis originated from endosymbiosis of cyanobacteria-like organisms, which ultimately became chloroplasts (Margulis, 1992).[4] These ancient life forms developed powerful pigments for harnessing the sun's energy, protecting against deadly solar radiations and handling "toxic" oxygen reactive molecules.

In the process of photosynthesis, some free-radical intermediates – the same as those produced by radiation - are also produced as water is stripped of electrons to form oxygen. These "free-radicals" need to be molecularly "caged" to prevent a breakdown of nearby molecules. Moreover, the final step of molecular oxygen is also released into the cell in large quantities. It is estimated that the oxygen concentration within cyanobacteria may reach three times atmospheric levels.

Microalgae represent one of the earliest photosystems, biological solar panels that capture light and convert photons into useable molecular energy. Their photosystem is a pigment-protein complex consisting of a) reaction center (RC) and b) a light-harvesting complex (LHC). Cyanobacteria have two reaction center complexes (type I and II), while all other photosynthetic prokaryotes have only a single RC. Within the light-harvesting complex are several pigment families (tetrapyrrole pigment molecules), including chlorophyll, phycocyanin, phycobilins and carotenoids.

Tetrapyrrole pigment molecules are found in virtually all living organisms on Earth. Tetrapyrrole biomolecules are commonly linked to light-harvesting, oxygen metabolism and their colorful pigments (e.g., chlorophyll and phycocyanobilin) act as redox signaling molecules and antioxidants. Tetrapyrrole pigments appear to play a key role in modulating inflammation.

Figure 10. Tetrapyrrole Light-Sensitive Pigment Molecules Are Found throughout Nature

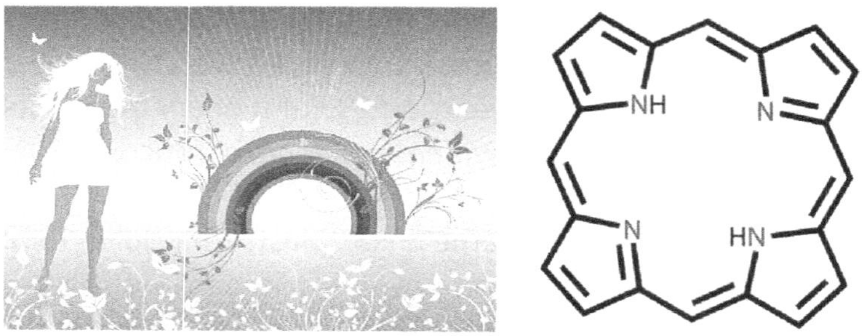

***Heme *Chlorophyll *Phycocyanin *Bilirubin *Biliverdin *Cytochrome C *Catalase *Nitric oxide synthase *Myoglobin**

Tetrapyrrole products allow algae, plants, and mammals to rapidly capture and exchange electrons. These colorful photo-sensitive molecules evolved to capture sunlight and thus played essential roles in the origin of life, and organisms' evolutionary ability to adapt to oxygen associated changes during times of stress and inflammation. Thus a "diverse colorfulness that the inflammatory response triggers during its evolution would reflect the major pathophysiological importance of these pigments in each one of its [healing recovery] phases. Therefore, the need of exploiting this color resource could be considered for both the diagnosis and treatment of the inflammation."[5]

Multi-colored pigments associated with living organisms exist, chiefly because of a relationship with sunlight. On a practical level, we would be unable to see most colored pigments, if solar wavelengths did not expose them. The only exceptions are phosphorescent pigments, which emit their own light sources. The chief pigments used by all photosynthetic reactions in microalgae and plants are chlorophylls; which converts solar photons into usable energy. Chlorophyll a originated in cyanobacteria (blue green algae) and absorbs only certain ends of the spectrum of visible light (especially blue, followed by red and violet). Chlorophyll looks green because it reflects back the middle part (green to yellow) of the light spectrum. A family of carotene pigment molecules also captures sunlight and their color can ranges from pale yellow through bright orange to deep red, is related to their structure. So-called "accessory pigments" like phycobilins and xanthophylls (a variation of carotene) absorb and reflect different visible colors of light that chlorophyll cannot.

Carotenoids are lipid-soluble pigments. They are mainly known for their antioxidant properties and there color properties. In plants and algae carotenoids display orange and red colors, and in aquatic animals may include blues, purples and reddish colors. Algae, plants, phytoplankton, as well as particular bacteria and fungi are the only organisms that can synthesize carotenoids. Humans as well as other animal species cannot produce carotenoids and must obtain them as part of their diet. Carotenoids with molecules containing oxygen, such as lutein and zeaxanthin, are known as xanthophylls.

<u>There are 5 main xanthophylls</u>:
* Astaxanthin * Canthaxanthin
* Cryptoxanthin * Lutein
* Zeaxanthin

Carotenoids, such as beta carotene, are good at absorbing blue spectrum light. The oxygen-free carotenoids, such as α carotene, β carotene, and lycopene, are known as carotenes.

Phycobilins (from Greek *phykos* meaning "alga," and from Latin: *bilis* meaning "bile") are light-capturing pigments found in cyanobacteria and in the chloroplasts of red algae.

<u>There are 3 main phycobilins</u>:
* Phycoerythrins * Phycocyanins * Allophycocyanin

Phycobilins are accessory pigments to chlorophyll. They transfer light energy to and possibly protect chlorophyll as well. Phycobilins, especially allophycocyanin, are better at harvesting light than chlorophyll a. Phycobilins are unique photosynthetic pigments as they are bonded to water-soluble proteins, known as phycobiliproteins. All phycobiliproteins are water-soluble, unable to exist within the cell lipid membranes, as do fat-loving carotenoids. Phycobiliproteins exist inside of larger structures called phycobilisomes, attached to special thylakoid membranes.

Different microalga can have distinct arrangements of chlorophylls, carotenes, xanthophylls, and phycobilins to maximize their ability to utilize light. Thus, a broad "antennae" of accessory light-harvesting pigments in microalgae help to capture photons, which then funnel these photons towards chlorophyll molecules at the "reaction center." When chlorophyll pigment molecules achieve an excited state, the photosynthetic process becomes activated. A long chain of electron-transfer steps finally results in the transfer of electrons between water and carbon dioxide molecules, using light (photon) energy.

The complex arrangements of light-harvesting pigments in microalgae likely evolved to capture different light color ratios and irradiance levels found at different depths in the water. Endowed with diverse combinations of "photoreceptor" pigments, some cyanobacteria have an ability to respond to four different light colors: blue, green, red and far-red. Microalgae have a greater diversity of light-sensitive pigment molecules than most land plants.

Most of us become aware of accessory light-harvesting pigments chiefly during the fall harvest season, especially in certain climate zones. The beautiful fall colors of autumn leaves emerge (yellow, orange and red carotenoid pigments), after trees decrease production of green colored chlorophyll. For similar reasons, as fruit ripens these colorful carotenoid colors become more dominate, eventually giving off aromatic compounds, smells, results from carotenoid pigment breakdown.

In conclusion, a broad "antennae" of accessory light-harvesting pigments help to capture photons, then funnel these photons towards chlorophyll molecules at the "reaction center." When chlorophyll pigments molecules achieve an excited state, the photosynthetic process becomes activated. A long chain of electron-transfer steps, finally results in the transfer of electrons between water and carbon dioxide molecules. Water and carbon dioxide are used in photosynthesis, and sugar and oxygen is made, using light (photon) energy.

Phycocyanin: The "Blue" in Blue-Green Algae

Phycocyanin is from the Greek *phyco* meaning "algae" and *cyanin* is from the English word "cyan," which means a shade of blue-green. Phycocyanin (PC) is a light-harvesting pigment-protein complex from the phycobilin family, which includes allophycocyanin (APC) and phycoerythrocyanin (PEC).

"A fundamental difference between [*Aphanizomenon flos aquae*] *AFA* and *Spirulina* rests on the different structure of the phycobilisomes. As opposed to *Spirulina*, the phycobilisome of *AFA* Klamath does not contain the pigment allophycocyanin, but only the pigment c-phycocyanin bound to a structural component which is missing in *Spirulina*, namely phycoerythrocyanin (PEC). PEC is a photosynthetic pigment which as of today has been identified only in a limited number of cyanobacterial species. PEC has a chemical structure very similar to the [phycocyanin] PC . . . Nevertheless, while every monomer of PC binds 3 molecules of PCB, PEC possess the unique characteristic of binding two molecules of PCB . . . and one molecule of phycoviolobilin (PVB) . . . which is responsible of the purple color of the pigment."[6]

According to the Italian University scientists who discovered this key difference between *Spirulina* and *AFA*, "This absolutely is the first time that the phycobilisome of Klamath algae is defined as peculiarly constituted by the union of c-phycocyanin and phycoerithrocyanin, and this different qualitative structure of the phycobilisome of *AFA* Klamath algae adds a further decisive factor distinguishing *AFA* from *Spirulina*."[7] Notably distributors of *AFA* microalgae had an early and long tradition of creating light displays of dissolved microalgae powder in columns of water to illuminate the color differences between *AFA* algae with its distinctive color hue range of violet, indigo, blue, green, yellow, orange and red pigments compared to other kinds of microalgae at conferences and trade-shows.

Phycobiliproteins have fluorescent properties that are used in medical and biotechnological immunoassay kits. Allophycocyanin absorbs and emits at longer wavelengths than phycocyanin C or phycocyanin R. Phycocyanin of blue green algae is a light blue color, absorbing orange and red light, particularly near 620 nm (depending on which specific type it is), and emits fluorescence at about 650 nm (depending on which type it is). As will be further explained in additional chapters, the light-harvesting and biliprotein pigments of microalgae contain some of the greatest health benefits.

CHAPTER

5

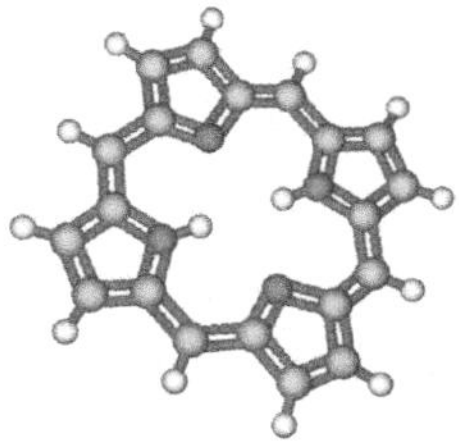

A Brief History of Heme Oxygenase (HO)

"The story of heme oxygenase takes us back to our beginnings. As oxygen gathered in the primordial nitrogen-containing atmosphere, our distant evolutionary ancestors were confronted simultaneously with a new opportunity and a new danger . . . the oxygen necessary to sustain multicellular life is inherently toxic. To make use of oxygen cells had to evolve mechanisms for surviving oxidative stress. One of the most ubiquitous such mechanisms is the enzyme heme oxygenase. From algae to humans, this enzyme is remarkably preserved; this preservation is an indication of its success as a strategy for protection and its importance for survival."[1]

— Morse D, and Choi A, 2005

"Chlorophylls are themselves the product of a long evolutionary development . . . The early part of the pathway is identical to heme biosynthesis in almost all steps and has clearly been recruited from that older pathway."[2]

— Blankenship R, 2010

This chapter will present a conceptual challenge for most readers. (Some readers may decide to skip this information). Fortunately after this chapter the remaining material is generally easier to understand.

However, the author believes an understanding of the cellular homeostatic relationships between cellular REDOX reactions, heme oxygenase pathways and phycobiliproteins (like the blue pigment in blue-green algae) helps to explain the numerous benefits of microalgae. While a complete understanding of heme oxygenase activity does not currently exist and is beyond the grasp of the author, this brief chapter will introduce some of the important concepts.

Light-absorbing heme proteins are found inside our cells in many different locations as cell pigments (cytochromes); visible in mitochondria, chloroplasts, and the endoplasmic reticulum. Cytochromes are capable of assisting REDOX reactions involving the exchange of electrons, which help to regulate cellular functions. REDOX reactions happen with reduction (the gain of electrons) and oxidation (the loss of electrons). Combustion reactions illustrate redox reactions whereby one molecule gains oxygen (is oxidized) and another molecule gains hydrogen (is reduced). Cellular respiration and photosynthesis are chief biological examples of redox reactions.

Heme oxygenase (HO) is regarded as a protective system for maintaining homeostasis. HO are dynamic sensors of cellular oxidative stress and modulators of redox balance. HO can act as cellular messengers.

Heme light-sensitive pigment proteins are central to photosynthesis (chloroplasts) and cellular respiration (hemoglobin). Heme is involved in both of these processes facilitating the exchange of energy, oxygen, hydrogen, CO2, iron, and bili pigment proteins. Scientists are discovering the unique ways that heme is exported from the mitochondrion to travel throughout organisms to do essential work. For example, heme protein groups interact with the nucleus, the fluids inside of the cell (via catalases), out into the extracellular compartments (via peroxidases) and are transported through our blood streams, in mammals.

The various biological mechanisms that create, transport and regulate heme at the intra- and extra-cellular levels are complex. In higher plants there is a tetrapyrrole molecule (called a protogen) that is shared between chloroplasts and mitochondria, which serves as precursor for both heme and chlorophyll.

Heme oxygenase (HO) enzymes catalyze the oxidation of heme to bili pigment proteins, carbon monoxide, and iron. HO enzymes perform a wide variety of roles in nature, from the recycling of hemoglobin to phycobilin biosynthesis in microalgae. HO is now increasingly regarded as a protective system for maintaining cellular homeostasis. HO are dynamic sensors of cellular oxidative stress and modulators of redox balance. HO and its various byproducts, like biliproteins and CO, are increasingly viewed as dynamic cellular messengers.

Heme oxygenase activity has been well documented in cyanobacteria, green algae, and red algae. The enzymatic properties of algal heme oxygenase are a little different from those of higher plants and somewhat different from those of animals. Heme oxygenase catabolizes free heme to yield iron (Fe^2), carbon monoxide, and biliverdin.[3] Yet, an analysis of the crystal structure of HO-1 from pea plants to rats "are found to be almost similar to that of human HO-1, but [some] difference also exists in structure due to evolution."[4] So universal are hemoproteins, that some scientists have hypothesized that the original evolutionary use of hemoproteins in cyanobacteria-like organisms' was for electron transfer, along primitive sulfur-based photosynthesis pathways, long before the appearance of molecular oxygen.[5]

Heme performs many different roles in biology across countless number of diverse organisms. Heme molecules are an "integral component of many proteins, so-called hemoproteins, in bacteria, plants and animals. It is the component of hemoglobin that gives blood its red color. [And chloroplasts their green color] Heme proteins display a wide range of biological functions; for example, hemoglobin and myoglobin are able to maintain a balanced supply of oxygen by functioning as oxygen transport and binding proteins."[6]

Bilirubin is to Hemoglobin as Phycocyanin is to Chlorophyll

The similarity between "bili" pigment production in microalgae and in mammalian systems is well documented.[8] Essentially what bilirubin is to blood heme, phycocyanin (blue pigment in microalgae) is to chlorophyll. Both of these bile pigment molecules converted from heme (cyanobacteria and human) and are produced during the carbon and oxygen exchange cycles. Blue-green algae synthesize phycocyanobilin and carbon monoxide (CO) at identical rates. Comparably, in mammals the only reactions known to produce CO likewise yields biliverdin.

McCarty (2007) first proposed that phycocyanorubin, a reduced form of phycocyanobilin, is a potent antioxidant species based on its molecular similarity to bilirubin. Remarkably human biliverdin reductase also reduces phycocyanobilin to phycocyanorubin, which contributes to the antioxidant effect after ingesting blue-green algae. Furthermore the scavenging of oxidative radicals revert phycocyanorubin to phycocyanobilin, which can then be re-oxidized to phycocyanorubin and have its reducing capacity regenerated. Human biliverdin and algal phycocyanin, linear tetrapyrrole molecules, both act as cellular protective molecules and powerful antioxidants. Mutually these bile pigment molecules have major health implications for cardiovascular, neurological, immune, and cellular wellness. Oral consumption of algal phycocyanobilins may well provide a therapeutic way to shift heme REDOX sensitive pathways.

Heme Molecules and By-products Support Life

In plants heme is made with the aid of chloroplasts. While, in animals heme synthesis is restricted to the mitochondria and then transported to other cells. For example, red blood cells are filled with hemoglobin, a protein which carries oxygen to tissues and carbon dioxide away from tissues. Hemoglobin carries oxygen by containing heme, a large ring-like molecule, which has at its center an atom of iron, which binds to oxygen.

In humans the iron in heme allows hemoglobin in red blood cells to carry the oxygen needed for life. "Humans survive by constantly recycling iron, a metal that is an essential component of red blood cells, but which is toxic outside of those cells. More than 90 percent of the iron in an adult human's 25 trillion life sustaining red blood cells is recycled from worn-out cells . . . Humans [are able] to recycle over 5 million red blood cells per second in our spleen and liver."[7]

Heme oxygenase (HO-1) is an enzyme which breaks down heme, the iron-containing oxygen-carrying part of red blood cells. Heme is both synthesized and degraded within an individual nucleated cell. Heme plays an essential role helping to support cellular homeostasis by sensing or using oxygen. Blood heme is so essential that little is ever wasted.

Figure 11. Heme Oxygenase Catalyzes Heme Recycling

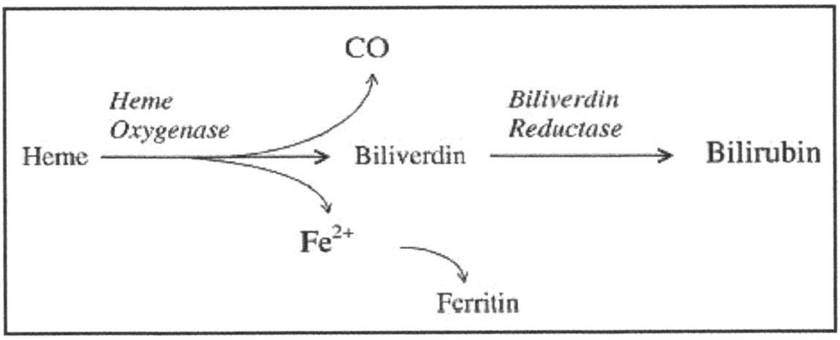

Heme oxygenase catalyzes the oxidative cleavage of heme to yield equimolar amounts of CO, iron (Fe^2), and biliverdin. Biliverdin is converted to bilirubin through the action of biliverdin reductase. Iron brings on ferritin synthesis

To better understand the function of heme, it is necessary to examine the products of its activity. When broken-down heme releases carbon monoxide, iron, and biliverdin, which is subsequently converted to bilirubin (see Figure 11). Heme degradation is one of the few natural sources of carbon monoxide in the human body. One molecule of carbon monoxide (CO) and one molecule of biliverdin occur for every molecule of heme degraded. Increasingly CO is recognized to play important roles in health, even as a neurotransmitter. The other major product of heme degradation is the release of iron (Ferritin). Each of these metabolic products plays various cellular protective roles or potentially damaging roles within the human body.

Heme oxygenase-1 (HO-1) is a potent anti-inflammatory, antioxidant and anticancer agent, providing protection against oxidative stress. HO-1 is increasingly viewed as a "general protective mechanism" that protects cells, tissues and organs from stressful insults, especially those involving oxidative or inflammatory factors.

Furthermore, accumulating studies over the past decade have shown that the use of HO end-products, when given at low concentrations, offer protective benefits for inflammatory stress or tissue injury. Cellular and tissue protection has also been reported for the use of bile pigments, biliverdin-IXa, and bilirubin-IXa, which represent the end products of the heme degradation pathway. Thus, an emerging consensus is the multiple benefits of HO-1 are likely due to the down-stream production and distribution of bioactive molecules resulting from the transformation of heme. Potentially similar health benefits might be achieved by the oral intake of cyanobacterial heme products, like phycocyanin.

The Different Colors of Tissue Injury Reflect Healing Responses

When cells are damaged colorful changes involving heme indicate different healing responses. "Heme oxygenase reactions can occur in virtually every cell; the classic example is the formation of a bruise, which goes through different colors as it gradually heals: red heme to green biliverdin to yellow bilirubin. HO-1 represents a highly evolutionarily conserved cellular defense system."[9]

Biliverdin is the pigment responsible for a greenish color sometimes seen in bruises. A similar linear tetrapyrolle bile molecule to biliverdin is phycocyanobilin, which is a blue-green colored pigment produced by blue- green algae.

The various colors that injured tissue undergoes during phases of recovery may mirror evolutionary adaptions that ancestral cells used to survive the change from an anoxic to oxygen-filled environment. As a bruise heals, it changes color until it disappears. Bruises typically change color in a progression from a reddish blush, to a gradually darkening blue or purple, followed by green, and finally a yellow-brown color. These colors relate to heme-related reactions, which help enable cells to stay alive, repair or might speed-up cellular death, when necessary. Depending on the level of cellular tissue damage and changes to blood and oxygen availability, different shifts among heme and bile light-pigment molecules occurs. For example tissues may appear a blue to purple color, "cyanosis," during acute hypoxia, lack of oxygen.

Bilins are derived from the closed tetrapyrrole ring of heme by an oxidative reaction catalyzed by heme oxygenase to yield their characteristic open chain. Biliverdin and bilirubin as well as algae phycocyanobilin help to protect cells against oxidative stress. Chlorophyll is associated with heme as well. Botanical chlorophyll carries a magnesium metal atom in the center of the ring, whereas human heme and biliverdin have iron (Fe), at their center. A similar tetrapyrrole molecular structure is shared across microalgae phycobiliproteins (e.g., phycocyanin), chlorophyll, biliverdin, and bilirubin.

Figure 12. Comparison of Similar Tetrapyrrole Structures

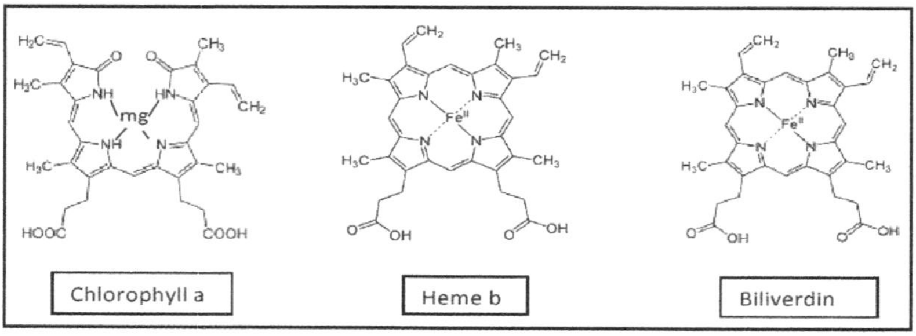

The green pigments (chlorophyll) in plants and algae, the human bile pigments (biliverdin, bilirubin), and the colorful pigments in blue-green algae (phycocyanobilin), and all share similar (tetrapyrrole ring) molecular forms

Heme b consists of a molecular ring similar to that of chlorophyll, expect for different metal ions at their centers. Heme is bright red, the pigment that give blood its characteristic red color. Chlorophyll's heme (in chloroplast) provides the green pigment color seen in plants. Also biliverdin is a green tetrapyrrole bile pigment, a product of heme. Heme breaks down into Biliverdin, which is further transformed into Bilirubin.

Biliverdin is the pigment responsible for a greenish color sometimes seen in bruises. A similar linear tetrapyrrole bile molecule to biliverdin is phycocyanobilin, which is the blue-green to violet colored pigment produced by blue-green algae.

Distinctively, all these "bili" like structures are shifted by light exposure, a fact exploited in the phototherapy of jaundiced newborns or the use of certain algal pigments for medical imaging. The center structure within biliproteins consists of a very stable network of alternating single and double bonds, a structure that readily conveys electrons, making them excellent photo receptors and light transducers.

Phytochromes are light-sensitive pigments that help shape algae and plant growth

Plants, algae, and cyanobacteria, organisms that rely on light as an energy source, possess multiple photo perception and signaling systems to better adapt to changing light environments. Phytochromes are light-sensitive pigment molecules that possess tetrapyrrole (bilin) prosthetic groups that enable them to absorb visible light. Through their ability to cause a reversible structural change by absorbing different light colors, "phytochromes function as molecular rheostats to regulate numerous responses to light quality, quantity, duration, and direction. Although the biochemical mechanism of phytochrome action is not fully defined, phytochromes are bilin-regulated light-modulated protein kinases that function to mediate gene expression."[10] For example, phytochromes help to signal when seeds germinate, the size, shape and number of leaves or even when plants flower.

Phytochromes, and potentially other phytochrome-related biliproteins, make up some of the most critical light-signaling systems of oxygenic photosynthetic organisms. The bilin prosthetic groups of phytochromes and phycobiliproteins arise from the oxygen-dependent ring opening of heme. In cyanobacteria, heme oxygenase (HO-1) catalyzes the conversion of heme to biliverdin IX (BV). Subsequently, BV is reduced to the chromophore phycocyanobilin (PCB) and ferredoxin oxidoreductase (PcyA). In cyanobacteria PCB is the immediate precursor of the chromophore of the phycobiliproteins, phycocyanin and allophycocyanin.[11] The diversity of light-harvesting pigments provide microalgae species a variety of unique colors.

Scientists have puzzled as to why organisms have devoted so much energy to the recycling of hemoproteins. "The spleen is required to remove a harmful compound such as heme from the circulation and to facilitate the recovery of iron from senescent red blood cells; hence the spleen expresses a high level of HO-1. However, it is not obvious why heme oxygenase should be present in other organs or in organisms without circulating blood — such as algae —where the need for hemoprotein turnover is less clear. Moreover, the ability of

HO-1 to be induced by stimuli other than free heme makes little sense if the sole object of activation is heme catabolism. Indeed, HO-1 is induced by a remarkable array of cellular insults and this is a key to understanding its function."[12]

Most importantly, HO-1 is part of a regulatory, homeostatic, system that attempts to achieve balance between the organisms and its environment. There is a complex set of relationships of REDOX feedback systems that involve mitochondria, nuclear DNA, heme and heat shock protiens, along with light-harvesting bili pigments. A complex set of relationships, which are essential to maintain optimal cellular energy balance.

Thus, our cells respond to oxidative stress and environmental insults may be shifted by the production or the utilization of photo-sensitive, heme-associated "bili" molecules. Shifts in biliprotein molecules may alter genetic expression. Scientists have reported that heat shock protein response can shift RNA methylation patterns, which influence epigenetic expression.[13] The word "epigenetics" refers to how "the development and maintenance of an organism is orchestrated by a set of chemical reactions that switch parts of the genome off and on at strategic times and locations."[14]

Color Pigments, DNA Methylation and Epigenetics

The word "epigenetics" refers to how "the development and maintenance of an organism is orchestrated by a set of chemical reactions that switch parts of the genome off and on at strategic times and locations.

Nearly a half century has passed since Watson and Crick first published their three-dimensional structure of the DNA double helix. The discovery that DNA encodes hereditary characteristics (like eye color) has captured the scientific and popular imagination. Yet "we are coming to realize that gene-centric theories of evolution are limited in their scope. The genetic blueprint, like a complex musical score, remains lifeless without an orchestra of cells (players) and epigenotypes (instruments) to express it . . . with some 30,000 genes in the human genome, the importance of silence, as with any orchestral performance, must not be underestimated. As cells develop, their fate is governed by the selective use and silencing of genes. This process is subject to epigenetic factors. DNA methylation patterns play a role in all sorts of phenomena where genes are switched on or off from the splash of purple on a petunia petal to the growth of cancerous tumors."[15]

Simply, evolution is first and foremost about our relationship to the sun and the various solar-ways living things gather energy. For instance, most plants grow towards the sun, but too much or too little sunlight will shape its phonotypical genetic expression (e.g., the shape or color of its leaves). Again, there is a complex set of relationships between REDOX states, influences of light sensitive pigment molecules, DNA methylation, and gene expression. Epigenetics explores how genetic codes can respond to changing environmental conditions. The flowering of a plant is a beautiful example of epigenetics at work, linked to colorful seasonal, solar, changes.

Color pigments are increasingly used as markers for epigenetic changes. For example, "Recent laboratory studies on inbred mice demonstrated how changes to their diet might influence their offspring. Their fur can be brown, yellow or mottled depending on how the agouti gene is methylated during embryonic growth. When pregnant mothers were fed methyl-rich supplements such as folic acid and vitamin B12, their young developed mainly brown fur. Most of the babies born to control mice (not given the supplements) had yellow fur."[16] Could eating colorful microalgae pigments change our genetic expression? As you read further in this book, you will find a resounding answer: yes. Edible plant and microalgae pigments need to be considered as essential food nutrients. Microalgae with their colorful, light-harvesting, bilin-pigments offer an excellent primer to study epigenetic mechanisms.

Heat Shock Proteins (HSPs)

Heme oxygenase-1 (HO-1) is one of the most common heat shock proteins found in humans and other mammals. Heat shock proteins

(HSPs) are well expressed under normal conditions, but rapidly increase in numbers when organism experience distress. The heat shock response, which involves HSPs, is a physiological, transient cellular response to injury, repair and defense. HSPs also known as "stress proteins" are part of a general stress response to a multitude of environmental changes; including sudden temperature changes, oxygen or water deprivation, starvation, exposure of cells to toxins, radiation, infection, fever, inflammation, and even exercise. HSPs are highly conserved throughout nature and found in virtually all living organisms, from bacteria to humans.

Heat Shock Proteins (HSPs) are proteins activated by cells in response to stressors such as radiation, toxins, infections, or fever. HSPs protect cells against danger posed by an immediate environmental threat and fortify cells to withstand future stresses, even from a different source.

The heat shock response not only protect cells against danger posed by an immediate environmental threat, but also fortify cells to withstand future stresses, even from a different source. HO-1 as one of the most universal heat shock proteins, part of a larger adaptive response to cellular stress. Learn how colorful heme associated pigment molecules indicate various states of inflammation as well as adaptive cellular responses to oxidative levels (see page 74). Moreover, some heat-shock proteins have the ability to bind to prokaryotic molecules that support immune memory and regulation.[17]

Microbiology scientists have observed a dynamic relationship that exists between certain HSPs along with bili- pigments like chlorophyll, carotenoids and components of phycobilisomes, including phycocyanins.[18] Several studies have suggested a link between the heat shock response and oxidative injury. For example, "Inhibition of NADPH oxidase by heat shock appeared related to the synthesis of heat shock proteins and may respresent an intrinsic cellular mechanism to down regulate superoxide production."[19] The pivitol role of NADPH oxidase and superoxide production in mitochondrial health and age-related diseases will become defined in later chapters.

While is still unknown how biliproteins like "phycocyanin exerts biological activities *in vivo* since proteins are generally broken down into individual amino acids or small oligopeptides before absorption. Yet, the phycocyanobilin chromophore structurally resembles the bile pigment bilirubin, which can be absorbed from any part of the small or large intestine as long as it remains unconjugated. This would suggest that the phycocyanobilin part of phycocyanin is mainly responsible for the antioxidant and anti-inflammatory effects seen after oral administration of phycocyanin."[20] However, bili- and heat shock proteins do not work in isolation, but interact within larger intra- and intercellular feedback networks, especially attuned to oxidative states.

Numerous health benefits are well documented by microalgae researchers (book part II chapters). Below Figure 13 lists major areas in researched health benefits of microalgae. One medical hypothesis is that the biliproteins pigments found within microalgae shift larger systemic mechanisms (e.g. heat shock response and REDOX states) to improve cellular signaling, repair and defense mechanisms.

Figure 13. Microalgae Health Benefits

Antioxidant/radical scavenging
Metallo (heavy metal) protection
Lipid peroxidation inhibitor
Hepatic (liver) protection
Anti-inflammation
Cox-2 inhibitor
Induces hematopoiesis (creation of new blood cells)
Anticancer
Immunomodulation
Antiviral/Antibacterial
Precursor of prostaglandin
Lowers Diabetic Complications
Stimulates stem cell production
Enhances brain function

Over the past decade, there is growing interest to identify natural and pharmacological agents with broad cellular protective properties associated with the heat shock response; including heme and microalgae (biliprotein, light-sensitive pigments) that are known to provide a wide array of health benefits. HSPs offer a wide range of functions that include assisting with the proper folding and unfolding of other proteins; acting as chaperone molecules, aiding transportation of proteins across membranes within the cell; and monitoring the cell for proteins degraded by stress or toxic insult, which may require immune system involvement or genetic repair. HSPs are in effect rapid epigenetic response molecules.

While there are a variety of different HSPs, heme oxygenase (HO-1) is one of the most widely studied. The similarity between bile pigment production in microalgae and in mammalian systems has already been mentioned. Both cyanobacteria and human bile molecules are converted from their heme sources. Human biliverdin and algal phycocyanin, both linear tetrapyrrole bile molecules, have major health implications for cardiovascular, neurological, immune, and adaptive cellular wellness and are associated with at least several HSPs.

Microalgae to the Rescue

Our planet has long shown that microalgae are essential for life, but the scientific community is only lately catching on. In the past two decades, hundreds of studies documenting the health benefits of microalgae have been published in peer-reviewed journals. This review is a compilation of many studies, most published. The primary health research on microalgae—presented in Part II of this book—may be categorized into 12 main areas, from antioxidant to safety concerns. How microalgae might contribute to planetary medicine, helping to restore or maintain the balance of our ecosystem is addressed in Part III.

Microalgae exhibit numerous therapeutic functions, including antioxidant, antibacterial, antiviral, anticancer, anti-inflammatory, antiallergic, and antidiabetic functions. Consuming certain microalgae appears to promote the growth of intestinal micro flora, as well. Increasingly, edible microalgae products are being considered important functional foods in the role of health care management. Additionally, microalgae compounds hold particular commercial interests in the food, cosmetic, agricultural, and horticultural industries and in environmental restoration, as well as in the health care industry.

While many of the studies mentioned here have been published in scientific journals, I have also included some as yet unpublished research results in cases typically where I've spoken to the researchers. In pursuit of the truth, a variety of research evidence is offered so that readers may arrive at their own conclusions. Still, it is important to recognize that even when we know something exists (e.g., gravity), we often lack a complete understanding of the causal mechanisms. For instance, Richard Smith, editor of the *British Medical Journal,* reported "only about 15% of medical interventions are supported by solid scientific evidence."[21]

PART

II

Algae to the Rescue: Microalgae and Human Health

CHAPTER

6

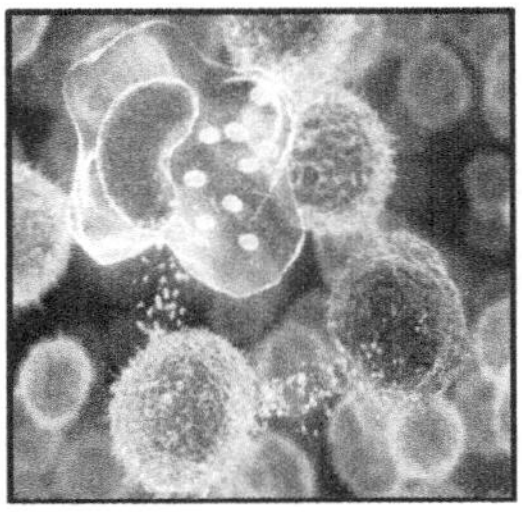

Oxygen, Antioxidants, and Inflammation

"Many of the elderly in the United States — and quite a few of the not-so elderly — experience terrible pain in their joints. Their fingers may become twisted and swollen, and they may be unable to button a coat without large doses of anti-inflammatory drugs . . . Many come to feel crippled and useless. By the age of 35, 35% of Americans have diagnosable arthritis in their knees. At least 85% of those over the age of 70 have it, and many have it severely."[1]

— Robbins J, 1987

"Inflammation is the cornerstone of the body's healing response. It is the process by which the immune system delivers more nourishment and more defensive activity to an area that is injured or under attack. But inflammation is so powerful and so potentially destructive that it must stay where it is supposed to be and end when it is supposed to end; otherwise it damages the body and causes disease."[2]

— Weil A, 2011

Microalgae Hypothesis I:
Anti-inflammatory and Antioxidant Effects

Chronic inflammation contributes to many health-related problems.

Microalgae offer a wide spectrum of potent antioxidants.

Evolutionary older antioxidants offer more protection than more recent plant or synthetic manufactured forms.

Microalgae antioxidants protect sensitive lipid membranes, DNA and amino acids from damage.

Algae oils reduce inflammatory arachidonic acid in blood and inhibit release of histamine.

Microalgae components can protect skin from UV rays.

Time Magazine (Feb. 23, 2004) ran a cover story titled, "The Secret Killer: The surprising link between inflammation and heart attacks, cancer, Alzheimer's and other disease." The inflammation-disease hypothesis proposes that inflammation happens as part of the body's innate response to stress. Stress occurs when our body, including our mind, reacts to potentially harmful inner or outer events (stimuli), which trigger our autonomic nervous system and release of stress hormones into our bloodstream.

Hans Selye (1907 – 1982) a pioneer of stress research, defined stress as "the nonspecific response of the body to any demand placed upon it." Nonspecific responses refer to our body's innate defense systems, which evolved over millions of years. Inflammation is a key mechanism within our body's stress response arsenal. For example, a wood sliver stuck into our big toe, elicits a variety of stress responses both locally at the site of the injury (e.g., inflammation) as well as larger physiological and emotional reactions (e.g., pain, changes in blood flow and even heart rate). Who is not familiar with experiencing inflammation when they experience a bruise, cut or illness?

Inflammation is the first response of our immune system to infection or tissue injury, attempting to protect the human body against a variety of insults. Chronic inflammation and stress often becomes far more deadly than short-term, acute, stress episodes. A negative spiral may occur, whereby parts of the body are experienced as "foreign invaders." Then inflammation is extremely difficult to stop. Consider someone who develops a bone spur and the formation itself provokes further inflammation, pain, immobility and further aggravated immune system attacks. Whenever a cluster of cells die, they automatically trigger an inflammatory response. Inflammation is our body's equivalent to setting off a burglar alarm. All sorts of physiological changes are initiated.

"An inflammatory response begins when cells of the immune system and/or cells involved in metabolic pathways sense pathogens, irritants and cellular damage, triggering the release of inflammatory substances, including cytokines, free radicals, hormones and other small molecules."[3] Also, prolonged metabolic influences such as lipid (fat) accumulation, elevated levels of glucose (sugar), excessive amounts of cytokines/hormones and even neurotransmitters can trigger inflammatory responses. Like a growing flame, the release of inflammatory factors further stimulates the cells that secreted them; thereby altering our cellular physiology, particularly immune, metabolic and nervous system pathways. All this impacts our ability to heal and protect ourselves.

Only recently have scientists begun to identify the role of inflammation to psychological stress and its association with greater risk for depression, heart disease, autoimmune illness and even infectious diseases. Researchers at Carnegie Mellon University showed "for the first time [that] the effects of psychological stress on the body's ability to regulate inflammation can promote the development and progression of disease." Likewise chronic psychological stress interferes with our body's ability to regulate the inflammatory response.[4] The same hormones and neuropeptides, such as catechol amines, glucocorticoids and cytokines mediate stress, mood, and inflammation.[5] Increasingly, researchers are linking stress to the activation of inflammatory responses, which can contribute to the development of depression or anxiety.[6]

Figure 14. Inflammation Connection to Disease States

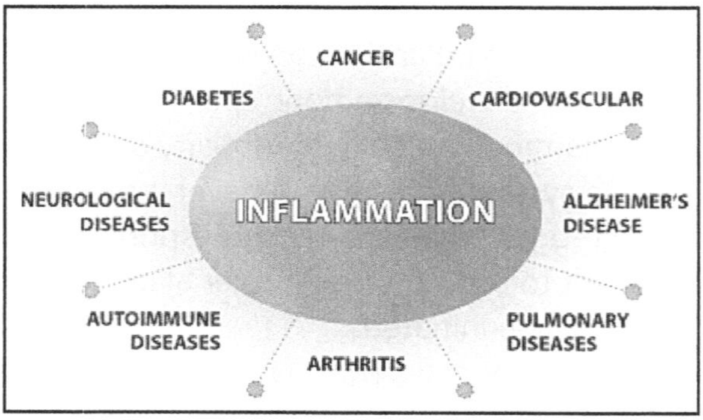

Substantial evidence suggests that cellular stress and excessive inflammation are causally linked to numerous illness conditions, such as arthritis, diabetes, chronic pain, cancer, high blood pressure, osteoporosis, heart disease, autoimmune diseases, neurodegenerative disease, digestion, and even skin problems. Chronic inflammation leads to (1) escalating immune system reactions; (2) impaired communication between the immune system and the brain; (3) amplification of pain perception; (4) delayed wound healing, with reduced stem cell production, along with accelerated aging; and (5) increased damage to lipids, proteins, and genetic material - all contributing to cellular, tissue, and organ dysfunction.

Beyond "Single-Bullet" Cures to Evolutionary Medicine

"Because many mediators of cellular stress and inflammation are regulated simultaneously, it is unlikely that a single response that integrates inflammation and stress signaling is responsible for the pathogenesis of a particular disease. Given this complexity, an effective approach would be to seek to re-establish functional homeostasis by modifying integrated biological outcomes rather than targeting single pathways."[7]

— Zhang, Kaufman, 2008

"Tetrapyrrole molecules [e.g., bilin pigments] are distributed in virtually all living organisms on Earth. In mammals, tetrapyrrole end products are closely linked to oxygen metabolism. Since increasingly complex trophic functional systems for using oxygen are considered in the post-traumatic inflammatory response, it can be suggested that tetrapyrrole molecules and, particularly their derived pigments, play a key role in modulating inflammation."[8]

The use of the sun's light energy by photosynthetic organisms offered the central blueprint for virtually all (cellular) life to exist on Earth. Photosynthesis efficiently converts light energy to electrochemical energy. Reduction-oxidation (redox) reactions are at the core of cellular biology, such as photosynthesis and cellular respiration. Photosynthesis involves the reduction of carbon dioxide into sugars and the oxidation of water into molecular oxygen. The reverse, but not identical, process of respiration oxidizes sugars to produce carbon dioxide and water. In animal cells, mitochondria perform similar functions yet are still dependent upon photosynthetic organisms to capture solar energy.

Photosynthetic pigments are categorized in three chemical groups: chlorophylls, carotenoids, and phycobilins. All three pigment groups are essential molecules of cyanobacteria, but not all or present in other green algae or land plants. These light-harvesting pigments not only capture sunlight for energy but impact other biological processes as well; from the function of DNA, influencing growth rates to cell death. Simply stated, without these essential pigments chlorophylls, carotenoids, and phycobilins (like C-phycocyanin), life as we know it could not exist on Earth. At a cellular level animals and plants start to look surprisingly similar. Though animal cells share a common ancestral architecture with plant cells, animals are ultimately dependent upon plants' photosynthetic pigments.

Colorful pigment molecules indicate various states of inflammation as well as adaptive cellular responses to oxidative levels. . . . the need of exploiting this color resource could be considered for both the diagnosis and treatment of the inflammation.

Photosynthetic pigments evolved to serve multiple purposes, to capture energy from the sun, and to protect cells from the damaging effects of oxygen, radiation, heat, toxic chemicals, and injury. Antioxidant pigments occur naturally in microalgae and plants. Antioxidants protect all life on earth - plants, animals and humans - against the damaging effects of living in an oxygenated environment. Over vast eons of time many living organisms coevolved and are now depend on each other for survival. So it happened that microalgae, fruit and vegetables, rich in antioxidants, are needed by humans and animals to preserve their health.

As such, we should start to consider photosynthetic pigments as essential nutrients. An essential nutrient is a nutrient required for healthy body functioning that cannot be synthesized by our body and must be obtained from a dietary source. Traditionally, vitamins, dietary minerals, essential fatty acids and essential amino acids have been reported. However, with the exception of bile, humans do not synthesize optimal amounts of photosynthetic pigments necessary for good health. The numerous health benefits of eating adequate amounts of fruits, vegetables and microalgae is connected to our need for photosynthetic pigments.

Colorful Pigments reflect Inflammation as well as Adaptive Cellular Responses to Oxidative Levels

According to Jose-Ignacio Arias et al. (2009) in their article, "Surgical inflammation: a pathophysiological rainbow," inflammatory processes reveals a color-coded healing pathway in response to injury, trauma, or infection.[9] Colorful pigment molecules indicate various states of inflammation as well as adaptive cellular responses to oxidative levels. For example, deeply injured tissues typically shift from colder colors (purple, blue, or greenish) to warmer colors (yellow, orange, or red). These color shifts represent the body's use of antioxidant, tetrapyrrole molecules (largely bilin proteins) to combat inflammation, initiate cellular repair, and their adaptive roles for changing oxygen tissue levels.

Jose-Ignacio Arias and his team proposed that tetrapyrrole molecules, such as heme, which "contribute a large variety of colors to the tissues, are employed through the evolutive process of acute inflammation" reflecting earlier phylogenetic cellular "adaption(s) to the metabolic and biochemical changes imposed by the development in different environments, from an unbreathable atmosphere to an environment fully enriched by oxygen." [10] They conclude, "Therefore, the need of exploiting this color resource could be considered for both the diagnosis and treatment of the inflammation."

For example, C-phycocyanin (Cpc) the blue pigment in cyanobacteria is the pigment most recognized as contributing to the many health benefits reported for blue-green algae. Cpc is used in a variety of industries for health food, antioxidant effects, natural colorings, and in medical fluorescent diagnostic usages. According to Danish scientists, "our understanding of the physiological functions of Cpc in humans has been improved by a mechanistic hypothesis that links the chemical properties of the phycocyanobilin chromophore of Cpc to the natural antioxidant, bilirubin, and may explain the observed health benefits of Cpc intake."[11]

Microalgae Offers Natural, Synergistic, Antioxidants

Scientists describe blue-green algae as having reached a sort of "evolutionary perfection," as some species show little evidence of mutation or genetic damage. Fossilized samples of cyanobacteria show that certain of these organisms have remained virtually unchanged for several billion years. The success of ancient microalgae depended upon their development of powerful antioxidant pigment shields. Four billion years ago Earth was subject to intense solar radiation. Before there was oxygen, there was no ozone layer and little atmospheric protection from the harmful rays of the sun. Therefore, antioxidant shields were vitally important to the protection of sensitive molecules— especially exposed nucleic acids and lipids—from the damaging effects of solar radiation.

The light-harvesting pigments and antioxidant defenses that protect against radiation also protect against oxygen toxicity and are the same kinds of biomolecules that protect against other types of physical challenges such as heat, infection, heavy metals or toxins. "In people, the genes activated by radiation also protect against oxygen poisoning, malaria or lead poisoning. The reason for this cross-protection is that many different physical stresses all funnel in to a single common damage-causing process in the cell, so all can be withstood through common protective mechanisms. This shared pathology is a rise in oxidative stress. Oxidative stress is defined as an imbalance between free-radical production and antioxidant protection."[12] This key understanding helps to explain why microalgae pigments can have beneficial effects across a wide range of health-related conditions.

> *Scientific studies have shown that the synergistic action of a wide spectrum of colorful antioxidant pigments is better than the activity of a single antioxidant. Natural sources of antioxidants have greater efficacy than synthetic antioxidants, and evolutionarily, ancestral, antioxidants appear more powerful than more recent phylogenetic forms.*

Primordial microalgae produced some of Earth's first and most effective antioxidants. The synergistic action of a wide spectrum of colorful antioxidant pigments and cofactors is better than the activity of a single antioxidant. Natural sources of antioxidants have greater efficacy than synthetic antioxidants; and evolutionary, ancestral, antioxidants are more powerful than recent phylogenetic antioxidant forms. Antioxidant and anti-inflammatory microalgae-based therapies may prove beneficial in the prevention and treatment of a wide range of illnesses, including various cancers, heart disease, diabetes, arthritis, depression, and premature aging, among others.

Ancient microalgae by producing oxygen and converting inorganic carbon and nitrogen into organic forms began the process of terraforming an inhospitable world into a life-sustaining world, a slow-motion process lasting hundreds of millions of years. So effectively did microalgae develop antioxidant and genetic survival codes that they provided a foundational blueprint copied by many higher species that came later.[13]

Within edible microalgae we have the full rainbow of antioxidant tetrapyrrole pigment molecules, including bilin proteins. Microalgae store the genetic evolutionary history of adapting from an anaerobic (oxygen deficient) to an oxygen-rich environment. Microalgae retain a rainbow spectrum of antioxidant pigments, which first captured and then harnessed the deadly rays of the sun, giving food to the world.

Prokaryotes, organisms without a nuclear membrane (e.g., blue-green algae), display a more diverse array of antioxidant pigments and a broader selection of carotenoids than terrestrial plants and most green algae. Scientists at the University of Wisconsin, Department of Food Microbiology, report that because of the remarkable health benefits of algal and microbial carotenes, there will be a substantial increase in the world-wide demand for a full range of this important family of antioxidants.[14]

Carotenoids make up one of the most widely distributed and structurally diverse classes of natural pigments, especially rich in microalgae, with important functions in photosynthesis and protection against oxidative damage. Along with bilin pigments, such as bilirubin in humans and phycocyanobilin in algae, light-harvesting pigment complexes help to protect cells against oxidative stress.

Natural Healthy Sunscreen Needed

For years, the media have reported on the dangers of sun exposure. A recent analysis of U.S. Medicare claims found that treatment performed for nonmelanoma skin cancer, most likely caused by UV ray exposure from the sun, nearly doubled from 1994 to 2004. "While the American Cancer Society estimates more than 2 million new skin cancers will be diagnosed this year, our research shows that the annual incidence in 2008 could actually have been 3.7 million," said Dr. Coldiron (2011), who presented his findings at the American Academy of Dermatology.[15]

Exposure to the shortest wavelengths in sunlight, i.e., ultraviolet light, constitutes a harmful environmental factor for most living creatures, from microorganisms to humans. The sun emits three forms of light: infrared (heat), visible light, and ultraviolet (UV) light. The UV light radiation is mostly responsible for skin damage. Prolonged exposure to UV rays can damage and kill skin cells, which then release chemicals that trigger the body's pain receptors. The reddening of sunburned skin is the result of increased blood flow; an attempt to remove dead skin cells and help heal the damaged skin areas. Energy from UV light also promotes the production of melanin, a pigment that causes the skin to darken, or tan.

> *Photoprotective pigment molecules found in various microorganisms such as cyanobacteria can be used as sunscreens, either taken internally or used topically.*

The conventional wisdom is to rely upon sunscreen to protect against UV exposure. However, sunscreen may not be so safe after all. Growing concerns over certain ingredients in today's sunscreens has arisen. Joseph Laszlo, Ph.D. works with the U.S. Department of Agriculture's Agricultural Research Service. He contends that sunscreens, a multi-billion dollar consumer driven product, are among substances that make up a relatively new family of water contaminants with potential adverse health effects on wildlife and people. "One, for instance, is a substance known as oxybenzone that is a suspected hormone disruptor that could contribute to the disruption of aquatic species reproduction."[16] Dr. Laszlo advocates the development of "nature-inspired skin-care materials that avoid such health concerns and at the same time have fewer adverse environmental impacts."

Likewise, cell toxicity studies by Dr. Yinfa Ma, Professor of chemistry at Missouri S&T, and his graduate student Qingbo Yang, suggest "that when exposed to sunlight, zinc oxide, a common ingredient in sunscreens, undergoes a chemical reaction that may release unstable molecules known as free radicals. Free radicals seek to bond with other molecules, but in the process, they can damage cells or the DNA contained within those cells. This in turn could increase the risk of skin cancer. Ma also found that the longer zinc oxide is exposed to sunlight, the greater the potential damage to human cells."[17]

Dr. Ma discovered that zinc oxide, an ingredient used in sunscreen, actually absorbs the UV rays. As a result, zinc oxide then releases electrons, which in turn may produce unstable free radical molecules in the zinc oxide solution. "Those free radical molecules then bond with other molecules and act as parasites, damaging the other molecules in the process." In his studies, the cells covered by zinc oxide and then exposed to UVA (ultraviolet A) were significantly worse off (dead or injured) than cells exposed to UVA with no zinc oxide solution, at all. Dr. Ma's research on zinc oxide's effects on cells is still in the early stages. Despite his findings thus far, he still advises sunbathers to use sunscreen and to limit their exposure to the sun. Instead of zinc oxide and petroleum-derived ingredients, might healthier and more effective sunblock pigments, made from microalgae, become available in the future?

Photo protective pigment molecules found in various microorganisms such as cyanobacteria can be used as sunscreens, either taken internally or used topically. UV light causes chemical changes in the body, which can suppress the body's immune system and generate toxic substances (known as free radicals) that stimulate the growth of cancer cells. Free radicals, which attack and damage our DNA, are believed to be one of the factors that cause skin cancer. Certain pigments from microalgae can absorb UV-A radiation. One such pigment, scytonemin, found in cyanobacteria sheaths can act as an effective UV protectant. Results from several studies found that the microalgae pigment scytonemin has multiple protective roles, including acting as an effective UV sunscreen and antioxidant protector, and could become incorporated into sun-protective lotions.[18-19]

Another remarkable microalgae reddish colored pigment, astaxanthin, "acts like a sponge absorbing UV rays. It also reduces pain and inflammation from sunburn. It's a powerful antioxidant, more than 500 times stronger than vitamin E and 10 times stronger than vitamin A."[20] Reportedly, patients who already had skin cancer saw a marked improvement after taking astaxanthin supplement.

> *A microalgae reddish colored pigment, Astaxanthin, "acts like a sponge absorbing UV rays. It also reduces pain and inflammation from sunburn. It's a powerful antioxidant, more than 500 times stronger than vitamin E and 10 times stronger than vitamin A."*

Skin pigmentation changes are an essential protective mechanism against oxidative stress or UV irradiation. However, abnormal hyperpigmentation in human skin may pose aesthetic as well as skin cancer problems. Blue-green algae have been reported to reduce hyperpigmentation. Taiwanese scientists investigated the effectiveness of C-phycocyanin (Cpc) extracted from *Spirulina* on melanin synthesis in a type of melanoma cell. Cpc significantly reduced melanin production in a dose-dependent manner. Microscopic analysis demonstrated that Cpc successfully entered melanoma cells. Furthermore, "Cps negatively regulated tyrosinase gene expression resulting in the suppression of melanin synthesis."[21]

Additional bioactive compounds that may offer protection to sun damage on the skin are algal-derived sulfated polysaccharides derived from red microalgae, which possess anti-inflammatory properties. Researchers at the Estee Lauder Companies have been searching to find natural compounds that decrease human skin inflammation. For example, the pain people commonly experience after a bad sun-burn is inflammation of the skin. "Human subjects were recruited to participate in a study in which the polysaccharide material was applied topically and shown to inhibit cutaneous erythema induced by a known irritant . . . The data obtained strongly suggest that sulfated polysaccharides derived from red microalgae have significant beneficial potential for use in topical products. In addition, the data suggested that the anti-inflammatory mechanism for the polysaccharide was, at least in part, due to inhibition of circulating immune cell recruitment toward inflammatory stimuli."[22]

Oxidative Damage Following Recovery from Surgery

Not only skin on the outside of the body but skin inside our body experiences the harmful effects of inflammation. Often these undesirable inflammatory effects occur following surgical operations. For example, restenosis is the name given to new blockages which occur in an artery, after a patient has been treated with angioplasty or "stenting." Restenosis literally means the recurrence of stenosis, a narrowing of a blood vessel, which leads to restricted blood flow. Post-operative, balloon- or stent-induced, injuries are a major drawback for managing atherosclerotic cardiovascular diseases using invasive vascular procedures, like stenting. These injuries elicit an overactive healing response and oxidative stress to the blood vessel wall. "The result is formation of a neo-intima over the site of injury. An exuberant healing response leads to intimal hyperplasia (thickening) that encroaches on the vessel lumen and causes stenosis. This process is analogous to keloid formation in epithelial wounds. It seems to be accelerated by the presence of prosthetic material in the vessel." It has been called "the bane of endovascular intervention and vascular surgery."[23]

The AFA extract was able to effectively slow down the rupturing of red blood cells and reduce oxidative stress in a dose-dependent manner. Also, the RBC loss of glutathione, the most important naturally occurring antioxidant, was significantly delayed when AFA extract was used to combat oxidative toxins.

Taiwanese medical researchers studied the preventative benefits of *Dunaliella salina* (EDS), a green micro-algae extract, on balloon catheter injury-induced neointimal formation. First the scientists demonstrated that an extract of the green microalgae was able to "markedly reduced the human aortic smooth muscle cell proliferation," a primary mechanism, of neointimal formation, in restenosis. The microalgae extract inhibited aortic smooth cell proliferation in a dose-dependent manner. The scientists were able to determine that "the apoptotic pathway may play an important role in the regulatory effects of EDS on cell growth.

These observations provide a mechanism of EDS in attenuating cell proliferation, thus as a potential intervention for restenosis."[24] An additional protective factor is microalgae's well-established antioxidant and anti-inflammatory effects, which likely reduce the risk of a stent implant failure as well.

Microalgae Antioxidant Protective Properties

The powerful antioxidant properties of microalgae, packed with free-radical-quenching molecules, have drawn the attention of numerous researchers. Early studies (Manoj et al.; 1992) reported an alcohol extract of a blue-green algae (*Spirulina*) inhibited lipid peroxidation significantly (65% inhibition) more than: alpha-tocopherol (35% inhibition); BHA, widely used antioxidant preservative (45% inhibition) or beta-carotene (48% inhibition). A water extract of *Spirulina* was shown to have greater antioxidant effects (76%) than gallic acid (56%). Gallic acid is the antioxidant property found in foods like blueberries, walnuts, and watercress. Also, blue-green algae's antioxidant effects were superior to those of cholorogenic acid (56%), major antioxidant compounds found in green tea and coffee.[25]

Benedetti and colleagues (2004), a team of Italian scientists, researched *Aphanizomenon flos-aquae* (*AFA*), a microalga rich in C-phycocyanin (Cpc) and phycocyanobilin (PCB). Cpc and PCB are pigments within the *AFA* that have strong antioxidant activity. Cpc had previously been tested in vivo and in vitro for its antioxidant activity, and was found to scavenge all the free-radical forms tested, with preference towards peroxyl radicals.[26] Still the Italian investigators decided to induce oxidative stress in red blood cells, with toxic chemicals, even up to the point of rupturing the red blood cells. The *AFA* extract was able to effectively slow down the rupturing of red blood cells and reduce oxidative stress in a dose-dependent manner. Also, the RBC loss of glutathione, the most important naturally occurring antioxidant, was significantly delayed when *AFA* extract was used to combat oxidative chemical toxins. Dramatically, the blue color from Cpc gradually faded, as the antioxidant donated its electrons to the free radical so the RBC could remain undamaged. Antioxidants work to protect living cells against oxidation by becoming oxidized themselves.[27]

Despite these results the Italian scientists continued their investigation. Cpc and PCB derived from *AFA* microalgae had never been tested using the oxygen radical absorbance capacity (ORAC) assay. The ORAC assay is a test that specifically measures the capacity of a substance to scavenge free radicals. A free radical is any atom or molecule that has an unpaired electron in its outer shell. The formation of these radicals is known as oxidation. Free radicals are constantly being formed as part of normal metabolism and they play crucial roles in our immune response. But because free radicals are inherently unstable, they can cause chain reactions that can lead to cell damage or death.

Antioxidants help by neutralizing free radicals and preventing more oxidation reactions. The ORAC assay is used to test the ability of an antioxidant to neutralize free radicals and thus make the molecules stable again.

The Cpc and PCB derived from *AFA* was tested against other well-known, potent, antioxidants such as Trolox (a vitamin E analog), ascorbic acid (AA = vitamin C), catechins (antioxidant found in green tea, red wine, and chocolate), and reduced glutathione (considered one of our bodies' master antioxidants). The researchers discovered that Cpc and PCB had considerably more, a higher, ORAC value than any antioxidant tested. Cpc was three times more powerful than catechins, a well-regarded antioxidant. Finally, "PCB was able to reduce the steady-state concentrations of the peroxyl radical by one-half, indicating a high antioxidant activity for this compound."[28]

Figure 15. Linear Regression of ORAC Values

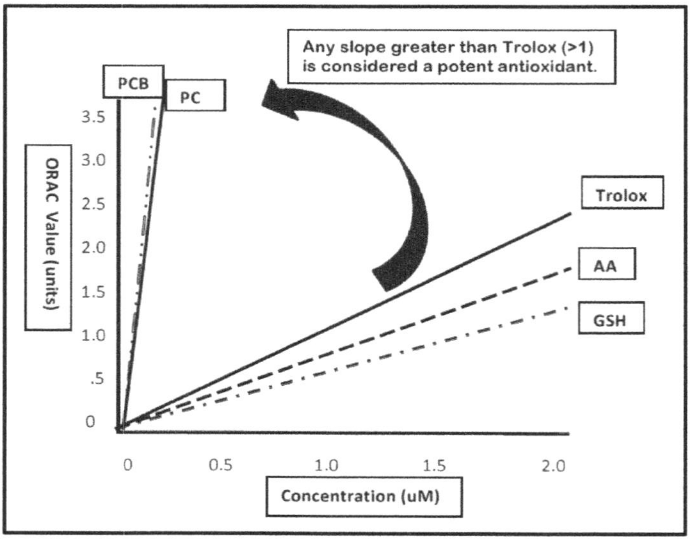

Graph adapted from: Benedetti S, et al. Oxygen Radical Absorbance Capacity of Phycocyanin and Phycocyanobilin from the Food Supplement *Aphanizomenon flos-aquae. Journal of Medicinal Food.* 2010; 13 (1): 223-227

Trolox a water-soluble vitamin E analog is considered the reference for antioxidants to determine ORAC value. Trolox has an ORAC value of 1.0. Anything greater than 1.0 is considered a potent antioxidant. Reduce glutathione (GSH) our body's chief antioxidants exhibited an ORAC value of .57. Vitamin C, ascorbic acid (AA), a widely regarded antioxidant had an ORAC value of .75. *AFA* microalgae pigments, phycocyanin (PC) and phycocyanobilin (PCB) had remarkable ORAC values of 20.33 and 22.18, respectively! The antioxidant activity of PC, which is a blue-green colored light-harvesting pigment, was nearly the same as the PCB bilin complex, which contains PC. The antioxidant activity of PC was significantly higher than other well-known antioxidant molecules, such as ascorbic acid.

Based on their results, the Italian scientists concluded that, "Altogether, these data further sustain the use of *A. flos-aquae*-based dietary supplements as helpful antioxidants in the treatment of clinical conditions related to oxidative stress . . . confirming the digestion and absorption of the protein per orally and making its chromophore PCB a functional compound that can prevent free radical-derived damage to biomolecules."[29] In other words, by eating *AFA* one can decrease oxidative damage due to free radicals produced by our body or accumulated from our external environment.

Overcoming the Dark Side of Pain Relief Medications

Oxidative stress can be a deadly side-effect of pain relief drugs, such as acetaminophen (Tylenol). Liver damage resulting from overdoses, chronic use, or combined use with alcohol or other drugs. Acetaminophen-induced liver injury (AILI) is the second leading cause of surgical liver transplants. Acetaminophen-related overdoses accounts for more than 56,000 emergency room visits, 2,600 hospitalizations, and an estimated 458 deaths due to acute liver failure each year in the United States.[30] Excessive amounts of Tylenol can deplete intracellular glutathione, resulting in a significant increase of reactive oxygen species (ROS). ROS are small molecules that are highly reactive as a result of the presence of unpaired electrons. ROS are important mediators of inflammation.

Kuriakose and Kurup (2010) compared experimental animals given an excessive dose of Tylenol, plus an extract of A. flos-aquae (*AFA*) versus control animals (given Tylenol only).[31] Concentrations of all animals' total lipids, phospholipids, triglycerides and cholesterol were measured. Also, a variety of endogenous antioxidants; superoxide dismutase (SOD), catalase (CAT), reduced glutathione (GSH), glutathione peroxidase (GPX), and glutathione transferase (GST) were assayed in all groups. Moreover, liver sections from each group were evaluated microscopically. The rats given Tylenol plus the *AFA* extract showed minimal liver necrosis (dead cells). *AFA*-treated rats had a significant reduction in the serum levels of enzyme markers of liver damage, along with evidence of liver cell regeneration. The *AFA* extract effectively prevented Tylenol-induced lipid peroxidation in liver cells, and glutathione (GSH) levels in the liver subsequently returned to near normal in the *AFA* treatment group. Pretreatment of experimental animals with *AFA* extract maintained membrane fluidity and resulted in a decrease in the cholesterol to phospholipid ratio, which were significantly elevated in the rats given with Tylenol alone.

Impressively, the animals' antioxidant defenses against free radicals, including all antioxidant enzymes studied (SOD, CAT, GPX and GST), were normal in the *AFA*-pretreated animals but were significantly reduced in the Tylenol-intoxicated animals. Overall, the *AFA* extract effectively prevented Tylenol-induced liver damage, protected antioxidants stores, maintained normal lipid and amino acid amounts.

The Orange Protective Pigment—Carotenoids

Carotenoids are pigments (e.g., orange color in carrots), which evolved alongside chlorophyll and work together with other light-harvesting pigments to capture the sun's rays and harness energy. Carotenoids represent a large family of biomolecules. Fucoxanthin, a carotenoid pigment found in some macro- and microalgae, has been "reported to exhibit various beneficial biological activities such as antioxidant, anticancer, anti-inflammatory, antiobesity, and neuroprotective activities."[32-33]

> *Natural algal betacarotene appears superior to synthetic betacarotene supplement in raising lipophilic antioxidants in human serum, with greater antioxidant potency compared to synthetic forms.*

Numerous studies have demonstrated that carotenoids protect against oxidative stress (e.g., protecting against lipid peroxidation), either through decreasing the production of free-radical derivatives or through the antioxidant activity of the protective compounds themselves. Rats and chickens fed a natural algal form of beta carotene showed at least a ten-fold higher accumulation of overall beta carotene in their livers than control animals fed equivalent amounts of synthetic all-trans beta carotene supplement. The higher accumulation of the natural algal carotenoids, over the synthetic isolated beta carotene, likely indicates a greater therapeutic value.[34]

Researchers have reported that natural algal beta carotene is superior to a synthetic beta carotene supplement in terms of raising lipophilic antioxidants (protecting PUFAs) in human serum.[35] What is more natural algal extracts of 9-cis beta carotene are shown to have a higher antioxidant potency compared to synthetic all-trans beta carotene with *in vitro* experiments.[36]

Antioxidants are biomolecules that protect organisms from the damaging effects of reactive oxygen species (free radicals) that are constantly formed in biological systems. Free radicals are molecules that lack one electron—a highly energetic particle—in what is usually a pair. To stabilize, free radicals randomly grab electrons from normal (i.e., healthy) molecules. This creates new free radicals, and like a row of falling dominoes, the damaging free radical cascade continues to spread. Free radicals are by-products of normal metabolism and can form as a result of exposure to radiation and chemical pollutants. Because they are highly reactive, they can damage cellular components, a particular concern with brain, vascular, and connective tissues and nucleic acids. They have been implicated in a variety of diseases and even in the acceleration of the aging process itself.

As life evolved and shifted from an anaerobic to an oxygen-rich atmosphere, multi-cellular organisms developed more sophisticated ways of handling oxygen molecules. Then more complex life forms, started to use various toxic oxygen molecules (free radicals) as weapons against the more primitive pathogens. Yet, the more sophisticated host species also needed to be able to defend themselves from these toxic oxygen species. Thus antioxidant defenses out of necessity were developed. Especially our immune system developed sophisticated cells capable of handling oxygen. For example, leukocytes can generate and launch toxic free-radical attacks against primitive pathogens. Yet leukocytes require their own effective antioxidant defense system to protect themselves.

Reactive Oxygen Species and Mitochondrial Health

Reactive oxygen species (ROS) are oxidants that, if left unchecked, can cause oxidative damage to the mitochondria, cellular proteins, lipids, and nucleic acids. ROS are the normal byproducts of cellular metabolism in the mitochondrion and may serve an essential component in cellular signaling (Redox Signaling Molecules). However, free radicals are a chemical species with a single unpaired electron, which is highly reactive. The majority of free radicals that damage our cells are oxygen radicals, which under normal circumstances our natural antioxidant defense systems detoxify. However, with aging and particular disease conditions, there is a significant decline in the function of our normal antioxidant defense system and therefore an increasing amount of cellular and organ damage.

While many antioxidants do not appear to operate at the level of mitochondria, several studies have documented the effectiveness of blue-green algae components to change superoxide radicals, which are primarily produced by mitochondria. German scientists investigated the use of blue-green algae as an "antioxidant and anti-inflammatory agent" and its effect upon neutrophils. Neutrophils are the most abundant type of white blood cell in mammals and form an essential part of the innate, nonspecific, immune system. "The results showed a dose-dependent inactivation of free superoxide radicals (antioxidant effect) as well as an anti-inflammatory effect characterized by a dose-dependent reduction of the metabolic activity of functional neutrophils and a dose-dependent inactivation of superoxide radicals generated during an oxidative burst. The results demonstrate that the tested *Spirulina* preparations have a high antioxidant and anti-inflammatory potential."[37]

Researchers at Oregon State University (OSU) in Corvallis are starting to recognize that pigment particles in blue-green algae may have an antioxidant effect in humans, preventing unstable compounds from damaging healthy cells. Balz Frei, director of OSU's Linus Pauling Institute, speculates that perhaps algae might one day be commonly eaten to help lower the risk of cancer, perhaps becoming as popular as taking vitamin C.[38]

Pigments, phytochemicals, vitamins, and trace elements from algae and higher plants can help boost the human body's antioxidant defenses.[39-40] *AFA* has an unusually wide variety of antioxidants, such as tocopherols, C-phycocyanin, carotenoids, zeaxanthin, polyphenols, flavonoids, superoxide dismutase, glutathione, polysaccharides, taurine, tryptophan, phenolic acid, and vitamins C, E, B5, and B2. Scientific studies have shown that the synergistic action of a wide spectrum of antioxidants is better than the activity of a single antioxidant. Furthermore, antioxidants from natural sources have a higher bioavailability and therefore higher protective efficacy than synthetic antioxidants.

When experimental animals were given blue-green algae (*Spirulina*) at two different dosages and then tested for levels of endogenous antioxidant enzymes, specifically superoxide dismutase, catalase, glutathione reductase, glutathione peroxidase, and reduced glutathione, researchers found that the levels were "increased significantly by both chosen doses of *Spirulina*."[41]

Blue-green algae contain a wide range of antioxidants in the form of specific trace minerals, amino acids, vitamins, and especially pigment—an impressive variety of carotenes along with potent green and blue pigments. Depending on the source of blue-green algae, the C-phycocyanin can account for up to 15% of its dry weight. C-phycocyanin a biliprotein has well-established antioxidant, anti-inflammatory and neuroprotective properties.[42-46]

C-phycocyanin—the Blue-Green in Cyanobacteria

A major antioxidant found in blue-green algae, C-phycocyanin is a type of phycobiliprotein. "Phycobiliproteins are a small group of highly conserved chromo proteins that constitute the phycobilisome, a macromolecular protein complex whose main function is to serve as a light harvesting complex for the photosynthetic apparatus of cyanobacteria and eukaryotic groups. The most common classes of phycobiliproteins are allophycocyanin, phycocyanin (Pcp), and phycoerythrin, all of which . . . carry different isomeric linear tetrapyrrole prosthetic groups (bilin chromophore) that differ in the arrangement of their double bonds."[47]

"The chemical structure of the bilin chromophores in phycocyanin is very similar to bilirubin, a heme degradative product." The authors note that bilirubin is considered to be a "physiologically important antioxidant against reactive species" and that it has been shown to protect serum albumin and other biological targets of free radicals by its action as a scavenger of oxygen radicals. They draw a parallel between bilirubin and phycocyanin, noting that phycocyanin inhibits the generation of reactive oxygen species as well as scavenging them "in a variety of test systems."[48]

Replicated studies with a range of experimental animal models have established the potent antioxidant and anti-inflammatory effects of C-phycocyanin. In rodents, experimentally induced colitis as well as edemas of the paw and ear all responded positively to C-phycocyanin.[49-50] C-phycocyanin extracted from microalgae has been demonstrated to effectively scavenge free radicals resulting from the administration of various chemicals and their metabolites toxic to the liver, kidney, and brain.[51] In studies showing that C-phycocyanin is a potent peroxyl radical scavenger, researchers found that C-phycocyanin proved to be almost 16 times more efficient as the antioxidant Trolox (a vitamin E analog) and nearly 20 times more efficient than ascorbic acid (vitamin C).[52]

"Bi-indoles isolated from a number of blue-green algae have anti-inflammatory and antiallergenic properties."[53] Additionally, a chlorophyll-related compound, pheophytin, derived from an edible green alga, has been shown to have potent anti-inflammatory effects in both *in vitro* and *in vivo* experiments. Both human and mouse immune cells showed an inhibited response to experimentally induced inflammation. Pheophytin has exhibited a significant suppression against edema formation in a mouse ear that was inflamed by toxic means.[54]

Gitte Jensen, an immunologist from McGill University, and her team at the Royal Victoria Hospital in Montreal reported that *AFA* algae may help to inhibit and to reverse inflammatory conditions. The researchers observed that small dilutions of *AFA* algae tend to dampen the release of reactive oxygen species from certain phagocytic cells in human blood.[55]

Chorella vulgaris, a green edible microalgae, has been found to exhibit antioxidant activity by suppressing inflammatory mediators and reducing the effects of oxidative stress.[56-57] However, in one study where the antioxidant effectiveness of blue-green algae (*Spirulina*) was compared to that of the phylogenetically younger, green algae (*Chorella*); the antioxidant activity of the blue green alga was superior to that of the green alga. The total phenolic content of *Spirulina* was almost five times greater than that of *Chlorella*, and a similar trend was evident in the ascorbic acid amounts.[58] Dietary phenolic contents of red wines, teas, and blue-berries are linked to their antioxidant properties.

Blue-Green Algae Treatment for Chronic Periodontitis

Blue green algae are associated with a wide range of antioxidant and anti-inflammatory health benefits. Periodontitis, gum disease, is an inflammatory condition induced by bacterial biofilms that accumulate in under the gums (gingiva). The gingiva, gums, are part of the soft tissue lining of the mouth, which surround each tooth and creates a seal around them. Our gums when health provides an effective barrier against periodontal insults to deeper tissue. However, when there is an accumulation of bacterial plaque, our gums have an increased tendency to bleed, which suggests a process of inflammation. When the gingival tissue is unhealthy, it provides a pathway for periodontal disease, which can eventually lead to loss of teeth.

Forty subjects were recruited from a sample of patients at the Department of Periodontology, at a Meenakshi Ammal Dental College in Chennai, India. The subjects were randomly assigned to two groups, a control and experimental group. Over thirty dental sites (in the mandibular anterior region were treated in both groups. The experimental group received Scaling and Root Planing (SRP) along with sub-gingival application of a *Spirulina* gel to periodontal pockets greater than 5 mm. The control group of patients only received the

Scaling and Root Planing (SRP) treatment. Clinical measures of Probing Pocket Depth (PPD) and Clinical Attachment Level (CAL) were assessed at baseline before treatments and then again at 120th day after the treatment protocol.

The dental researchers found that a significant reduction in both periodontal pockets, along with increased evidence of clinical attachment levels was significantly improved in the experimental treatment group with blue-green algae treatment. The resulting decrease in probing depth and gain in CAL are a major clinical signs used to assess for success in any periodontal treatment program. The researchers noted that, "The soft tissue walls changes caused by resolution of gingival inflammation explained the improvement in clinical parameters in this study . . . On comparing the post-treatment values of experimental and control groups with regards to PPD and CAL, it was found that there was greater improvement in the clinical parameters in the group exposed to *Spirulina* gel as compared to control group."[59]

Mud Baths and Microalgae—A Blessed Union

Bathing in Iceland's famous geothermal Blue Lagoon, has been known for many years to be beneficial for human skin in general and for patients with psoriasis and atopic dermatitis in particular. Yet, the scientific rationale for this observation remained elusive. German researchers carefully examined extracts prepared from the silica-rich mud, along with two microalgae species common to the Blue Lagoon. The algae demonstrated the ability to support beneficial gene expressions in skin cells, such as increased collagen, along with inhibiting UVA radiation-induced changes. The researchers concluded that, "In aggregate, these results suggest that the bioactives [of algae and mud] in Blue Lagoon have the capacity to improve skin barrier function and to prevent premature skin ageing."[60]

Scientists at the University of Padova Italy found that diatoms, golden-brown unicellular algae, produce anti-inflammatory chemicals that are the main active ingredients in mud-pack treatments. In European health spas, the use of mud-packs for the treatment of rheumatic and osteoarthritic patients has a long and relatively successful history. The maturation of thermal mud is dependent upon the full colonization of the mud by thermophilic microorganisms, with diatoms producing anti-inflammatory sulfoglycolipids (SGL), similar to those in blue-green algae. A typical cycle of treatments requires 12 packs of thermal mud.

"On this basis we calculated the amount of SGL taken up by each patient in a cycle of treatments, and found a figure not far from the recommended dose of nonsteroid anti-inflammatory drugs utilized for the same pathology. However, unlike pharmaceutical preparations, the amount of SGL taken up by the patients after the mud-packs does not exert any adverse gastrointestinal effect on these patients," reported the scientists.[61]

Additionally, the anti-inflammatory action of SGL is consistent with the decrease of serum interleukin-1 observed in arthritic patients treated with mud-packs.[62]

Microalgae Can Help Regulate Cytokines

Cytokines are immunoregulatory molecules secreted by cells, and they have various different effects. One type, the interleukins, that was significantly reduced in the microalgae mud-pack study controls inflammation and produces fever. As a result of biotechnology and advances in production of interferon from synthetic sources, there has been a widespread increase in interferon therapy for fighting several forms of cancer, chronic viral hepatitis, and multiple sclerosis. However, a commonly reported side effect of interferon therapy is severe depression. Dr. Andrew Weil has written about the "depression-inflammation connection." He believes the connection is linked to cytokines. Dr. Weil writes, "Cytokines are the principal chemical mediators of the inflammatory response. Anything you can do to keep them within their proper bounds will reduce your risks of chronic disease and also, it now appears, protect you from depression."[63]

According to Dr. Weil, a commonly reported side effect of interferon therapy is severe depression. Long-term activation of the immune system, as in autoimmune disease, also seems to go along with depression. Recently, evidence has accumulated linking inflammatory cytokines to increased risk for anger, as well. According to Emil Coccaro, M.D. (2013), chair of psychiatry at the University of Chicago and an expert on intermittent explosive disorder (IED), "These data suggest a direct relationship between plasma inflammatory processes and aggression in humans."[64]

> *Antidotal reports from pet owners suggest that feeding blue-green algae has reduced aggressive behaviors in animals. Microalgae nutrients are helpful in a variety of inflammatory conditions ranging from rheumatoid arthritis to depression.*

Antidotal reports from pet owners suggest that feeding blue-green algae has reduced aggressive behaviors in animals. Microalgae nutrients including omega-3 fatty acids have been found helpful in a variety of inflammatory conditions ranging from rheumatoid arthritis to depression.[65] Biologists at Israel's National Institute of Oceanography fed blue-green algae to animals and discovered that the high levels of the important fatty acids EPA and DHA in blue-green algae reduced the levels of arachidonic acid in the blood and liver.[66] Arachidonic acid increases brain cell oxidative damage and impairs the flexibility of cell membranes. It is a precursor of the inflammatory prostaglandins—the molecules people are trying to suppress when they take anti-inflammatory drugs.

Fish oils are essentially derived from the consumption of aquatic algae-derived PUFAs, either directly or indirectly. Wild cold-water fatty fish are good sources of the health-promoting omega-3fatty acids because of the cold-water algae in their food chain.

In a retrospective study of 208 documented cases, medical researchers found positive evidence suggesting that *AFA* (a blue-green algae) may be helpful in the treatment of fibromyalgia, a painful inflammatory condition with multiple etiologies. [67]

Overall, evidence suggests that microalgae demonstrate at least five antioxidant properties:

1. Supports Redox Signaling mechanisms
2. Scavenges reactive oxygen species (free radicals)
3. Regenerates endogenous antioxidants, such as SOD and glutathione reductase
4. Chelates heavy metals and toxic chemicals
5. Repairs oxidation-damaged proteins and lipids

During cellular metabolism, reactive oxygen species (ROS) are generated spontaneously. ROS are implicated in the etiology of a variety of acute and degenerative illnesses, including heart disease, stroke, rheumatoid arthritis, diabetes, and cancer. A major health benefit of microalgae can be attributed to its assortment of powerful antioxidant compounds that help to neutralize ROS.

CHAPTER

7

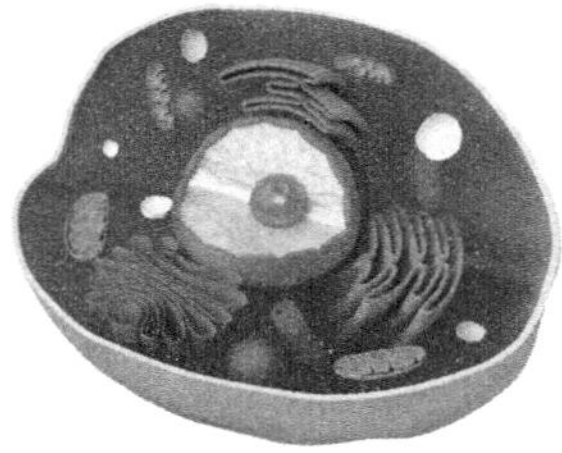

All Life Begins and Stays Alive Because of Cells

"The cell is immortal. It is merely the fluid in which it floats that degenerates. Renew this fluid at intervals, give the cells what they require for nutrition and, as far as we know, the pulsation of life may go on forever."

— Carrel A, 1924

[Alexis Carrel, winner of two Nobel Prizes, kept the cells of a chicken alive for 38 years. The cells eventually died because of a custodial error made by his lab technician.]

"On the one hand, immortality releases us from the biological push to survive, which is the basis of egotistical struggle. On the other hand, mortality lends life a time-limited sweetness and preciousness. How interesting that we mortal organisms are composed of 75 trillion cells which have some connection with immortality? How possible is it to connect to both sides of this polarity by embodying both the human and the cellular levels of consciousness?"[1]

— Aposhyan S, 1999

Microalgae Hypothesis II: Cellular Regeneration

All life is dependent upon the well-being of cells.

Microalgae safeguards mitochondria from oxidative stress.

Microalgae provide rich sources of REDOX supportive nutrients.

Microalgae can stimulate adult stem cell production.

Microalgae can boost bone marrow cell recovery, muscle regeneration and even brain neuroplasticity.

Microalgae provide rich sources of nucleotides, which may have regenerative effects.

Our Life is Dependent upon the Well-being of our Cells

Cells are the basic essential building blocks for life. Each individual cell is a complete, independent, living entity. Complex assemblies of cells unite, to create tissue and organ systems. We are a composite of vast quantities of living cells. It stands to reason then, that our health is directly linked to the well-being of our cells. Even complex multi-cellular organisms, such as humans, can reveal surprising pairings or uniformity at the cellular level. For example, despite the extreme external differentiation between a human brain and a bucket of urine; cell biologists have pioneered techniques that may actually allow human urine cells to be converted into brain cells.[2] While the science behind this type of cell transformation is extremely complex; this illustrates that at the deepest evolutionary level all cells share a common, universal, heritage.

A growing recognition of the universal ancestral origin of cells across cell types and across species, akin to the biblical story of Adam, is creating a revolution in biological and medical science. Evolutionary biologists refer to the last universal common ancestor (LUCA), as the ultimate cellular form from which the current diversity of life on Earth originated. When the genetic code was deciphered in the 1960s and was found to be universal across all species, most scientists became convinced in the origins of LUCA. "The unearthing of the genetic code and the subsequent demonstration that it is common to all life (a gene from a human can be read by the translation machinery of a bacterium) is without a doubt a key piece of evidence in establishing that there was a LUCA."[3] Today, genetic engineers combine genes from across the most unlikely organisms, interchanging genes from bacteria, plants and animals. The very reason this is possible is due to the common origin of cellular life.

Blue-green Algae, Evolutionary Dynamism, and Necessary Protection from the Sun

Paleobiologists, such as J. William Schop, describe how some blue-green algal species have shown minimal changes over billions of years. Fossils of blue-green algae from central Australia, dating back more than 3.5 billion years reveal early life forms that are remarkably similar to contemporary species. It appears that blue-green algae achieved a biological apex—with perhaps little need for alteration—accompanied by strong protective mechanisms that minimized genetic mutations.[4] A surprising variety of important biological mechanisms found in microalgae are characteristic throughout the entire animal kingdom. For example, the energy-producing mitochondria of animal cells are hypothesized to share common characteristics with cyanobacteria. Furthermore, primitive molecular mechanisms used for vision, movement, circadian rhythms, and even human polypeptide hormone-like substances, such as acetylcholine and melanin, and even forms of insulin are found in microalgae species.

Solar radiation in the visible range is essential to the process of photosynthesis. However, solar UV radiation is damaging, especially to unprotected DNA close to cell membrane surfaces. One of the most vulnerable areas of a cell is its lipid membrane. "The polyunsaturated fatty acids in biological membranes serve as both the target and source of oxidative damage and can be regarded as the most unstable class of biomolecules in the body. Lipid peroxides arising from both spontaneous and enzymatic oxidation of polyunsaturated fatty acids are the major source of endogenous DNA damage linked to various age-related pathologies and initiating carcinogenesis."[5]

Micro algal antioxidant pigment solutions have proven to very effective in protecting against lipid peroxide oxidative damage. (Learn about micro algal sunscreen pigments, see pg. 68.) The lipid, protein and DNA protective benefits that certain microalgae have acquired over hundreds of millions of years of adaptation, can be incorporated by humans and animals simply by oral consumption.

A Host of Antioxidant Pigments and Protective Agents

All contemporary plants species can trace their chloroplast genetic structure, their ability to photosynthesize, back to cyanobacteria.[6] Likewise, the origins of antioxidant defense systems are likely connected to the early adaptations made by these cyanobacteria. As Earth's original photo synthesizers, cyanobacteria depended on sunlight for food production, yet were susceptible to damage from that same energy source. Two and a half to 3.8 billion years ago — when blue-green algae was one of Earth's principle life forms — solar radiation was especially formidable because the protective atmospheric ozone levels were at less than 1% of present levels. Absent the planet's current protective "ozone shield," blue-green algae developed a host of potent antioxidant pigments and protective agents, including carotenoids, and phycobilins in order to survive.

Blue-green algae, scientifically called "cyanobacteria," are similar to other bacterial forms, in that it lacks a nucleus. Its genetic material is not surrounded by a nuclear membrane, like those present in all higher plants and animals.

"Ultraviolet (UV) radiation has provided a challenge for the evolution of life on Earth. On the one hand, UV radiation is a mutagen, arguably the most important naturally occurring mutagen. Genetic novelties, the result of mutations, are the raw material of evolution. On the other, UV radiation is a selective agent because it affects metabolic processes from photosynthesis to vitamin D formation, and causes DNA damage."[7]

Notably, just enough nonlethal change to the correct part of the DNA is required for evolution to occur. Too much damage to the DNA and the organism simply dies. Therefore, blue green algae out of necessity developed a family of potent antioxidants to protect unguarded DNA and membrane fats. A wider range of antioxidants is available in cyanobacteria than is found in land plants. Furthermore, the antioxidants pigments in cyanobacteria are both water and fat soluble, as well as readily available for assimilation.

Mitochondria the Cell's Powerhouse and Oxidative Stress

Cells are one of the smallest living entities that compose all the tissues and organs of our body. As our body is constructed of organs, our cells likewise have specialized membrane-bound structures, called organelles, which both support and are dependent upon the whole cell for survival. The label used to describe these intracellular components "organelles,' literally means "little organs." One of the most essential cellular organelles is mitochondria, tiny energy factories found in nearly all cells, including fungi, plants, and animals. These spherical, rod-shaped, organelles are referred to as the "powerhouse of the cell" since they assist in the production of high-energy, electron-rich, carrier compounds (e.g., ATP), during cellular respiration. ATP, an abbreviation for adenosine triphosphate, is the major energy currency molecule of the cell.

The coupling and uncoupling of the ATP atomic molecular structure (covalent bonds) helps to power the activity necessary for the vast majority of our body cells. By comparison, most nutrients typically contain low-energy covalent bonds, which are not very useful to power the work of cells. However food nutrients, especially glucose, are necessary to build ATP within the mitochondria. ATP is like a battery for the cell that builds up and releases energy. Because the total human body content of ATP is only about 50 grams; ATP is continually recycled in the mitochondria numerous times. Remarkably, the average adult daily intake of 2,500 calories of food results in the bewildering turnover of about 400 pounds of ATP.[8] At any single moment one cell contains nearly one billion fleeting ATP molecules, which last only a few minutes and then are recycled. This amazing energy transformation occurs only in the mitochondria, for multicellular organisms. This steady recycling of ATP is so crucial that anything that interrupts ATP transformation, like a loss of oxygen, rapidly kills an organism in minutes.

Mitochondria are the Powerhouse for Cells

Mitochondria produce ATP, which powers most cellular activity. These energy factories are especially plentiful in cells and tissues that have high energy demands, such as those found in the brain, heart and the skeletal muscles. Mitochondria are responsible for 90% of the cellular energy production that fuels tissues and organ systems and are the main producers of reactive oxygen species. During cellular respiration about two to five percent of oxygen use is converted into reactive oxygen species (ROS).

Due to their high energy demands, accompanying ROS production, the half-life of most mitochondria is estimated to be only a few weeks. The precise mechanisms by which the mitochondrial synthesis occurs are not completely understood. There is evidence to suggest that mitochondrial DNA (mtDNA) divide, facilitate mitochondrial replication and complete the synthesis of new mitochondria. Mitochondria can develop from preexisting mitochondria due to mitochondrial fission, which occurs in all eukaryotic cells, including neurons.[9] Impressively, mitochondrial DNA outnumbers nuclear DNA by several hundred times within the cell. But why does mitochondria have different DNA than the rest of the cell?

"Endosymbiosis" Bacterial Origins of Mitochondria

Most evolutionary biologists agree that the prokaryote origin of mitochondria, through a process called "endosymbiosis," gave rise to later complex life-forms. A fundamental difference exists between prokaryotes (e.g., bacteria and cyanobacteria) and eukaryotes (multicellular life forms). Prokaryotes lack organelles, like mitochondria. Therefore, in prokaryotes most of the cell's energy-conversion (e.g., ATP) occurs within the cell membrane. Accordingly, as a precursor to mitochondria, prokaryote cells became incorporated into primordial, ancestral eukaryote, host cells.

Figure 16. Endosymbiosis of Cyanobacterium into Host Cell

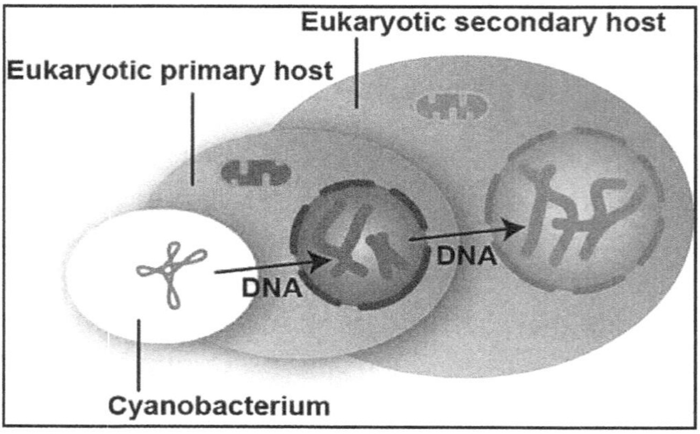

Over time, this mitochondria-type oval bacterium became dependent upon their cell hosts and conversely the larger host cells became dependent upon them (mitochondria) for survival. Now neither can survive independently. Yet mitochondria keep their own separate bacterial-form of DNA, very different from the cell's nucleus, and reproduce independently as well.

"The complex biogenesis of mitochondria as endosymbiotic organelles and their manifold biological functions render these organelles particularly vulnerable to accumulating damage during cell life. . . . Mitochondria are the sites of ATP production through oxidative phosphorylation, but as a by-product they are exposed to high concentrations of reactive oxygen species (ROS), which can induce protein modifications, lipid peroxidation and DNA damage."[10]

According to Rugarlie and Langer (2012), "Mitochondria contain their own genome . . . yet, the vast majority of mitochondrial proteins are encoded by the nuclear genome, synthesized in the cytoplasm and subsequently imported in the organelle. Although nuclear-mitochondria communication systems coordinate the expression of the two genomes, a possible imbalance between nuclear-encoded and mitochondrial-encoded respiratory chain subunits and the accumulation of misfolded polypeptides is a risk that mitochondria continuously face."[11] Ultimately, mitochondria represent the critical "weak link" in living organisms over time.

Mitochondrial Theory of Aging and Disease

"Only one human genetic system is known to be present in thousands of copies per cell to exhibit regional genetic variation, and to be directly involved in calorie utilization: the mitochondrial genetic system." - Douglas C. Wallace (2005)[12]

In 1972, Denham Harman proposed a *Mitochondrial Theory of Aging,*[13] which represents a refinement of his earlier Free Radical Theory of Aging. According to Harman's theory, aging and illness are accelerated by the cumulative damage wrought by free radicals to mitochondrial DNA plus harming nearby macromolecules particularly lipids and proteins. Mitochondria are central to many theories of aging, as they are essential to cell survival, as well as producers of deadly reactive oxygen species (ROS). Harmful ROS are also termed "free radicals." Mitochondria are major sites of premature electron leakage to oxygen, which generate superoxide and thereby contribute to oxidative stress. Mitochondrial ROS, mainly in the form of superoxide radicals (0^2), along with other free-radical variants, act as redox signaling molecules, supporting cellular balance or in excessive amounts damage cellular functions.

The formation of excess superoxide radicals (0^2) by the mitochondria can lead to cell death, apoptosis. Thus, the delayed-onset of most age-related diseases is believed due to the cumulative and excessive amounts of ROS – produced by mitochondrial activity – decaying mitochondrial DNA (mtDNA), damaging cellular functions, resulting in declining energy levels over a lifespan. Due to rapid mitochondrial duplication, cells are particularly vulnerable to any defects or loss of mitochondrial functions. This makes mitochondria especially vulnerable to injury and genetic mutations.

Evolutionary medicine researchers (Wallace, 2005; Reddy, 2008)[14-15] describe a "Mitochondrial Paradigm of Metabolic and Degenerative Disease, Aging, and Cancer." Their theory goes something like this;

mitochondrial dysfunction, especially vulnerable mitochondrial DNA (mtDNA), plays a central role in a wide range of age-related disorders. Mitochondria are a) essential to convert dietary calories into usable energy, and b) are primary generators of reactive oxygen species (ROS), which damages as well as helps regulates cellular functions, and c) mitochondrial DNA (mtDNA) interact more with their environment than cellular DNA, and d) combined with mtDNA higher rates of production, leads to greater mutation risks than typical cellular DNA.

Mitochondrial DNA (mtDNA) is typically inherited unchanged from the mother, yet its mutation rate is far higher than that of nuclear DNA.[16] Due to these characteristics, mitochondrial DNA is commonly used to trace ancestral lineage maps. Whether your ancestors came from Northern or Southern Europe stock can be determined from your mtDNA.

> *Mitochondria replicate far more often than the cells they live in: every couple of days in liver cells, and every few weeks in post-mitotic brain cells.*

While the majority of mitochondrial genes long ago migrated into the cell nucleus, those remaining as mtDNA are primarily involved in the process of cell metabolism. Mitochondria replicate far more often than the cells they live in: every couple of days in liver cells, and every few weeks in post-mitotic brain cells. Generally, the faster the mitochondria turn over, the better – it's a good idea to replace mitochondrial populations before too much mtDNA damage accumulates."[17]

Naturally, this vigorous strategy of rapid mitochondrial replication works more efficiently in younger organisms than older organisms. Over time small amounts of genetic errors and environmental damage accumulates. Aging organisms have more difficulty maintaining perfect patterns of mitochondrial DNA replication, and their ROS defenses are typically weaker than younger organisms. Thus, the delayed-onset and progressive course of most common age-related diseases, such as type II diabetes, heart disease, various cancers, and neurodegenerative diseases may be due to the buildup of mtDNA mutations during tissue renewal, in conjunction with the accumulated effects of ROS production. Age-related decreases in cellular energy production, the body's equivalent of a "brownout," ultimately lead to tissue and organ failures with the development of clinical disease or illness.

Individual and regional variations for major degenerative diseases, such as heart disease, may result from multiple complex interactions between: 1). Dietary intakes and environmental toxic exposures, 2). Physical activity levels, 3). Population variants of inherited genes (involving both mtDNA and nuclear DNA interactions); 4). Acquired and inherited ROS defenses and 4). Longer-term impacts on mitochondria from aging.

According to Dr. Douglas Wallace, the director of the Center for Molecular and Mitochondrial Medicine and Genetics University of California, Irvine (2008) a single change in the DNA of mitochondria has been found to cause degenerative heart and muscle disease. "While these diseases traditionally have been assumed to result from mutations in the genes encoded by DNA in the cell's nucleus," he said, "most common degenerative diseases frequently do not exhibit inheritance patterns wholly consistent with our understanding of these nuclear DNA genetics. Our demonstration that mutations in the mitochondrial DNA can also cause the same diseases means that both nuclear and mitochondrial DNA genes that affect mitochondrial function can contribute to disease risk."[18]

How else to explain the rapid shifts in modern illnesses like diabetes, heart disease and various cancers within a span of a single generation, closely associated with diet, lifestyle, and environmental changes? Typically most genetic changes would take thousands of years to effect large human populations. Moreover, our most vital energetic organs like the heart and the brain have the highest concentrations of mitochondria; also represent the most vulnerable organ systems, the leading causes of death and illness.[19] Ultimately aging associated with dysfunctions of cellular energy is the strongest predictor for leading health risks, excluding accidents or environmentally acquired illnesses.

Redox Comes from Two Opposite Reactions—REDuction and OXidation

Although most scientists have focused on reactive oxygen species (ROS) as toxic by-products of cellular metabolism, more recent studies discovered that cells harness the power of ROS for cell signaling, homeostatic, purposes. The term "redox signaling" describes the beneficial ways that cells utilize ROS as communication signals or feedback loops to better regulate the demands of both the intra- and the extracellular functions. The term Redox comes from two opposite chemical reactions, REDuction and OXidation. REDOX is the transfer of electrons from an electron donor (reducing agent) to an electron receiver (oxidizing agent). "Life is an extraordinary REDOX program, designed to capture, store and expend energy through the reorganization of chemical bonds. Thus, REDOX potential controls the fate of every cellular organism – its survival, differentiation, multiplication, death and beyond. REDOX signaling is a precise modulator of gene expression, transcriptional factors, protein structure, enzyme functions and various cellular responses. . . any derangement in REDOX homeostasis can result in oxidative stress, a detrimental condition that plays a role in the pathogenesis of every dysfunction or disease."[20] Currently several strategies are being considered to treat mitochondrial diseases, including targeting mitochondria with antioxidants.

> *Microalgae pigments may be considered REDOX signaling molecules. Cyanobacteria alone and in combination with other nutrients have demonstrated the ability "to promote mitochondrial respiration and/or proliferation of these cells in culture."*

In this regard, edible microalgae offer a rich variety of antioxidants, some of which work at the mitochondrial level. Mitochondria are ancient bacterial symbionts with their own mitochondrial DNA (mtDNA), RNA, and protein synthesizing systems, separate from the DNA of the host cell. Most antioxidant pigments are unable to work at the cellular mitochondrial level. However, animal and human studies suggest that microalgae nutrients may be absorbed and protective of human mitochondria.

Microalgae pigments may be considered REDOX signaling molecules. Cyanobacteria alone and in combination with other nutrients have demonstrated the ability "to promote mitochondrial respiration and/or proliferation of these cells in culture"; as well as protecting against inflammatory insults and oxidative stress.[21] Mitochondrial protection and support may account for the numerous health benefits associated with eating edible microalgae.

Microalgae can Enhance Circulation of Adult Stem Cells

Modern medicine is characterized by the use of drugs (pharma), cutting (surgery) or burning (radiation and chemo). However, advances in biotechnological, especially stem-cell research offer the potential avenue whereby our bodies may be supported to dramatically repair or renew themselves. Stem cells have the potential to develop into many different types of cells. As such, they serve an essential role within our body's repair system. One alternative to growing stem-cells outside of the body for transplantation is to activate our body's own adult stem cells (human hematopoietic progenitors) to multiply and disperse to areas of need. Many people remain unaware that every day our body naturally produces stem cells that assist in the replacement or repair of damaged tissues.

Scientists at the University of South Florida, College of Medicine's Center for Excellence for Aging and Brain Repair, have investigated a number of nutritional compounds that activate adult stem cell production. A compound of blueberry, green tea, catechin, carosine, and vitamin D(3) was found to increase the proliferation of human bone marrow and human stem cell production in a dose-related manner. These researchers completed a series of studies to demonstrate that a combination of these "nutrients produce a synergistic effect to promote proliferation of human hematopoietic progenitors [stem cells]."[22] The resulting nutritional compound was called "NT-020." NT-020 has successfully increased the proliferation of adult stem cells in a number of human and animal studies, in conjunction with decreasing harmful oxidation effects. Adult stem cells appear particularly vulnerable to oxidative damage, which can reduce the number of stem cells and thus reduce the body's ability to heal itself.

Like NT-020, research with extracts of cyanobacteria (*Aphanizomenon flos-aquae*), in combination with NT-020 or AFA by itself, have demonstrated the ability to proliferate adult stem cells and inhibit oxidation. The strongest adult stem cell activation effect was observed when *AFA* was combined with NT-020.[23] Blue Green algae increased stem cell genesis and protected animals against LPS induced declines in neural stem cell proliferation. Blue Green algae was able to increase proliferation of human hematopoietic stem cells *in vitro*.[24]

Brain Cells are Vulnerable to Oxidative Stress, Microglial Activation, and Inflammation

One area where cellular protection and repair is most critical, especially in addressing the myriad of insults and challenges of aging, is in the brain. The neurons in brain cells are especially susceptible to damage from oxidative stress, microglial activation, and to proinflammatory insults (which are known to decrease brain cell proliferation and limit neuroplasticity). "Neuroplasticity" is the ability of the brain to heal and "re-wire" itself. Injury or trauma to the brain provokes a "cascade of molecular and cellular events, which manifests in the form of microglial activation and release of various proinflammatory molecules. This eventually culminates in a profound neuroinflammatory reaction which characterizes the brain's response to stress, injury, or infection."[25] Proinflammatory molecules, like chemokines, released during this process will damage and reduce the production of neural and other adult, progenitor stem cells. Two areas of the brain responsible for neurogenesis, the production of new neural stem cells are the sub ventricular zone and the sub granular zone of the hippocampus both of which can be especially disrupted by the release of chemokines following inflammation.

> *Fortunately for aging or injured brains, diets supplemented with edible microalgae, which are high in oxygen radical absorption capacity (ORAC) have been shown to significantly reverse increased levels of proinflammatory cytokines in the aged cerebellum.*

Fortunately for aging or injured brains, diets supplemented with edible microalgae, which are high in oxygen radical absorption capacity (ORAC) have been shown to significantly reverse increased levels of proinflammatory cytokines in the aged cerebellum. Only five mg of blue-green algae supplied in a rat's food or in its' drinking in water for two weeks, significantly reduced a biomarker of oxidative damage, malandialdehyde (MDA); and improved cerebellar beta-adrenergic receptor function in aged rats.[26]

Dr. Paula Bickford at the Center of Excellence for Aging and Brain Repair at the University of South Florida have demonstrated through a series of four rodent studies the effectiveness of blue-green algae (*Spirulina*) in supporting a variety of neurological function. Blue-green algae were effective in assisting to:

- Reducing brain damage from experimentally induced strokes by up to 70 % (Yasuhara et al.; 2008).[27]
- Reducing the loss of spatial memory function typical of aging brains (Acosta; 2010).[28]
- Enhancing the dopamine neuronal recovery and may be protective against Parkinson's disease (Stromberg, 2005; Gemma et al., 2002; Wang et al., 2005).[29-31]
- Promoting neurogenesis, enhanced brain stem cell production (Bachstetter et al, 2010; Acosta; 2010).[32-33]

- Providing neuroprotection to brain cells from oxidative stress (Pabon et al., 2012; Yasuhara, 2008).[34-35]

AFA Boosts Bone Marrow Cell Recovery and Muscle Regeneration

Myeloablation is the severe or complete depletion of bone marrow cells typically resulting from irradiation. A research team demonstrated the effectiveness of "a water-soluble extract of *Aphanizomenon flos-aquae* (*AFA*), on hematopoietic recovery after myeloablation as well as recovery from cardiotoxin-induced injury of the anterior tibialis muscle in mice."[36] After the control and experimental *AFA* extract treated female mice were irradiated, they were transplanted with bone marrow stem cells, embedded with a green fluorescent protein (GFP) and allowed to recover. GFT allowed the scientists to track how GFP bone marrow derived stem cells were incorporated into muscle, using fluorescence imaging. After hematopoietic (bone marrow cell) recovery (23 days), mice were then injected with toxin, which damaged a section of their leg muscle. The *AFA* extract significantly improved muscle regeneration, by inducing mobilization of stem cells from the bone marrow to the damaged muscle, compared to the control group.

> *Using animal models they discovered that the rate of protein breakdown is significantly lower when optimal levels of the amino acid leucine are present in people's diets.*

Exercise has an acute effect on muscle protein metabolism. After exercise both muscle protein breakdown and muscle protein synthesis are increased. Researchers at the University of Texas Branch in Galveston, Texas (1999) reported "improving of muscle protein anabolism in humans with an oral amino acid supplement after exercise."[37] They found that "an oral dose of amino acids after resistance exercise will result in positive muscle protein balance similar to seen when the amino acids are [intravenously] infused." Therefore, oral intake of amino acids, like those found in high ratios within certain microalgae can accelerate muscle growth after exercise.

Typically once adults reach the age of 40, they begin to lose between 0.5 and 2% of their muscle each year. A French team (2005) from the Human Nutrition Research Center of Auvergne believes they have found a way to help the elderly decrease the loss of muscle mass. Using animal models they discovered that the rate of protein breakdown is significantly lower when optimal levels of the amino acid leucine are present in people's diets. Dr. Didier Attaix who led the team's efforts said: "Preventing muscle wasting is a major socio-economic and public health issue that we may be able to combat with a leucine-rich diet."[38]

Subsequent research (2006) supports the idea that the amino acid leucine is responsible for the stimulatory effect of dietary protein on protein synthesis.[39] Leucine is readily available in dairy, eggs, seafood and meat, nearly 8 to 9 % of the total protein. Yet leucine is found at even higher levels in blue green algae. For example, leucine is more than 11 % of the total protein in *AFA* algae; one of most abundant essential amino acids of blue green algae.

Italian researchers (2011) demonstrated that BCAA-enriched amino acid mixtures can increase the lifespan of male mice, increased mitochondria biogenesis, and support ROS defense systems, with reduced oxidative damage, both in cardiac and skeletal muscles of middle aged mice. They concluded that, "A body of recent evidence suggest that amino acids, and in particular BCAAs, behave as evolutionary conserved modulators of lifespan of different organisms, ranging from yeast to mammals. Our data demonstrates that oral intake of a BCAA-enriched balanced amino acid mixture improves motor coordination and endurance and promotes longevity of male mice . . . A number of preclinical and clinical reports, here reviewed, supports the use of dietary supplementation with balanced amino acid formulas containing BCAAs to prevent disability and prolong healthy life expectation of elderly subjects."[40] Again, microalgae can provide a complete source of BCAAs, including high levels of leucine.

Microalgae Flavonoids and Chromosomal Protection

> *New research provides evidence that dietary flavonoids (i.e., pigments) help repair a range of free radical damage in DNA and offer protection against strand breaks and base alterations in our cells' genetic material.*

New research provides evidence that dietary flavonoids (i.e., pigments) help repair a range of free radical damage in DNA and offer protection against strand breaks and base alterations in our cells' genetic material. Scientists at the University of Auckland in New Zealand demonstrated that antioxidant flavonoids can reduce the incidence of single-strand breaks in irradiated solutions of double-stranded DNA, *in vitro.* Using advanced pulse radiolysis measurements, scientists found that electron transfer from the flavonoids to free radical attack sites on DNA appear to result in faster chemical repair, lessening the oxidative damage to DNA.[41]

Beta carotene, a flavonoid compound, derived from the microalga *Dunaliella* demonstrates antimutagenic effects on human lymphocytes. Chinese researchers used *in vitro* micronucleus and chromosomal aberration tests on human white blood cells. The inhibitory effect of microalgae-extracted beta carotene on mutagenesis induced by both gamma-rays and mitomycin, a known mutagenic agent, was apparent.[42]

Microalgae are Masters of Regeneration—A Vital Rescue Food

Steve Gagne, a macrobiotic counselor and author of *The Energetics of Food* (1990), reports that "Algae are the masters of regeneration—they probably are the most highly regenerative foods on the planet."[43] In support of this empirical observation, it is noteworthy that micro algal extracts added to culture mediums dramatically increase human cell survival rates. In 1984, a U.S. Patent (no. 4,468,460) was granted to S. Kumamoto for "A Method of Human Cell Culture," described as follows: "A method of culture of human cells is disclosed which comprises effecting the cultivation in a culture medium containing an extract of microalgae...said method permitting the normal successive cultivation of human cells to be maintained efficiently without any morphological and genetic mutations over a greater number of successive generations than has hitherto been possible."[44]

Dr. John Apsley states, "I have spent nearly 20 years focusing on the places within biology where the Regeneration Effect appears the strongest. In my opinion, the most rewarding places to study surround undenatured chlorophyll and raw nucleoproteins." Drawing from the Nobel Prize-winning work of Dr. Alexis Carrel, Apsley notes that regenerative food is best obtained from young, "embryonic" cell cultures. Microalgae can be considered an embryonic cell source, since given a suitable environment within a single day or two; they undergo rapid and massive cellular replication. Furthermore, according to Carrel, avoiding excessive heat is essential as most "regenerative effectors" are destroyed when heated above the body's physiological temperatures. Finally, optimal regenerative foods contain a wide and balanced spectrum of trace minerals, vitamins, all ten essential amino acids, and essential fatty acids, as found in freeze-dried *AFA* microalgae.[45]

Chlorophyll is considered to be a cell regenerator because its central magnesium atom plays an important role in so many (325) different enzyme systems. Also, as an antioxidant, chlorophyll may help to protect DNA during cell division, a very vulnerable time in the life of any cell. Unfortunately, most of the published research on the regenerative properties of chlorophyll is found in older or foreign medical journals, which rely more upon natural therapies.[46-49] Nevertheless, microalgae offers the richest sources of chlorophyll.

Dietary Nucleic Acids are Unrecognized Promoters of Health

"Perhaps the single most underestimated or under-appreciated deficiency that results in premature aging, death and an excess of various aberrant health conditions. That single most overlooked nutritional component is the provision of dietary nucleic acids, the building blocks of DNA and RNA."[50]

—Ovokaitys T, 2010

Nucleic acids are large biological molecules essential for all life. Nucleotides are the individual components that combine to form the structure of nucleic acids, including DNA (deoxyribonucleic acid) and RNA (ribonucleic acid). Along with proteins, nucleic acids are the most vital biological macromolecules. Nucleic acids are essential for encoding, transmitting and expressing genetic information. Even the smaller components of nucleotides function as cellular signaling agents and enzymatic cofactors, for in nature little is wasted.
As in all bacteria, the DNA of blue-green algae forms a simple loop, without chromosomes or a nuclear membrane. Thus, unlike nucleic acids in more advanced plant and animal species, cyanobacteria nucleic acids constitute from 4 to 7% of the microalgae's dry weight and without a nuclear membrane, are more orally available.

Dietary Nucleic Acids, Including MicroRNAs can be Absorbed and Utilized by Cells

The absorption and metabolism of dietary nucleic acids has received little attention compared to that of other organic nutrients, largely because of challenges in tracking their cellular utilization.
Conventional wisdom was that nucleic acid and nucleotides played no useful nutritional role and were simply destroyed within the digestive tract. "Yet, it has been demonstrated that DNA is not broken down in the gastro-intestinal tract to the extent formerly believed. Ingested DNA sequences large enough to contain whole genes have remained intact and entered the blood and tissues. Gene fragments from the

food are circulated and sometimes integrated into the chromosomes of the predator."[51]

Research conducted at Nanjing University in China, has revealed that strands of RNA, containing a sequence of nucleotides from consuming vegetables flow into our bloodstream, and can regulate the expression of our genes once they're ingested. The Chinese research team first discovered that certain microRNAs (miRNAs) found in human blood originated from the rice in their subject's diet. Intrigued, the scientists next confirmed dietary miRNAs from plants could be transported into the blood stream and be circulated via microvesicles (MVs). MVs are "shed from almost all cell types under both normal and pathological conditions . . . and that many types of MVs in human plasma contain miRNAS."[52] Carried by the blood-stream plant-based miRNAs are capable of shifting gene expression in animals, thus demonstrating a "cross-kingdom regulation of microRNAs." The team concluded that "miRNAs may be a new class of functional components in food, like vitamins or minerals—even in an animal that's pretty far removed from their home organism; they can manipulate gene expression and have an effect on nutrition."[53]

Can various other nucleic acids also make their way into the blood stream? Scientists at the University of Goettingen in Germany have found that significant quantities of orally ingested nucleic acids are capable of surviving the digestive process and are absorbed intact in specific cell tissues, as demonstrated in radioactive-tagging studies.[54] Most importantly orally consumed nucleic acids display a strong tissue-regenerative effect. Might algal nucleic acids offer a sort of "bio regeneration effect" that operates at the genetic level?

Insufficiency of Dietary Nucleic Acids Results in Premature Aging and Illnesses

According to Todd Ovokaitys, M.D. "The absorption studies and metabolic studies indicate a special sensitivity of the pituitary, thymus, thyroid, adrenals, salivary glands and intestinal lining to receive additional nucleic acids for optimum function and regeneration . . . An insufficiency of dietary nucleic acids – the body not having enough nucleic acid bases to repair and regenerate the immune system, the intestinal lining and major endocrine glands – is an unsuspected, undeclared and undiagnosed nutritional deficiency that is resulting in premature aging, unnecessary demise, and unnecessary health conditions."[55]

While nucleotides are present in high concentrations in mother's milk and are utilized to fortify some infant formulas, easily absorbed and high concentrations of nucleotides are not readily available in our food supply. Rich sources of nucleotides found in blue green algae offer foundational nutritional support that assists the protection and repair of cells at even a genetic level.

Additional evidence is found in experiments conducted with chickens and mice at the Stable Isotope Laboratory in Baylor College of Medicine in Houston Texas. Poultry and rodent food rations were supplemented with radioactively labeled blue-green algae. Labeled algal dietary nucleosides, pyrimidine and purine, were detected in the experimental animals' isolated hepatic RNA. The researchers observed that large quantities of dietary pyrimidine nucleosides and minimal quantities of dietary purine nucleotides were incorporated into the animals' hepatic nucleic acids, without hydrolytic removal of the ribose moiety.[56] In other words, these laboratory results support the potential nutritional role for nucleosides and suggest that pyrimidines are "essential organic nutrients" that can be genetically incorporated at a cellular level. Pyrimidines are nitrogenous bases that are critical building blocks of nucleic acids (DNA and RNA).

Algal Diets Used to Stimulate the Regeneration of Blood Serum and Liver Proteins

Two studies by Devi and his team demonstrated the ability of algal diets to stimulate the regeneration of blood serum and liver proteins in rats.[57-58] Because micro algal protein is composed of shorter and less complex polypeptide chains - with an abundance of all essential amino acids—it can be more readily utilized at the cellular level. One can think of it as supplying the foundational building blocks for cellular repair in easily usable form. Even more than just added protein benefits, Brazilian scientists reported that blue-green algae ingestion increased the "muscle protein synthesis rates" in experimental young rats, over and beyond what was expected from simply an "adequate protein quality" source.[59]

Might a Diet Supplemented with Algae Increase Life-span?

Might a diet supplemented with algae increase life-span? "In extensive traveling, beginning in 1927, Professor S. Kondo, of Tohoku University, discovered that geographic region and especially diet play a determining part in life-spans of the Japanese people...On islands and in fishing villages, people eat less rice and salty food and more sea vegetables, with the result that they live longer. The village of Oki Island, in Shi mane Prefecture, where the people eat plain food and soybeans and sea vegetables—of which they are very fond—has the highest incidence rate of longevity of any...place in the nation..." and one of the highest median lifespans in the world.[60]

Feeding microalgae to elderly people or animals was shown to be effective in providing protection from a variety of age related diseases, such as cardiac hypertension, hyperlipidemia, risk of cancer, Alzheimer's disease and inflammatory conditions. Edible microalgae may well be the ultimate in antiaging food!

CHAPTER

8

Radiation Protection for a Modern World

"The transfer of energy that is produced by radiation is similar to that caused by other forms of acute injury such as an automobile crash or a bullet wound...The difference between a bullet and an X ray lies principally in the size of the particle. While a bullet destroys tissues and entire organs, a particle of radiation collides with single atoms or molecules deep within the cells."[1]

— *Needleman H, Landrigan P, 1994*

Microalgae Hypothesis III: Radiation Protection

Microalgae can protect cells from radiation and oxidative damage.

Microalgae can support immune function (T cells and macrophages).

Microalgae can help to protect endocrine glands (e.g., thyroid).

Microalgae can reduce cellular damage from radiation exposure.

Microalgae can ease radiation sickness and urine radioactivity levels.

Microalgae antimutagenic effects may include DNA repair mechanisms and help to fix single strand breaks.

Antioxidants have Antimutagenic Properties

High doses of radiation destroy cells completely, while lower doses of radiation can damage cellular DNA replication and thus lead to cell mutations. A mutation occurs when a DNA gene is damaged or changed in such a way as to alter the genetic message carried by the gene. Radiation effects may be countered by increasing the cell's protective antioxidant defenses, by enhancing the repair mechanisms that fix genetic base code damage or by increasing the body's ability to identify and eliminate mutant or damaged cells. Many natural antioxidants have antimutagenic properties that may prove useful in reducing long-term radiation effects. Furthermore, nutrients that help modulate endogenous antioxidants, such as superoxide dismutase, may be useful in some radiotherapy protocols.[2]

Dr. Nick Lane (2002) in his book "Oxygen: The Molecule that made the World" proposes that microalgae developed protective mechanism against oxidative stress, long before there was any oxygen in the atmosphere. He describes how early life must have developed effective defenses against ionizing solar radiation early on, during an evolutionary "radiation bottleneck" that happened nearly 4 billion years ago. Dr. Lane addresses the vital question, "As oxygen is undoubtedly toxic, life might somehow have adapted to the threat in advance. Could it be that life adapted to radiation early on, and that this formed the basis of its response to other kinds of threats? . . . New evidence suggests that resistance to oxygen and radiation alike was built into the very earliest organisms."[3] Accordingly, cyanobacterial antioxidant and radiation defenses took on a multi-purposed adaptive function evident within colorful light-harvesting pigments.

Humans are not immune to the challenges of oxidation either. Dr. Lane portrays that "The continual production of superoxide radicals by cells means that, despite the emotive associations of radiation sickness, breathing oxygen carries a qualitatively similar risk." He explains "the fact that hydroxyl radicals are produced from both breathing and radiation allows a direct comparison to be made. As a typical chest x-ray delivers 50 micro-sieverts of radiation, breathing for a year would seem to be 1000 times more dangerous than a chest x-ray, or 50 times as dangerous as all the radiation that we normally receive from all sources in the course of an entire lifetime."[4] Again, Dr. Lane's point is not for us to stop breathing, which would be impossible while alive, but to recognize that antioxidant defenses developed by life's most primitive organisms (e.g., cyanobacteria) might offer protection against radiation, as well as antioxidant benefits.

Japanese Survived Nuclear Radiation and Attack

The release of radiation from Japan's Fukushima Daiichi Nuclear Plant disaster (2011) has compelled many people to seek supplements in an attempt to reduce the risk of the resulting environmental radiation. This has resulted in the widespread use of iodine *kelp* supplements, complete with rich antioxidant pigments found in marine algae. These may block the uptake of radioactive iodine isotopes released from the nuclear reactors, by preventing their binding to the thyroid glands iodine receptors.[5]

Historically, Japan is the only country that has experienced the dreadful consequences of a nuclear attack. Tatsuichiro Akizuki, M.D. was Director of the Department of Internal Medicine at St. Francis's Hospital in Nagasaki, located near to the center of the atomic blast zone. He instructed his staff and patients to eat a strict diet of brown rice, miso soup, lots of seaweeds, and prohibited the consumption of sugar and sweets. Reportedly, due to a sea vegetable abundant diet, Dr. Akizuki saved everyone in his hospital while many other survivors perished from the radiation sickness.

According to the Linus Pauling Institute at Oregon State University, "Sea vegetables, commonly known as sea weeds, such as kelp, purple dulse, wakame, and sea lettuce are high in iodine. . . One quarter teaspoon of kelp provides 3 mg of iodine. Consuming one-half teaspoon, or two grams, of purple dulse on a daily basis is enough to meet the recommended 150 mcg dosage of iodine. [Blue green algae] is also effective in binding plutonium."[6]

Scientists at the Gastro-Intestinal Laboratory at McGill University of Montreal demonstrated that alginates derived from seaweed could reduce by 50-80% the amount of radioactive strontium absorbed through the intestine.[7] Sodium alginate (Alginic acid) is an important intercellular polysaccharide found in large brown algae like *kelp*. Sodium alginate is claimed to bind tightly to many radioactive substances such as strontium, calcium, barium, cadmium and radium.

The Largest Nuclear Disaster in History was Ukraine's Chernobyl Disaster

The largest nuclear disaster in world history was Ukraine's Chernobyl disaster of 1986. In comparison, the more recent Japanese Fukushima nuclear disaster, released only an estimated 10 to 30% of the radiation unleashed by the Chernobyl accident.[8] Radioactive contamination discharged by the Chernobyl disaster was described as exceeded that of the 1945 atomic bombings of Hiroshima and Nagasaki. Hundreds of cases of thyroid cancer in children and adolescents were directly linked to the Chernobyl radiation poisoning.

Belarus is a region neighboring the Chernobyl plant and lay within the most severe radiation contamination zone. At the Institute of Radiation Medicine in Minsk, Belarus, five grams a day of blue green algae (*Spirulina*) was ingested by children (n = 100) for one month and was found to reduce urine radioactivity levels by 50 percent. The Institute's team described beneficial health effects of blue green algae on children with radiation sickness from the disastrous Chernobyl melt-down. "Use of *Spirulina* decreases radiation dose load received from food contaminated with radionuclides, Cesium-137 and Strontium-90. *Spirulina* is favorable for normalizing the adaptive potential of children's bodies in conditions of long-lived low dose radiation."[9]

Reportedly, the blue-green algae supplementation helped with regeneration of bone marrow, white blood cells, spinal fluids, blood and liver tissues. The children who received the microalgae treatment at the Institute made rapid improvements in their overall health compared to those children who did not receive the microalgae. Based on the Institute of Radiation Medicine in Minsk results, the Belarus Ministry of Health concluded that *Spirulina* accelerates the evacuation of radionuclides from the human body.

Based on these findings and other successes in treating radiation poisoning, a Russian (1994) patent was awarded for *Spirulina* as a medicine to reduce allergic reactions from radiation sickness. Seven hundred and nine children who had long-term exposure to radiation during and after the Chernobyl accident were subsequently relocated to Israel. Of these children, 99 were provided twice daily with a beta carotene powder of *Dunaliella* microalgae. The researchers found that irradiation increased the susceptibility of lipids to oxidation overall in the Chernobyl children. Moreover, they reported that micro algal beta carotene may act as an *in vivo* lipophilic antioxidant or radio protector in the supplemented group of children.[10] Victims of the Chernobyl incident and those medical relief workers exposed to the long-term effects of radiation poisoning reported positive benefits after eating *AFA* blue-green algae.[11]

Furthermore, the application for microalgae in radioactive environmental clean-up operations is promising. Bio absorption of radioactive cadmium by cyanobacteria (*Spirulina platensis*) immobilized on alginate gel and silica gel was studied. The results showed that cyanobacteria cells could absorb up to 70 micrograms of cadmium per gram of algae biomass. Temperature did not influence radiation or metal absorption, which occurred even at the widest range of pH (from 3 – 8), with the highest absorption levels occurring at a pH of 6. Impressively, algae cells could be repeatedly used in the radioactive absorption process up to five times.[12]

Multiple CT Scans Can Expose Patients to Harmful Radiation Levels

Numerous patients are needlessly exposed to radiation when they receive X-rays, especially when this diagnostic imaging is coupled with a computed tomography (CT) scan. A CT scan can expose patients to potentially harmful levels of radiation, up to the equivalent amount of radiation received by atomic bomb survivors in the low-dose range. "Epidemiologists estimate that as many as 29,000 new cancers have been related to CT scans performed in the US in 2007 alone. There are now approximately 70 million CT scans performed every year, up from a mere 3 million in 1980."[13]

States radiologist, Shella Farooi, M.D., director of research for Columbus Radiology Corp, "Radiation safety is, without a doubt, a large concern for practicing radiologists today . . . However, the current focus does not account for the possibility of harm to future generations from radiation delivered today . . . to consider potential harm to the patient's offspring and their offspring's offspring." She describes how, "ionizing radiation effects have been shown to span generations, resulting in heritable defects in mice. Moreover, Epigenetics may ultimately turn out to have a greater role in health and disease and treatment of genetics itself; and given this knowledge, one wonders if future trends in diseases will be linked to today's utilization of CT."[14]

In 2008, the National Institutes of Health (NIH) announced that $190 million would be directed toward epigenetics research over the next five years. Epigenetics may work in two possible directions: 1) replicating today's environmental exposures or 2) lifestyle habits to be inherited by future generations in potentially positive or negative ways. Are there ways to protect oneself from the harmful effects of radiation and even to protect our future offspring? Given the potential disastrous alternatives, this is a question we must consider.

Carotenoids Are Potent Free-radical Quenchers and Lipid Antioxidants

Beta carotene and other carotenoids, found abundantly in microalgae, are known to be potent free-radical quenchers and lipid antioxidants. Natural beta carotene (50 mg/kg diet), obtained from the unicellular alga, *Dunaliella,* was fed to lab rats exposed to a single high dose of whole-body radiation (4 Gy). Radiated control animals, not fed algal carotenoids, suffered a significant loss of body weight and decreased liver concentrations of beta carotene and retinol, compared to algal beta carotene supplemented rats. Normal increase in body weight and the absence of ill effects were noted in the groups of rats whose diet was supplemented by beta carotene before and after irradiation. Furthermore the effect of irradiation was "partially cured" by supplementation with beta carotene after irradiation in sub-group controls. These results suggest that beta carotene (especially 9-cis isomers) and retinol protect *in vivo* against the cellular damage by free radicals induced after whole body irradiation.[15]

However one possible contraindication of using beta carotene extracts, versus giving the whole algae, is that high doses of beta carotene treatment may markedly lower concentrations of vitamin E in tissues.[16] The use of mega doses of beta carotene, like any isolated (pro-) vitamin extract, can contribute to other nutritional imbalances.

Chlorophyll Offers Radiation Protective Benefits

Foods rich in chlorophyll have been shown to offer radiation-protective benefits. "Tests performed by the U.S. Army showed that chlorophyll-rich foods may also be effective in decreasing the effects of radiation. In one controlled study it was found that a chlorophyll-rich diet doubled the life span of animals exposed to fatal doses of radiation."[17] Again, microalgae — in this instance, *Chlorella* —are cited as particularly rich sources of chlorophyll.

Chlorophyllin, a water-soluble derivative of chlorophyll, has demonstrated an ability to protect DNA against gamma-radiation-induced strand breaks exposing plasmid DNA *in vitro.* Researchers discovered that chlorophyllin effectively protected plasmid DNA against ionizing radiation, independent of DNA repair or other cellular defense mechanisms; revealing direct evidence of free radical-scavenging properties of chlorophyll.[18] Notably, *Aphanizomenon flos-aquae* (*AFA*) microalgae has one of the highest percentages of chlorophyll (2-3%) of any food. For example wheat grass, which is used by many health advocates due to its high chlorophyll content, contains only ¼ the amount of chlorophyll as *AFA*.

Microalgae Demonstrates Remarkable Radio protective Effects

Extracts of phycocyanin (the blue pigment) from blue-green algae helped to restore the efficiency of antioxidant defenses, dehydrogenase activity, and energy-rich phosphate levels in rats exposed to X-rays (dose of 5 Gy).[19]

When microalgae was administered orally to mice, radio-protective effects of microalgae were shown to occur both before and immediately after exposure to sub-lethal gamma-rays.

Several animal and *in vitro* studies using microalgae have demonstrated remarkable radio protective effects.[20-24] When microalgae was administered orally to mice, radio-protective effects of microalgae were shown to occur both before and immediately after exposure to sub-lethal gamma-rays.[25] Significant benefits were observed in the number of bone marrow cells and the spleen weight. According to Dr. John Apsley, "*AFA* has no less than 20 potent nutrients that limit and reverse the most common forms of radiation injury and may help to partly reverse genetic damage as well."[26] For example, "research has shown that salts of aspartic acid, which are present in *AFA* as potassium aspartate and magnesium aspartate, are useful for protection against radiation damage."[27] The cyanophycin storage granules in blue-green algae are made of aspartic acid and arginine, which act to stimulate the thymus gland, promoting white blood cell activity. Also, microalgae can increase glutathione levels in the body, making it an effective radio protective agent.

Water-Soluble Extracts of Blue Green Algae Promote Repair of Damaged DNA Strands

The influence of water soluble polysaccharide, from *Spirulina platensis* , on promoting nucleotide "excision" repair of DNA was investigated by means of endonuclease assay and using radioactively labeled isotopes, a process called "autoradiography." Nucleotide excision repair involves several important DNA repair pathways to fix single strand DNA damage; by first identifying damaged DNA strands, then repairing, removing, or creating new "patched up" DNA sequences. Chinese researchers showed that the presence of the microalgae polysaccharide significantly enhanced both the repair activity of radiation damaged DNA and the unscheduled DNA synthesis (UDS). During the nucleotide excision process, it was shown that the presence of a *Spirulina* polysaccharide extract both increased the initial rates of fixing damaged DNA strands, and also led to longer phases of nucleotide repair and DNA synthesis.[28]

Might algal diets be able to confer to other cells some aspect of protection from genetic mutations? Researchers at the Institute of Molecular and Subcellular Biology in Slovakia believe so after they found that freeze-dried *AFA* blue-green algae demonstrated antimutagenic effects on bacterial cells exposed to a mutagen, a substance that disrupts DNA/RNA transcription and causes mutations. When the algae powder was added to the cell culture at the same time as the chemical mutagen, no benefit was shown. However, if the algae powder was added to the cell culture medium 2 to 24 hours before exposure to the mutagenic agent, a significant antimutagenic effect was evident.[29] The maximum suppression of mutagenic activity was achieved when the algae powder was mixed in the cell culture medium 24 hours before the addition of the mutagen. This suggests that the algal phytochemicals were utilized by the cell culture as a protective cellular influence rather than neutralizing the chemical mutagen directly.

CHAPTER

9

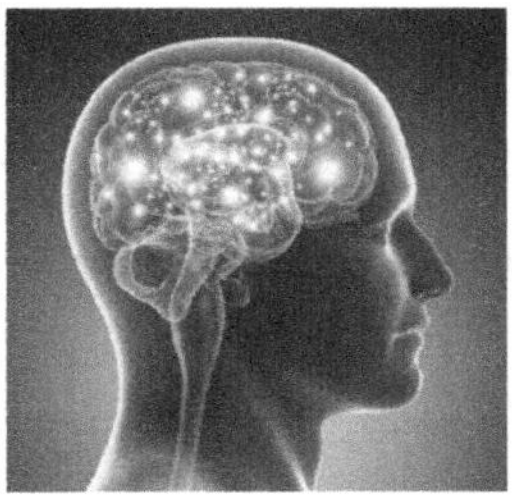

Eat Light and Feel Bight: Ultimate Brain Food

"Blue-green algae and my new diet have helped me focus and concentrate better in school and on my homework. I am more relaxed and I don't have stomachaches anymore. I have more friends now and my mom is happier too."

— Chelsea, 8 years old

"Our brains are unique among all species in their enormous metabolic demand. If we can explain how our brains sustain such a tremendous metabolic flow, we will have a much better chance to understand how the brain works and why it sometimes breaks."

— Khaitovich, 2008

Microalgae Hypothesis IV: Supports Mental Health

Microalgae can improve production of neurotransmitters.

Microalgae can support better focus, working memory and stamina.

Microalgae can provide a rich source of bioactive B-12 and decrease homocysteine.

Microalgae serve as an "adaptogen" to increase resistance to stress.

Microalgae can provide a rich source of omega-3 healthy brain fatty acids (esp. DHA).

Microalgae can protect neurons against toxic insults and lower inflammation.

Microalgae can help balance blood sugar levels and moods.

Microalgae can increase new brain cells and brain repair.

After the Heart, the Brain is the Most Voracious Consumer of Energy in the Body

The brain uses one of the highest concentrations of nutrients of any organ in the body. "After the heart, the brain is the most voracious consumer of energy in the body."[1] While an adult brain accounts for less than 2 percent of our total body weight, on average our at-rest brain metabolism accounts for a whopping 20 to 25 percent of adults total energy needs. This percentage is far more than the 8 to 10 percent found in nonhuman primates and way more than the meager 3 to 5 percent consumed by the brains of most other mammals.[2] The single best indicator of fuel use, metabolism, is oxygen consumption. That is because oxygen is required by our cellular mitochondria to change glucose (our primary body fuel) into ATP, the energy that powers all cellular life. A lack of oxygen results in rapid injury to the brain and heart, the most mitochondrial loaded organs of our body.

Because of multiple reasons our brain is highly susceptible to oxidative damage, including: a). the brain's high content of unsaturated fatty acids, which are susceptible to oxidation; b). high use of oxygen means greater risk of oxidization; c). greater mitochondrial activity, contributes to free radical formation; d). relatively high concentrations of Redox active metals, like iron, zinc and copper, contribute to oxidative stress and finally e). limited access, within brain tissues, to antioxidant defense systems. Altogether, our brains are highly susceptible to oxidative stress, far more than most tissues in the body.

Mitochondrial Dysfunctions in Brain Cells can lead to Learning Disabilities

All living cells, except red blood cells, have mitochondria. A typical animal cell may have on the order of 1000 to 2000 mitochondria. Surprisingly, up to one third of mitochondria found in our brain, are able to move along axons to help module neurotransmitter signal release. Healthy mitochondrial mobility along axons controls the strength of signals by modulating the way nerve cells release transmitters.[3] Neuroscientists at the Flanders Institute for Biotechnology (2013) have shown that mitochondrial dysfunctions in brain cells can lead to learning disabilities. Their results demonstrate the wider importance of mitochondrial health for the optimal function of the human brain, which impacts our ability to think, feel, perceive and even act.[4] Researchers from U.C. – Davis (2010) found that mitochondria from children with autism consume far less oxygen, a significant reduction of 66 percent less, than mitochondria from neurotypical children.[5]

Children have Increased Neurological Risk from Toxic Exposures

Children's brains are hungrier, more metabolically active, and proportionally larger than adults' brains. Unlike many organs (e.g., the liver) that have cellular fuel reserves, the brain is almost entirely dependent upon a continuous blood supply for fuel. Per pound of body weight, children eat more food, drink more fluids, and breathe more air than adults, thereby increasing their potential exposure to toxins. Also, younger children's blood-brain barriers (BBB) and intestinal linings are not as developed and are therefore less protective than those of most adults. This means that more incompletely processed food proteins and environmental toxins can leak into a child's bloodstream and brain. Furthermore, "This theoretically renders the [brain] CNS vulnerable to immune-mediated damage should the BBB be breached."[6] Finally, because of immature metabolic pathways, children are not able to synthesize all of the amino acids that adults can synthesize and must obtain a wider range of amino acids in their diets than adults. As a result, children are more likely to experience nutritional problems and toxic risks.

All these factors contribute to children's heightened susceptibility to dietary imbalances. The increased susceptibility of children to neurotoxins from what they eat, breathe, and drink, is likely contributing to the epidemic of neurobehavioral problems sweeping our country today.[7] For example, many pesticides used on food crops are specifically designed to attack the nervous systems of pests and can act as neurotoxins in humans. Vigorous industrialization and urbanization creates a greater discharge of hazardous environmental wastes. Given that our nervous systems are designed to be highly sensitive to environmental changes, brain cells, especially mitochondria organelles, are especially vulnerable to the effects of pollution and oxidative stress.

Most Children are Not Getting Enough Protective Antioxidants

Despite the slew of public health messages such as "eating 3 to 5 cups of fruits and vegetables daily," many children fall way short. According to the Center for Disease Control and Prevention (CDC, 2011) most teens "only eat fruits and vegetables an average of 1.2 times a day . . . about 30 percent of US high school students consumed fruit less than once a day. The same is true of vegetables, with 33 percent taking in leafy greens and carrots less than once daily."[8]

Basically children are not getting enough protective antioxidants, found in fruits and vegetables, or receiving the optimal macro- and micronutrients found in a healthy diet.[9] Studies done by the American Dietetic Association, the U.S. Department of Health and Human Services, and the National Cancer Institute indicate that a large percentage of children in the United States are not obtaining even the recommended daily allowance (RDA) of nutrients from their food.

Today's average restaurant meal is four times larger than the 1950s meal, and adults are, on average, 26 pounds heavier. Yet, despite an abundance of food, many Americans still unknowingly suffer from nutrient deficiencies. The CDC's Second Nutrition Report (2012), an assessment of diet and nutrition in the U.S. population, concludes that a variety of specific nutrients are lacking in a significant number of Americans' diet. Specifically, vitamin B6, iron, and vitamin D had the highest prevalence of deficiency, followed by vitamin C and vitamin B12.[10-11] Unfortunately other important food co-factors like antioxidants, enzymes, or many trace minerals were not measured. In 1997, Munoz et al. studied 3307 children in the U.S. (age 2–19) to determine how many were meeting national recommendations for food group intake and found that only 1% met all the recommendations.[12] Furthermore, 64% of children studied failed to meet the minimum RDA requirements for vegetable intake, and, for the 36% that actually met these requirements, one-fourth of all the vegetables they consumed were in the form of French fries! This data is startling, and alarmingly many children are not receiving the essential vitamins, mineral, and phytochemicals required for proper brain functioning.

Are Children the Canaries in the Mines?

Figure 17. Canary and Miner

Miners used to bring canaries into coal mines as an early warning system. If the air was toxic enough to harm the canary, soon it would become noxious enough to harm the miners. The canaries were sensitive to environmental toxins, such as deadly fumes, so their demise cast a warning for miners to leave the mines, before it was too late. Young children's immune and nervous systems are far more susceptible to environmental toxins and nutritional imbalances than are those of adults. So we must ask ourselves are our children the metaphorical "canaries in the mines" of today's high stress, high toxic environments?

A number of disturbing statistics indicate that children are becoming more at risk for brain-related problems. For example, during the past several decades the prevalence of autism and related pervasive developmental disorders has dramatically risen in the U.S. In 1970s and 1980s autism was a rarely diagnosed disorder affecting about one out of every 2,000 children. By 2008 the CDC estimated that nearly 1 in 88 children were diagnosed with autism. That statistic represented an overall increase of about 25 percent since the CDC's prior analysis in 2006 and a near-doubling of the rate reported in 2002. By 2013 the CDC estimated that nearly one in every fifty children is now affected by an autism spectrum disorder.

While autism spectrum disorders are part of a changing diagnostic entity, the severities of children's associated symptoms are easily noticeable by family members. Most parents do not seek out or desire a diagnosis of autism disorder for their children. Genetic factors are clearly important, as indicated by high concordance rates among twins and siblings. Yet, genetic factors alone cannot account for an epidemic that developed in the relatively short period of 10-20 years. Thus environmental factors are very likely to account for a major portion of the increased prevalence of autism. Recent studies of sulfur metabolism in children with autism reveal a pattern of abnormalities indicative of the presence of oxidative stress and impaired methylation (James et al., 2006).[13]

Major depression is the leading cause of disability in the U.S. and across the globe in this century[14] Studies of 39,000 people in nine different studies from five regions of the world found that young people in each country are more likely to experience depression than their elders' generation. Rates have risen in the past decades, doubling about every ten years in industrialized countries."[15-16] Alarmingly, preschoolers are among the fastest-growing market for sales of antidepressant drugs. At least four percent of preschoolers — over a million — were reportedly diagnosed as clinically depressed."[17]

Rates of depression and anxiety among young people in America have steadily increased over the past fifty years. "Today five to eight times as many high school and college students meet the criteria for diagnosis of major depression and/or an anxiety disorder as was true half a century or more ago."[18] According to Dr. Gray (2010), the increase in anxiety and depression disorders is not the result of changed diagnostic criteria; it holds even when similar questionnaire measures (e.g. MMPI) are held constant. According to the World Health Organization, major depression is the leading cause of disability among Americans age 15 to 44.

Not only depression, but rates of attention deficit hyperactivity disorder (ADHD) in children and adolescents continue to steadily rise over the past decades. The percentage of children diagnosed with ADHD increased from 5.5 percent in 1997 to 8.4 percent in 2010, an increase of more than 50 percent.[19] By 2010, more than one out of every ten males ages 3 to 17 (11 percent) received a diagnosed of

ADHD, based on parental reports. Among youth diagnosed with ADHD, 66.3 percent were taking medication for their "disorder."[20]

The term "children with special needs" includes children with a broad range of chronic health conditions, from major developmental disabilities like autism to less severe conditions such as asthma. In 2010, an estimated 18 percent of elementary-school-aged children and teenagers were identified by their parents as having "special health needs." Some estimates have suggested that "As many as one in six children nationwide has a neurodevelopmental disability, including autism, speech and language delays, and attention deficit hyperactivity disorder."[21] The percentage of boys with a specific learning disability, identified in kindergarten through grade 3 (between 2001 and 2007), increased from 8 to 13 percent. Of those children receiving special resource services, nearly three-quarters (72 percent) had some type of speech or language delay.[22] For parents, coordination of care for their "special needs" children, often involves schools, doctors, and community resources and requires a substantial commitment of time, money, and energy.

What is happening to the mental and physical health of our children? Could the dramatic rise in childhood immune and central nervous system (brain) problems, be related to environmental toxins and nutritional imbalances? If so, microalgae solutions offer a number of nutrient factors that can help to protect our children from harmful toxic exposures, help to detoxify contaminates, help to fill in missing nutritional components, and help to improve overall brain function.

The Evolution of the Human Brain and Microalgae Aquatic Nutrition

Louis Leakey, legendary British archaeologist and son of a missionary, helped establish Africa as the origin of human evolution. He declared that in order to find early human fossil remains, it was necessary to locate areas where water had been. Why? Water is essential for biological life, necessary for early humans to keep cool, and a rich source of aquatic plant and animal nutrition. Primarily, microalgae growing in water and eaten by fish provide the richest sources of "ready-made" DHA and other important omega 3 fatty acids. Animal life first originated in the sea. The fatty acids that provided essential components for brain cell membranes and for the photoreceptors in the eyes are most plentiful in the microalgae-fertile waters. Cold-water fish, which provide food sources for DHA and EPA omega-3 fatty acids, obtain these nutrients from microalgae.

The chemistry of the brain and nervous system is characterized by a heavy investment in lipid chemistry that accounts for up to 60 percent of its structural material.[23] One of the greatest evolutionary mysteries, the question of what spurred the change in hominid brain size remains a point of significant contention among anthropologists. Our modern brain weighs roughly twice as much as did the brain of our similarly sized primordial human relative, *Homo habilis* two million years ago.

Some brain researchers have suggested that the development of the human brain—which requires up to 10 times as much energy as that of other land-based mammals—depended on a rich source of fatty acids, especially DHA (docosahexanoic acid). DHA, the most abundant fat in the brain, is found in marine and coastal food chains and is not as easily obtainable on land.[24-25] The ultimate source of important fatty acids in the food chain, algae may have significantly contributed to the evolution of the human brain.

"Of necessity and convenience, early hominids would have made use of the aquatic food chain thereby making possible the spectacular evolution of the brain and brain size . . . Our brain, therefore, needs a direct dietary supply of DHA [docosahexaenoic acid], AA [arachidonic acid], and EPA [eicosapentaenoic acid]. It is important to note that much of the AA, EPA, and DHA in the food supply originate from microalgae which grow in water. Hence aquatic food chains are especially rich in these three important brain fatty acids,"[26] so states Dr. Stephen Cunnane and Dr. Kathy Stewart (2010).

Dr. Michael Crawford describes that "DHA is the one constant in the visual and nervous systems in all organisms, attesting to its singular importance for signal transmission in these systems. One of ... [the] most compelling arguments about the impact of DHA for brain evolution involves brain size vis-à-vis body size. In mammals consuming a terrestrial diet low in DHA and AA, as body size increased from very small rodents to massive savanna herbivores, with just one exception, relative to the rest of the body, brain size did the reverse—it shrank. The exception is humans. The question is why?"[27]

Recent fossil evidence indicates that the rapid expansion of our species' ancestral human brain took place in coastal areas, where aquatic food rich in long-chain polyunsaturated fatty acids (PUFAs), such as algae, mollusks, crustaceans, and fish, was abundant. "The shores gave us food security and higher nutrient density" says Dr. Stephen Cunnane, a professor of nutrition at the University of Toronto, Canada Research Chair in Brain Metabolism and Aging, and author (2005) of "Survival of the Fattest."[28]

Human babies are typically born fatter than the offspring of any other mammals; they are the only primate babies born with a built-in excess fat reserve. This fat reserve accounts for about 14 percent of their birth weight, a percentage similar to that of their brains. Dr. Cunnane hypothesizes that it's this baby fat, fortified with microalgae-oil rich food sources, which helped to fuel early hominids' evolutionary brain expansion. Human brains aren't just comparatively big, they're also very hungry. The average newborn's brain consumes an amazing 75 percent of an infant's daily energy needs.

Ancient Shoreline Environments Provided a Year-Round, Accessible and Microalgae Rich Foods

Brain development and function requires ample supplies of a particular polyunsaturated fatty acid: DHA. DHA is critical to proper neuron function. Human baby fat provides both an energy source for the rapidly growing infant grey matter, and also, says Dr. Cunnane, a greater concentration of DHA per pound needed than at any other time in life.

Ancient shoreline environments provided a year-round, accessible and rich food supply, unlike the prehistoric savannas or forests.

> *The particular physical size and shape of individual fatty acid molecules is what gives the cell membrane its structural flexibility and fluid-like properties. Altered "trans fatty acids" pack more tightly together, making the cell membrane more saturated, rigid and less flexible.*

Wetlands, rivers, and lakes dominated East Africa's prehistoric Rift Valley, in which early humans evolved. Aquatic, algae-oil-rich shorelines provided essential brain-boosting nutrients and minerals that launched *Homo sapiens'* brains past those of their primate peers, says Dr. Cunnane. Also, aquatic foods are also rich in iodine, a key brain nutrient. Iodine is present in much lower amounts in terrestrial food sources. It was this combination of abundant shoreline food and the "brain selective nutrients" that sparked the growth of the human brain, he says.

"Initially there wasn't selection for a larger brain," argues Dr. Cunnane. "The genetic possibility was there, but it remained silent until it was catalyzed by this shore-based diet." His ideas are supported by fossil evidence collected by Dr. Kathy Stewart from the Canadian Museum of Nature in Ottawa. Her study of fossil material excavated from numerous *Homo habilis* sites in eastern Africa revealed a bevy of chewed fish bones, particularly catfish bones. Other researchers discovered that the bones of early modern humans found in Britain, Russia, and the Czech Republic (dated 20,000 to 28,000 years ago), showed that fish and seafood accounted for 10% to 50% of their dietary protein.[29] In contrast, the inland Australopithecines did not have easy access to aquatic omega 3 fatty acids and retained a brain capacity that was not much bigger than a chimpanzee for three million years.

Yet, Cunnane is not without his critics, who claim that "Calculations indicate that a diet composed of 35% terrestrial animal matter and 65% terrestrial plant matter would have provided more than adequate raw material for brain-building purposes, not only sufficient amounts of DHA but also of AA and docosatetraenoic acid. . . The 'aquatic foods argument' also offers no real explanation for why these foods stimulated human brain expansion."[30]

Still the importance of key shoreline nutrients to brain development is still painfully with us. For example, iodine deficiency is one the world's most common nutrient deficiency. It affects more than 1.5 billion people, mostly in inland areas, and causes suboptimal brain function. Iodine, especially rich in seaweeds, was legally required to be added to salt in more than 100 countries. In fact salt, so abundant in the seas, was once considered more precious than gold by Europeans. Most terrestrial salt had long-ago washed into the seas or was deposited under the earth. Obtaining salt therefore was extremely costly, until Europeans discovered how to acquire salt from oceans and from underground mines.

Microalgae Offer a Number of Neurologically Active Substances

In addition to providing a vital source of essential fatty acids, microalgae potentially offer other neurologically active substances. For example, microalgae can offer good sources of B vitamins. Deficiencies in one or more B vitamins may contribute to fatigue, irritability, anxiety, mental slowness, memory loss, and confusion—many of the mood changes and behaviors associated with depression.[31] These changes in mood are not surprising, since the classical symptoms of some vitamin deficiencies appear as psychological problems.[32] Likewise, amino acid deficiencies can directly affect the production of key neurotransmitters.

That amino acids are essential and affect brain function was established over half century ago in a series of experiments led by William Cumming Rose. Healthy male graduate students were placed on diets consisting of cornstarch, sucrose, butterfat without protein, corn oil, inorganic salts, the known vitamins, a large brown "candy" made of liver extract flavored with peppermint oil (to supply any unknown vitamins), and mixtures of highly purified individual amino acids. Rose reported that the symptoms of nervousness, exhaustion, and dizziness were encountered to a greater or lesser extent whenever human subjects were deprived of even a single essential amino acid.[33] Given findings such as these, it is worth noting that blue-green algae contains all of the brain food nutrients below, including neuroactive amino acids.

Microalgae Offer a Rich Source of Amino Acids and Essential Co-Factors

<u>Neuroactive Nutrients in (1g) *AFA* microalgae</u>:
Glutamine 152 mg
Phenylalanine 51 mg
Tryptophan 15 mg
Glutamic acid 8 mg
Tyrosine 35 mg
Histidine 19 mg
Asparagine 95 mg
Valine 64 mg
Proline 57 mg
Threonine 65 mg
Glycine 59 mg
Methionine 15 mg
Gama Linolenic Acid
Mycosporine-like amino acids (MAAs)
C-Phycocyanin/phycoerithrocyanins complex (C-PC/PEC)

Blue-green algae offer an especially rich source of amino acids and essential co-factors which in turn supports the production of major neurotransmitters. Amino acids like tryptophan, threonine + glycine, and phenylalanine + tyrosine are essential building blocks for major neurotransmitters. B-vitamins, vitamin C, trace minerals (e.g., zinc and magnesium) and essential fatty acids are all involved in facilitating healthy neurotransmitter functions.

Tryptophan is an essential amino acid building block for the production of serotonin (a calming neurotransmitter) and melatonin (a sleep-inducing neurohormone). Tryptophan helps to improve sleep, mood (both depression and anxiety) and to decrease stress. Tryptophan has been reported to benefit patients with restless leg syndrome, help to control appetite (lower carbohydrate cravings), to improve behavioral impulse control, and to lessen suicidal thoughts.

Threonine works with glycine and increases glycine levels in the CNS. Other nutrients are better absorbed when threonine is present. Threonine may be helpful in treating some types of anxiety and depression. Symptoms from threonine deficiency include emotional agitation and confusion. Increased need for this amino acid occurs during times of stress. A deficiency of threonine can result in irritability in children. Threonine is used to treat a variety of nervous system disorders including spinal spasticity, multiple sclerosis and amyotrophic lateral sclerosis (ALS, Lou Gehrig's disease).

Glycine has been used to treat hyperactivity, mood disorders and to prevent epileptic seizures. Glycine is considered one of the most important inhibitory neurotransmitters in the central nervous system (CNS); an amino acid reported to calm the nervous system and have an anti-inflammatory effect. Glycine works in the brain to reduce unwanted muscle contractions (spasticity).

Phenylethylamine (PEA) is an amino acid linked to energy, mood, and attention. PEA exists naturally in the human central nervous system and is responsible for the feelings associated with pleasure and mental awareness. The most important action of PEA is promoting the neurotransmission of catechol amines: noradrenaline, dopamine, and epinephrine. Catecholamine helps our body respond to stress, such as "flight, fright, or freeze" survival responses.

Tyrosine is created from PEA or ingested directly as a dietary amino acid. The catecholamine metabolic pathway is as follows: PEA converts into L-Tyrosine, followed by L-dopa, which is metabolized into dopamine, to be converted into norepinephrine and finally changes into epinephrine. Thus tyrosine is essential in the production of excitatory neurotransmitters that boast mental energy, concentration, and arousal. Tyrosine has also been used to decrease hunger, for appetite control, and to improve mood.

Taurine and **Glutamine** both contribute to the synthesis of GABA in the brain. GABA is the major inhibitory neurotransmitter, occurring in 30 to 40 percent of all synapse, responsible for calming a distressed or overly excited mind and controlling impulsive or aggressive behaviors. Vitamin B6 is also essential for the production of GABA.

The list could go on. For example, **Methionine** increases the production of specific enzymes that help to increase the amount of memory-enhancing neurotransmitters, such as dopamine, and mood-elevating endorphins. **Proline** may help with learning ability. Oral consumption of nucleic acids may help to improve memory. Nutritional pioneer, Cheraksin Williams (1974) listed threonine, along with most of the B vitamins, magnesium, ascorbic acid, iodine, potassium, tryptophan, lysine, inositol, and glutamic acid as being essential in the prevention and treatment of mental illness.[34]

The Impact of Omega 3, DHA on Mental Health

"Comparative studies first demonstrated the universality of docosahexaenoic acid (DHA) in neural signaling systems of fish, amphibian, reptiles, birds, mammals and primates alike . . . Genomic research has explained that DHA is responsible for transcribing over 170 genes required in brain development."

— Crawford, M

DHA and EPA are absent from most vegetable fats and oils, and deficient in dairy products. The richest dietary sources are found in certain microalgae and seafood sources, though poultry and eggs can provide lower, but important, sources of EPA and DHA. Interestingly, microalgae-source DHA is being feed to chickens and cows to increase the DHA content of eggs and milk, respectively. A few organic dairy processors are even adding DHA algal oil directly into their organic milk products. While our body can convert DHA from its precursor fatty acids, this process varies from person to person and many factors hinder our bodies' ability to convert (ALA/LNA) essential fatty acids into DHA and EPA.

Some researchers believe an imbalance of omega-3 and omega-6 fatty acids leads to a variety of mental disorders, including hyperactivity, depression, brain allergies, and schizophrenia. A balanced ratio of these two fatty acid families is necessary for a healthy brain, which is structurally composed of a 1:1 ratio of omega-6 to omega-3 fatty acids. It has been estimated that the ratio of omega-6 to omega-3 fatty acids in the diet of early humans was 1:1, but the ratio in the typical Western diet is now almost 10:1 due to increased use of vegetable oils rich in LA and reduced fish consumption.[35-36] A number of medical experts have suggests that increasing the relative abundance of dietary omega-3 fatty acids will provide numerous health benefits.

In addition to the many stresses of modern life, scientists at the National Institutes of Health (NIH) associated the increase in depression in North America during the last century with the decline in consumption of DHA (docosahexaenoic acid) during that same period. Remember that DHA is the primary essential fatty acid within our brains and that microalga is the primary source of DHA in the food chain. Joseph Hibbeln, M.D., and Norman Salem, Jr., Ph.D., (1995) concluded that the "relative deficiencies in essential fatty acids may also intensify vulnerability to depression."[37]

Doctors in Australia who studied the blood of people with moderate to severe depression found the balance of essential fatty acids was significantly altered, with the level of the omega-3 fatty acids being too low.[38] Researchers of Purdue University found that individuals with symptoms of hyperactivity and attention deficit had lower levels of the omega-3 fatty acids, including DHA, in their blood.[39] All these clinical observation studies related imbalance in dietary intake of fatty acids to impaired brain performance and diseases in some genetically vulnerable individuals.

Evidence is rapidly accumulating to suggest that omega-3 fatty acid deficiency contributes to cognitive decline in older adults. An epidemiology team led by Kalmijn tested 1,613 subjects, ages 45-70, for various cognitive functions at baseline then again after five years and correlated the results with habitual food consumption reported on a self-administered food questionnaire.[40] Subjects exhibiting the most impaired cognitive function (lowest 10 percent of the group score) also had the lowest intake of DHA/EPA or fatty fish. Overall cognitive performance and psychomotor speed were positively correlated with DHA/EPA status. High intakes of cholesterol and saturated fat were also linked to increased cognitive impairment in this middle-aged population.

In the past decade, various epidemiological studies indicate relatively high DHA and EPA intake lowers the relative risk of dementia incidence or progression.[41-42] Recent studies have associated deficits in DHA with cognitive decline during aging and in neurodegenerative disease.[43-44] The importance of DHA-derived neuroprotection contributes to brain cell survival and repair; through nourishment of neurons, prevention of cell death, and anti-inflammatory (REDOX) signaling factors. Regarding young developing brains, many public health officials have recommended that DHA be added to infant formulas to support healthy brain growth in babies.

A study published in the journal *Behavioral Neuroscience* (2009) examined the relationship between omega-3 fatty acid intake and the nervous system's ability to handle sensory input – a function called "sensorimotor gating." Mice were fed two diets. One diet containing only ALA; the other diet had ALA plus EPA and DHA. All mice were evaluated on their response to a loud noise following a soft tone. The soft tone acts as a warning so that the animal is not startled by the loud noise. The mice fed ALA + EPA + DHA had a 12% increase in brain DHA content and displayed improved sensorimotor gating – they responded more calmly to the loud noise.[45] Inadequate sensorimotor gating is central to human disorders like schizophrenia, autism, and ADHD. Also, children with ADHD are reported to have reduced DHA levels, and DHA has been shown to improve the symptoms associated with ADHD.[46-47]

Microalgae is an Excellent Natural Source of Branched Chain Amino Acids (BCAAs)

More recent scientific research has demonstrated that dietary consumption of branched chain amino acids (BCAAs) leucine, isoleucine, and valine; "increased mitochondrial biogenesis . . . and up-regulated ROS defense system, with reduced oxidative damage." "Mitochondria are critical in providing metabolites for the de novo synthesis of nonessential amino acids."[48] Furthermore, these Italian researchers claim that a "balanced amino acid mixture with a high content of branched-chain and other essential amino acids. . . had been found to improve age-related disorders in animals and humans."

Amino acids are degraded into compounds that can be metabolized into CO_2 and H^2O, or used to make glucose from noncarbohydrate sources (gluconeogenesis). Thus, the oxidative degradation of amino acids produces an estimated 10 to 15% of total metabolic energy in animals. Given the heavy energy requirements of the brain, having optimal amino acids available is essential to support healthy brain function. Blue-green microalgae provide one of the highest and most balanced profiles of BCAAs, with a complete range of essential and nearly all nonessential amino acids.

Amino acid (BCAA) therapy has demonstrated "reversed injury-induced shifts in net synaptic efficacy, and led to reinstatement of cognitive performance after concussive brain injury," in controlled experiments with brain-injured mice. "These results suggest that dietary BCAA intervention could promote cognitive improvement by restoring hippocampal function after a traumatic brain injury."[49] Most importantly, human clinical data has likewise shown supplementation of BCAAs to enhance the cognitive recovery of patients with traumatic brain injury, even patients in a vegetative or minimally conscious state.[50-51]

Nutrition Strategies may Enhance Recovery for Ischemic Stroke Patients

Italian medical researchers (Aquilani, et al. 2011) reported that nutrition strategies enhance recovery for patients in rehabilitation after an ischemic stroke.[52] They report that "nutrition interventions can enhance the recovery of neurocognitive function in individuals with ischemic stroke. Experimental studies have shown that protein synthesis is suppressed in the ischemic penumbra [area surrounding the dead brain cells]." Patients given amino acid supplementation have an enhanced recovery of neurocognitive function. Cellular damage in cerebral ischemia is also partly caused by oxidative damage secondary to free radical formation and lipid peroxidation." Furthermore, "patients whose daily zinc intake was normalized had better recovery of neurological deficits than subjects given a placebo." The Italian's also described the neuroprotective benefits of B-vitamins. Microalgae especially rich in amino acids, antioxidant pigments, absorbent zinc and a family of B-vitamins offers a fairly complete nutritional strategy for patients recovering from stroke.

A Uniquely Discovered Light Pigment Complex in *AFA*

An evolutionary view of brain development and microalgae must lead to the conclusion that if our human brains were significantly shaped by the unique biochemical profile of microalgae, then microalgae has additional, even unknown properties, which may influence the neurotransmission properties of the brain.

For example, researchers in Spain described the effects of several species of algae on the central nervous system. Aqueous extracts of two species of microalgae showed antidopaminergic effects and anticholinergic properties.[53-54] Extracts of two other microalgae species showed promise as a central nervous system (CNS) depressant and a potential muscle relaxant.[55] Certainly the CNS effects of micro algal species will vary considerably. Yet microalgae offer promising, relatively unexplored and novel sources, of neuro- and biologically active molecules.

One of the most remarkable features of cyanobacteria, the most ancient of the microalgae, is their evolutionary ability to harness the rays of the sun. Like the Greek myth about Prometheus who gave humankind the gift of fire, cyanobacteria gave all plants and trees their original photosynthetic pigments to capture the rays of the sun. Pigments are chemical compounds which reflect only certain wavelengths of visible light. Pigments found within flowers, animal fur, corals, and even human skin makes things appear "colorful." Yet, even more important than their ability to reflect light, pigments have the ability to absorb certain wavelengths of light energy. However, since each pigment only captures a narrow range of the light spectrum, usually several pigments, each of a different color, are needed to catch more of the sun's energy.

Most people are familiar with the green light-harvesting pigment, chlorophyll, as taught in elementary science classes. However, few people know that there are a couple different chlorophyll pigments, and extra pigments that capture an even wider range of the light color spectrum. Thus nutritional scientists talk about a family of pigments, such as chlorophylls, carotenoids, or phycobilins (found in red and blue-green alga); all which have powerful antioxidant, as well as other health benefits. In microalgae research, the most studied pigment to date is phycocyanin, a photosynthetic blue "bilin" pigment, which along with chlorophyll gives cyanobacteria its characteristic "blue-green" color. Phycocyanin extracts have been shown to possess antioxidant and anti-inflammatory properties across a number of different physiological systems, such as the liver and the brain. They have been shown to assist in cell functioning, protect brain cells, support the liver, neutralize damage from free radicals and strengthen cell walls.

Yet, an interesting finding came to light when Italian University scientists were measuring the oxygen radical absorbance capacity (ORAC) values of *Aphanizomenon flos-aquae* (*AFA*). They discovered that the antioxidant properties found within the light pigment complexes, phycocyanin (PC) and phycocyanobilin (PCB), derived from *Aphanizomenon flos-aquae*, were both considerably higher than other tested pure antioxidant materials.[56] Because of their findings, this team continued to investigate the remarkable light pigment complexes and made further substantial discoveries.[57]

To start with, they were able to identify "a fundamental difference between *AFA* and *Spirulina*," based on a different structure found within their respective phycobilisomes. A phycobilisome is a light-harvesting complex containing various photosynthetic accessory pigments, found in all cyanobacteria and red algae. All phycobiliproteins are water soluble pigments complexes that adhere to membranes called phycobilisomes. As opposed to *Spirulina*, the phycobilisomes of *AFA Klamath* does not contain the pigment allophycocyanin, a pigment that absorbs and emits red light, but contains a pigment missing in *Spirulina*, namely phycoerythrocyanin. Phycoerythrocyanin is a magenta colored light-harvesting pigment, which contains both a violet pigment, phycoviolobilin, and a blue pigment, phycocyanobilin.

> *Phycoerythrocyanin a unique light pigment complex found within AFA is a powerful MAO-B inhibitor. MAO-B inhibitors support the effectiveness of dopamine, a powerful neurotransmitter associated with the "feel good" reward system within our brain.*

So far, only a few microalgae have been discovered to date that contain this pigment complex, phycoerythrocyanin. These scientists were able to demonstrate that the unique light pigment complex found within *AFA* was a powerful MAO-B inhibitor. MAO-B inhibitors support the effectiveness of dopamine, a powerful neurotransmitter associated with the "feel good" reward system within our brain.

Monoaminooxidase (MAO,) comes in two forms A and B. While both are found in *AFA*, by far the highest concentrations are MAO-B inhibitors, found within *AFA* microalgae. MAO-B has a beneficial effect on dopaminergic transmission which is involved in neurological disorders, where dopamine is essential, such as depression, mood disorders, problems with addictions, as well as both Parkinson and Alzheimer's disease.

In addition to the powerful dopamine enhancement of *AFA's* unique light pigment complexes, these scientists found other unique bioactive molecules; namely the amino acid, phenylehylamine (PEA), which supports dopamine production, this in turn is synergistically enhanced by powerful MAO-B inhibitors.

AFA is a uniquely concentrated food source of phenylethylamine (PEA), a monoamine alkaloid, found in certain microalgae and chocolate. In the 1980's phenylethylamine became popularized by Michael Liebowitz (1983) in his book the "*Chemistry of Love*," that reported chocolate was loaded with phenylethylamine.[58] His view led to the "chocolate theory of love," which caused a media sensation.

PEA is a member of a class of chemicals well known for their psychoactive and stimulant effects. Pharmaceutical versions of PEA have many applications in pharmacology, including use as stimulants and antidepressants. PEA can function as a neuromodulator of catecholamine neurotransmission in the brain. A deficit in phenylethylamine can be responsible for depression.[59] Studies show phenylethylamine can relieve depression, even in those that were unresponsive to standard treatments.[60-61] Also, PEA levels have been shown to be significantly lower in a group of children diagnosed with ADHD, as compared to a control group of children, with no elevated symptoms of ADHD.[62] A benefit in the use of phenylethylamine is that it improves mood without producing a tolerance often associated with amphetamines.[63] The mechanism of its effect probably is related to the stimulation of dopamine release.[64]

While phenyletheylamine is sold as a dietary supplement for reported mood and weight loss related therapeutic benefit, orally ingested phenethylamine is quickly inactived by monoamine oxidase (MAO-B), which turns PEA into phenylacetic acid. Therefore, MAO-B prevents orally consumed PEA from ever reaching therapeutic levels in the brain.[65-66]

> *AFA's unique light-pigment complex was described thus: "the phytochrome has proven to be the most powerful MAO-B inhibitor of all known substances to date.*

Remarkably, the Italian scientists' (2008) detection of *AFA's* unique light-pigment complex was described thus: "*the* phytochrome has proven to be the most powerful MAO-B inhibitor of all known substances to date. It's very high affinity for the MAO-B enzyme, and its effective inhibition at dosages of few nanomolars, makes this molecule not only a perfect therapeutic agent on its own, but the factor that seems to provide the most important contribution to the high neurological effectiveness of the *AFA* extract(s)."[67] Importantly, the multiple kinds of MAO-B inhibitors (*AFA* phytochrome, *AFA*-PC and MAAs) found within *AFA* were reported as having a "competitive and reversible character of the MAO-B inhibition [which] assures at the same time high efficacy and a physiological and side-effects free activity." Additionally, the MAO-B inhibitor molecules within *AFA* had a low molecular weight capable of easily crossing the blood-brain barrier and theoretically capable of protecting the PEA content in the algae to elevate therapeutic concentrations of both PEA and dopamine within the brain.

These scientists were so impressed by their University research findings they filed a 2008 patent claiming that *AFA* extracts could be effective therapeutic agents for the following neurological diseases conditions or disorders including: Alzheimer's disease, Parkinson disease, multiple sclerosis, hyperactivity and attention deficit disorders (ADHD), autism, depression, memory deficit and mood disturbances.

Can *AFA* Improve Neurological and Psychiatric Conditions?

Even earlier published research by a psychiatrist, Gabriel Cousens (1985) documented the clinical effectiveness of *AFA* blue-green algae to improve Alzheimer's disease symptoms in two patients. Dr. Cousens reported "some significant return of function" such as decreased hand tremors, better balance, and improved short term memory, attention span, judgment, and reasoning in one patient. In the second patient there was no significant return of previously lost function, but there was a halting of the typical "progressive degeneration associated with Alzheimer's" along with a corresponding improvement in the patient's marital relationship.[68]

A psychiatrist and perceptive observer, Dr. Cousens reports that compared to *Spirulina*—the other most popular species of blue-green algae—*AFA* acts more effectively on the central nervous system, making it clinically more useful to improving patients' "mental and emotional health." He advocates *Spirulina* for general detoxification and somatic (physical) purposes, while he endorses the use of *AFA* microalgae for psychiatric (mental, emotional, and behavioral) purposes within his medical practice.[69]

Dr. Cousens describes *AFA* as having "a unique energy that has an affinity for improving brain function. We now have clinical research showing *AFA* balances EEG readings and improves central nervous system functioning. . . I've had people who've been depressed for years and years, and literally, within a few days after receiving *AFA,* their depression lifts. This is because the *AFA* gets to the root of helping heal the addictive brain chemistry, which is behind a lot of depression. . . I've also had excellent results using the algae with children who've been autistic. . . I take *AFA* as a sort of mental adaptogen for stress."[70]

An article by Gabriel Cousens, M.D. (1995) titled "Microalgae: First and finest Superfood" reported that, "Perhaps the greatest value of *AFA* is not in its nutrient concentration or even its high concentration of neuropeptide precursors, but in its amazing positive effect on the nervous system in general, and specifically on the pituitary, pineal, and hypothalamic structures. *AFA* is an elixir for the brain-mind complex." He went on to describe how, "Over the years, people taking *AFA* have consistently reported an overall increase in mental alertness, mental stamina, short and long-term memory, problem-solving ability, creativity, dream recall, enhancement of the visualization process, and a greater sense of well-being and centeredness."[71]

Spiritual Nutrition and the Rainbow Diet

In his far-reaching book, Spiritual Nutrition and the Rainbow Diet (1986) Dr. Cousens advocates eating a "rainbow diet" based on the relationship between your food color choices and corresponding colors of the human chakra, energy, system.[72] He describes how the variously colored light-harvesting pigments in plants may interact with the human energy system. Dr. Cousens described an esoteric, spiritual, healing approach, while taking efforts to make connections with available scientific data. He describes how "Food is a dynamic force which interacts with humans on the physical body level, the mind-emotional level, and also the energetic and spiritual level." Dr. Cousens proposes a three tiered system of interaction between our circulatory, nervous, and meridian energy systems. Wherein he describes the interplay between light harvesting pigment molecules (e.g. chlorophyll), circulating blood heme, and electromagnetic, resonate, frequencies broadcast by crystalline-like biomolecules.

Optogenetics Controlling Brain Circuits with Light

Perhaps a day will come when certain colored light-pigments found within food will be found capable of shifting cellular function, including brain neurons. That day has already arrived in certain biotechnology laboratories under the name of "optogenetics," a new field of brain science. Optogenetics is the use of light, optics, combined with genetic biotechology to activate or silence specific cells, found in living tissue.

Finding effective mental health or psychiatric treatments is extremely difficult because the human brain is so complex. The standard paradigm of psychiatric disorders based on chiefly chemical imbalances, including altered levels of neurotransmitters, cannot adequately explain the speed of our brain's multi-tasking ability. Now neuroscientist have a radical new set of tools to precisely map and to control the firing of brain cells by the use of light sensitive microbial and microalgae, opsin genes.

"At least 40 years ago biologists knew that some microorganisms produce proteins that directly regulate the flow of electric charge across cell membranes in response to visible light. These proteins, which are produced by a characteristic set of "opsin" genes, help to extract energy and information from the light in the microbes' environments. In 1971 Walther Stoeckenius and Dieter Oesterhelt, both then at the University of California, San Francisco, discovered that one of these proteins, bacteriorhodopsin, acts as a single-component ion pump that can be briefly activated by photons of green light—a remarkable all-in-one molecular machine."[73]

"Using nothing more than safe pulses of visible light, we attained reliable, millisecond-precision control over the cells' patterns of firing of action potentials—the voltage blips, or impulses, that enable one neuron to convey information to another . . . by introducing a microbial opsin gene, we could make neurons precisely responsive to light. . . They worked in part because mammalian tissues contain naturally robust quantities of all-trans retinal—the one chemical cofactor essential for photons to activate microbial opsins—so nothing beyond an opsin gene needs to be added to targeted neurons."[74]

Certain Microalgae Appear to have an "Adaptogenic" Effect on the Nervous System

As previously reported by Dr. Cousen, some microalgae appear to have an "adaptogenic" effect on the nervous system. In herbal medicine, an "adaptogen" is a substance reported to help the body's ability to adapt to stress. For example, male rats exposed to water immersion-induced stress (i.e., experimental near-drowning) showed significantly fewer gastric mucosal lesions when fed *Dunaliella bardawil* (whole green microalgae) as compared to a control group given isolated synthetic betacarotene.[75]

Oral administration of unicellular green algae, *Chlorella vulgaris,* was found to prevent stress-induced ulcers in rats, as well.[76] Researchers hypothesized that microalgae prevent ulcer formation primarily through the "immune-brain-gut" axis and secondarily through possible gastric mucosa protective factors.

"A test reported in the Japan Medical News in 1965 indicated that two grams of supplemental *Spirulina* were able to cure all symptoms of gastric ulcers. In addition, seven out of nine cases of duodenal ulceration were completely cured, while the remaining two showed marked improvement."[77] The antipeptic ulcer effect of *Chlorella* was demonstrated in both human and animal studies.[78]

Over the past two decades or so, medical science has determined that most gastric ulcers involve a bacterial infection, and antibiotics are often the treatment of choice. Blue-green algae may be helping through a number of different possible mechanisms: an enhancement of the immune response; direct antibacterial effects; general support for the healing process (enhancing tissue regeneration and modulating inflammation); or possibly by providing the trace mineral, bismuth, known to be effective in healing ulcers caused by the bacterium *Helicobacter. pylori.* Anecdotal and experimental evidence from humans ingesting *AFA* blue-green algae is suggestive of an adaptogen, "antistress," effect.[79-80]

Antioxidants found abundantly in microalgae may contribute to protecting the central nervous system (CNS) from oxidative stress. Lipids, which constitute most of the brain tissue, are especially sensitive to oxidative damage. Researchers in Israel and Japan have demonstrated the protective effects of antioxidants in experimental animal brain trauma models.[81-82] Oxidative stress has been implicated in the pathogenesis of some disorders of the brain; hence antioxidants have become attractive therapeutic agents.[83] Furthermore, brain trauma and injury tend to increase whole-body oxidative stress.[84]

Scientists at the National Institute of Nutrition in Hyderabad, India, demonstrated that blue-green algae offer a cost-effective source of antioxidant carotenes for children.[85] Microalgae represent one of the richest natural food sources of carotenes, including beta carotene. Beta carotene offers powerful anticancer, antiaging, and antioxidant properties without the toxic risks of taking fat-soluble vitamin A.

Beta carotenes may act as "Tiny Molecular Shields" to Protect against Cognitive Declines

Beta-carotenes and antioxidants found abundantly in microalgae may contribute to protecting the central nervous system (CNS) from oxidative stress. Lipids, which constitute most of the brain tissue, are especially sensitive to oxidative damage.

Researchers at Erasmus University Medical School in the Netherlands conducted a three-year study of 5,100 people between 55 and 95 years of age and found that beta carotene molecules acted as "tiny molecular shields" and may provide dramatic protection against the ravages of aging, memory impairment, and general brain damage.[86] These findings suggest that beta carotene foods, like micro algal foods, need to be further investigated with our aging population. Microalgae might also provide a cost-effective way to reduce brain-related risks of aging. As "baby boomers" turn into "senior boomers," the number of Americans with Alzheimer's disease is projected to increase by more than 300%.[87]

Neuroprotection from ischemic brain damage is another area where edible microalgae solutions may be of benefit, as there is substantial evidence that oxidative stress and inflammation play a key role in the neurodegeneration following ischemic injury. Several studies have investigated the ability of blue-green algae to protect the brain before a focal ischemic stroke and reperfusion injuries.[88-90] Rats feed blue-green algae for one month prior to a surgically induced major stroke showed a close to 70% reduction in infarct size.[91]

Mild traumatic Brain Injury is one of the Most Common of All Neurological Disorders

Mild traumatic brain injury (MTBI)—also called concussion, minor head injury or minor brain injury, is one of the most common of all neurological disorders, accounting for "at least 75 percent of all traumatic brain injuries in the United States," according to a report from the Centers for Disease Control. Sports-related injuries account for an estimated 20 percent of MTBIs, with children and adolescents often involved. The report goes on to state, "However, it is clear that the consequences of MTBI are often not mild. . . These injuries may cause long-term or permanent impairments and disabilities. Many people with MTBI have difficulty returning to routine, daily activities and may be unable to return to work for many weeks or months."[92] Furthermore, twenty percent of combat veterans returning from Iraq and Afghanistan likely suffer MTBI.[93]

Researchers created a combination of nutrients (called NT-020) to which microalgae was added that promotes the proliferation of neural stem cells (neurogenesis).

Patients who survive head injuries often present with disabilities persisting up to decades after the injury.[94] Although the severity of disabilities varies, which may be directly associated with the severity of the injury itself.[95] The most common disabilities associated with head injury include sensory-motor problems, learning and memory deficits, anxiety, and depression.

Brain researchers identified an "overarching theme advances the concept that a massive neuro-inflammation after TBI represents a second wave of cell death that impairs the proliferative capacity of cells, and impedes the regenerative capacity of neurogenesis in chronic TBI . . . that chronic TBI-induced neuro-inflammation interfered [with the brain's] endogenous repair mechanisms."[96]

These findings support the potential benefits of using anti-inflammatory therapies, like microalgae antioxidants, following traumatic brain injury. While TBI is generally considered an acute injury, longer term effects of neuro-inflammation often result in further neuronal cell deaths, which surround the focal site of the acute injury, and decrease the ability of the brain to repair itself. The concept of untreated neuro-inflammation helps to explain the often devastating longer term, unexplained effects, of traumatic brain injuries, that may worsen or change over time.

Until recently, scientists believed that adult brains were incapable of producing new brain cells (neurons). However, a number of researchers have demonstrated that neurogenesis, the production of new brain cells, occurs during adulthood. There are two areas where neural stem cells reside in the brain, namely the sub ventricular zone (SVZ) and the sub granular layer (SB) of the dentate gyrus of the hippocampus.

Scientists at the University of South Florida have shown that natural compounds such as polyphenols from blueberry and green tea and blue green algae are high in antioxidant and anti-inflammatory activity that decreases the damaging effects of reactive oxygen species (ROS), in the blood, brain, and other tissues of the body. These researchers developed a combination of nutrients (called NT-020) to which microalgae has also been added that creates a synergistic effect that promotes the proliferation of neural stem cells (neurogenesis) *in vitro* and *in vivo*."[97]

In sum, NT-020 may promote health, proliferation, and maintenance of neurons in the age animals and exert anti-inflammatory actions that promote function in the aged stem cell niche. . . . After treatment with NT-020, there was a significant reduction in the numbers of aged impaired rats and a significantly reduced spread in the cognitive scores." Furthermore, through a series of animal experiments, these researchers were able to demonstrate that NT-020 provided significant levels of neuroprotection in an experimental treatment for stroke. They proposed that the "use of dietary supplementation as an adjunct therapy is likely to improve stroke outcome."[98]

Andrew Valencia and colleagues at the Neurolab Clinic associated with the University of New Mexico demonstrated that patients suffering from mild brain injury who ate *AFA* showed a 25% improvement in about half the time as patients who did not receive algae.[99] According to Valencia, in his study of more than 150 patients over two years, patients who ate *AFA* algae had improvements similar to those in a two-month hospital-based rehabilitation program. However, the best results were achieved when neuro-rehabilitation was combined with eating *AFA* algae, better than *AFA* alone or the hospital program alone. Valencia's research team hypothesized that *AFA* algae promotes reparative neuroplasticity—or, in lay terms, rewiring of the circuitry of the brain.

Valencia also conducted electrophysiological studies of brain waves and found that the ingestion of *AFA* algae was linked with pronounced improvements in brain function, notably in the ability to focus and discriminate between various auditory signals.

Benefits of *AFA* on Child Cognition, Mood, Behavior, and Academic Performance

In 1992 a program was initiated which provided 1 to 2 grams of AFA algae supplementation to 2,000 primary school children and 500 adults from Nandaime, Nicaragua. Due to the dramatic health benefits reported, the director of the Carmela Noguera School in Nandiame requested AFA algae be provided for their students, as well. After AFA supplementation was provided for all students during the 1993 academic year, the school's national standardized test scores soared and the school was awarded first place among all primary schools in Nicaragua, an outstanding achievement!

Two doctoral students, Irma Sevilla and Nereyda Aquirre impressed by the school's improvements, conducted a scientific, controlled evaluation of AFA microalgae's effects on children. They chose a different school, in the same town, which had not yet received the microalgae. From June through December, 1994, under the auspices of the Centro Comunitario in Nandaime and the Central American University in Nicaragua, these doctoral students assessed the health, behavior and academic status of first-, second-, and third-grade children attending the Monseñor Velez School in Nandaime, Nicaragua. The AFA supplemented students demonstrated an 81% increase in the average standardized test scores receiving only 1 gram of *AFA* a day over a six-month period. In contrast the control students (not receiving AFA) had a decrease in their standardized test scores over the same time period. Malnourished students given microalgae showed significantly increased classroom attendance and class participation, as well as marked improvement in overall health.[100]

Impressed by the Nicaragua findings, Claudia Jarratt, family therapist at the Center for Family Wellness in Harvard, Massachusetts, elected to see if behavioral improvements might occur in a sample of American children. She assessed 105 children given *AFA* and found significant improvement in their behavior evidenced by both parent and teacher ratings. The children, who displayed a variety of behavioral problems (e.g. from ADHD to depression), consumed between 0.5 and 1 gram of *AFA* daily and were observed over a 10-week period. Data from the Achenbach Child Behavior Checklists (ASEBA) and extensive case histories were collected for all participants. The ASEBA is a comprehensive evidence-based rating system developed through decades of research. Significant improvements were reported by parents across all 11 ASEBA behavioral rating scales in children receiving AFA. These findings were corroborated by teachers' ratings, which revealed significant improvements across seven of the ten behavioral symptom areas measured. Notably both inattention and hyperactivity symptoms were rated as significantly improved by teachers. The use of an expectancy scale revealed little correlation between parents' initial expectations of treatment benefits and final outcomes.[101] Subsequently, Jarratt continued to work with an additional 250 children, using an *AFA*-based nutrient program, with similar positive behavioral results.[102]

My own research team studied 26 students with reading difficulties, who participated in a three-month *AFA* supplement study.[103] All were enrolled in the Stilwell Learning Center, a reading tutorial program in Sierra Vista, Arizona. Participants included 18 boys and 14 girls, ranging from 6 to 17 years old, with a mean age of 11. The children were randomly assigned to one of two groups, (1) the low-*AFA* (1.5 g) treatment group or (2) the high-*AFA* (3 g) treatment group. There was also a non-*AFA* comparison group.

Both *AFA* treatment groups showed significant improvements on the following measures over the three-month trial period: attention and concentration indices, a sequential memory index, academic testing, parent and teacher behavioral ratings, health symptoms, tutorial attendance records, and flushing of toxic aluminum from the body. Both algae-supplemented treatment groups, consuming either 1.5 g or 3 g of *AFA* daily, demonstrated significant improvements compared to pre-test baseline measures and a small nonsupplemented comparison group.

This study, known as the Sierra Vista study, showed that all of the significantly elevated symptom scales (e.g., Hyperactivity, ADHD Index, and Cognitive Problems) as rated by parents at the start of the study were within a normal range by the conclusion of the three month study. Furthermore, teacher ratings revealed significant improvements on the two highest clinical scales (Hyperactivity and Cognitive Problems), as well. Neuropsychological measures showed a nearly 20 percent improvement on children's attention and concentration index and a whopping 45 percent improvement on children's sequential memory recall index. Moreover, AFA supplemented students showed significantly higher improvements in standardized test scores of reading accuracy, reading comprehension, and spelling compared to randomly selected control group of students who participated in the same tutorial program.

Children with learning challenges frequently report greater health problems. At the start of the Sierra Vista study, using a Metabolic Screening Inventory for Children, the overall group mean was suggestive of poor metabolic health, across a variety of physical symptoms. However, by the end of the study, both *AFA* treatment groups were significantly restored to a normal symptom range.

Figure 18. Pre- Post *AFA* Metabolic Symptom Changes

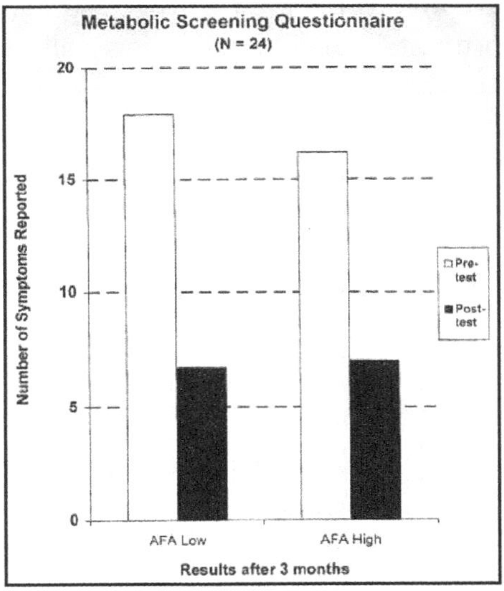

The health areas that showed the greatest improvements were symptoms related to the nervous system (e.g. less headaches and dizziness); improved muscles and joints; less gastrointestinal problems; and overall improvement in energy (e.g. less fatigue).

At the initiation of the study, hair assays from learning and behaviorally challenged students revealed highly elevated aluminum levels, exceeding toxic reference levels. By the end of the study, little aluminum was detectable, a significant reduction ($p < 0.0001$) in both *AFA* treatment groups. (Aluminum is a common mineral element, a suspected neurotoxicant, found in aluminum containers, baking powder, toothpaste, baby wipes, skin moisturizers, and vaccines).

Marnee Foldoe's Sonoma State University master's thesis documented that *AFA* improved attention deficit/hyperactivity disorder (ADHD) symptoms in three elementary school students. She examined the children's, teachers', and parents' attributions as to what caused the noticeable improvements in these diagnosed children. Interestingly, none of the teachers attributed the improvements to the dietary intervention, whereas the parents were convinced that eating the *AFA* made the biggest difference. Only the children reported that multiple factors were involved—their own volition, eating *AFA,* and social factors. Unfortunately, teachers in this small sample appeared to be completely unaware that diet could have any strong influence on ADHD.[104]

> *Attention Deficit Hyperactivity Disorder is our nation's leading mental health problem in children. Depression is one of the leading causes of disability in the world today. Both of these brain-related problems are associated with difficulties in arousal.*

A team of medical researchers headed by Dr. Krylov of the University of Illinois concluded, after examining hundreds of well-documented case histories, that *AFA* appears promising for the treatment of depression and ADHD as well as several other health challenges.[105]

A pilot study of several thousand *AFA* consumers indicated that symptoms of ADHD, depression, and memory difficulties improved with *AFA* consumption. This study was based on medical questionnaires similar to those used in a wide-scale health survey by the National Center for Health Statistics. In addition to specific improvements in a number of areas, the study found that people consuming *AFA* reported significantly better overall health than that reported by the general population in the larger survey.[106]

In an epidemiological study of 230 patients with multiple sclerosis in the Ukraine, it was reported that the intake of *Spirulina* increased the length of remission in patients with disseminated sclerosis.[107] Multiple sclerosis (MS) is a progressive disorder of the nervous system, where the lipid myelin sheath surrounding the nerves is gradually destroyed. Deficiencies of omega-3 fatty acids may contribute to the syndrome of MS.

British biochemists at the Institute of Brain Chemistry and Human Nutrition in London report that any deficiency in essential fatty acids (EFAs) during early brain development in childhood can greatly increase the risk of learning disorders. Nervous tissue contains 50% to 60% lipids on a dry weight basis. Lipids play an essential role in the structure, fluidity, and function of brain membranes.[108] *AFA* blue-green algae is especially rich in PUFAs, which are very important in maintaining membrane fluidity, comprising up to 10% of its dry weight.

Animal research at Massachusetts General Hospital and Harvard Medical School found that *AFA* algae dramatically raised blood levels of eicosapentaenoic acid (EPA) and docosahexaenoic acid (DHA). EPA and DHA are known to be important for optimal functioning of numerous organ systems, including the nervous system. They are extremely difficult to obtain in the modern American diet. The researchers found that *AFA* was far more effective than soybean oil, a good source of PUFAs, at raising blood levels of the important omega-3 fatty acids.[109]

A monumental report released from the Center for Science in the Public Interest (1999) acknowledged that food dyes and certain foods can adversely affect children's behavior. The report "Diet, ADHD & Behavior, A Quarter-Century Review" cites 17 controlled studies that detail the effects of diet on behavior. Michael F. Jacobson, the executive director of CSPI stated "It's high time that the government—as well as doctors—provided the public with accurate information that might help many children."[110]

In addition to EFA deficiencies, several studies have indicated specific vitamin and mineral deficiencies in children with ADHD.[111-112] For example, in 1997 Kozielec and Starobrat-Hermelin examined 116 children with ADHD and found that 95% of those examined were deficient in magnesium in blood serum and hair.[113] Magnesium has a profound influence in regulating the central and peripheral nervous systems. Chronic magnesium deficiency results in hyperactivity, impaired reaction to external stimuli, irritability, fatigue, difficulty sleeping, and poor mental concentration. Researchers also discovered a positive correlation between zinc deficiency and hyperactivity.[114]

Numerous studies suggest that heterogeneous biochemical etiologies for ADHD cluster around at least eight risk factors: food and additive allergies, heavy metal toxicity and other environmental toxins, low-protein/high-carbohydrate diets, mineral imbalances, essential fatty acid and phospholipid deficiencies, amino acid deficiencies, thyroid disorders, and B-vitamin deficiencies.

Children with AD/HD are more likely to show deficiencies in DHA than children without attention problems, and EFA supplementation has shown initial promise in the treatment of AD/HD.[115] Gamma-linolenic acid (GLA), a powerful fatty acid found in *AFA,* may be helpful as well.[116] GLA has been linked to the release of neurochemicals that improve mental attitude, increase alertness, and reduce depression.

Attention deficit/hyperactivity disorder is our nation's leading mental health problem in children. Depression is one of the leading causes of disability in the world today.[117] Both brain-related problems, depression and ADHD, are associated with difficulties in arousal. While ADHD children appear hyperactive, many scientists believe this is actually a problem of under-arousal, which increases symptoms of self-stimulation, risk-taking, and poor attention. This helps to explain why amphetamines (stimulant drugs) are prescribed to many children with ADHD. Depression in children may sometimes even appear like ADHD. Are there any links between the epidemic of hyperactivity in children (with acute arousal symptoms) and depression in adults (with chronic arousal symptoms)?

Children diagnosed with ADHD at ages 4 to 6 are more likely to suffer from depression as adolescents than those who did not have ADHD at that age, according to a long-term study published in the October 2010 issue of the *Archives of General Psychiatry.* The researchers found that 18 percent of children diagnosed early with ADHD suffered from depression as adolescents, about 10 times the rate among those without ADHD. Children with early ADHD were five times as likely to have considered suicide at least once and twice as likely to have made an attempt.[118]

Although depression and ADHD are usually diagnosed and treated as different "disorders," Larry Christensen in his survey of research on diet and behavior found evidence that these two brain-related problems are the two types of behavioral disorders most likely to respond to dietary intervention.[119] Interesting? Given that these two conditions appear to be increasing, what might this suggest about the standard American diet (SAD)?

Two Experts in this Area Render Their Opinions:

John Taylor, Ph.D., psychologist, author, and AD/HD expert states, "I have been in a position to talk with thousands of parents and professionals very frankly about AD/HD. And blue-green algae are consistently mentioned to me by parents as being of help for children with AD/HD."[120]

Further, Edward Hallowell, M.D., the author of the *New York Times* bestseller *Driven to Distraction* and a leading authority on learning disabilities, particularly AD/HD. In his appearance as a keynote speaker at the 1998 Pacific Region Learning Disabilities Drug Treatment Conference in Honolulu, Hawaii, Dr. Hallowell reported that blue-green algae appears to offer a promising nondrug alternative for people with AD/HD.[121]

CHAPTER

10

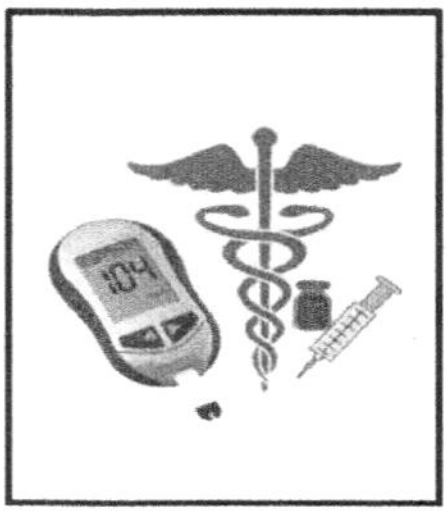

Microalgae Solutions for the Diabetic Epidemic

"Research on humans and animals shows algae components offer significant utility in the prevention and control of diabetes. Aligned studies have demonstrated algae's therapeutic value for the diseases common with diabetics; cholesterol management, blood pressure, heart disease and cancers. Algae can moderate chronic inflammation that often proceeds and accompany degenerative diseases. Algae compounds provide therapeutic value for diabetes and fat metabolism." [1]

— Edwards M, 2013

"Banting's discovery of insulin only allowed them [diabetic patients] to live just long enough to develop blindness, renal failure, and coronary disease." [2]

— Brownlee M, 2005

"Oxidative stress plays a pivotal role in the development of diabetes complications, both microvascular and cardiovascular . . . increased superoxide production causes the activation of 5 major pathways involved in the pathogenesis of [diabetic] complications." [3]

— Giacco F, Brownlee M, 2010

Microalgae Hypothesis V: Improves Blood Sugar Balance and Reduces Diabetic Complications

Microalgae can help improve hyperglycemia control.

Microalgae can balance redox potential and protect mitochondria damage from free radicals.

Microalgae omega-3 fatty acids support cardiovascular health.

Microalgae can decrease blood platelet aggregation.

Microalgae can protect against oxidation of lipids (fats).

Microalgae can lower triglyceride and "bad" cholesterol levels.

Microalgae can improve vasodilation and increase blood flow.

"Obesity itself is not the cause of our dramatic upswing in chronic disease. Rather, it's metabolic syndrome, which can strike those of "normal" weight as well as those who are obese. Metabolic syndrome is a result of insulin resistance, which appears to be a direct result of consumption of added sugars . . . obesity is a marker for metabolic syndrome, not a cause." [4]

— Bittman M, 2013

Diabetes mellitus (DM) is a metabolic disorder characterized by high amounts of sugars, fats (lipids), and amino acids in the bloodstream, along with impairments in insulin secretion and insulin effectiveness. A deadly "triopathy" of diabetic complications (retinopathy, nephropathy, and neuropathy) is observed in substantial numbers of diabetic patients. Among adults, diabetes is the leading cause of new cases of blindness, kidney failure, and the amputations of feet and legs unrelated to accidents or injury. People with diabetes have a far shorter life expectancy — their risk of death due to vascular complications is three-fold higher in patients with DM than in the remaining population. The Centers for Disease Control (CDC) estimates that for children diagnosed with diabetes, their life span could be shortened by as much as nineteen years. Tragically, we are reaching an alarming reversal of expected lifespan. Most children born today are likely to die earlier than their parents and grandparents, if current trends continue.

Nearly 26 million people in the United States, over eight percent of the population, have diabetes today. Yet the CDC predicts that, if American's don't change their lifestyle and dietary patterns (e.g., poor diets, high stress, lack of exercise), fully "1 of 3 U.S. adults will have diabetes by 2050." On average, the medical expenses for an individual diagnosed with diabetes are over twice the amount as a person without diabetes. Total cost (direct and indirect) of diabetes for the U.S.A. in 2012 was estimated at $245 billion.[5] "We know diabetes has been increasing for decades, but to see 18 states [from 1995 to 2010] having an increase of 100 percent was shocking," described Linda Geiss, a statistician at CDC's division of diabetes. The tragedy is that diabetes is largely preventable through lifestyle changes, she added.[6]

The Sugar and Diabetes Connection

The sugar industry, including the world's largest food and beverage companies, have claimed for decades that there is no relationship between the consumption of sugar and the risk of developing diabetes. Using tactics similar to those employed by the tobacco industry to stall public awareness and governmental legislation, the sugar lobby has lied, selectively funded research, and influenced groups like the American Dietetic Association to look the other way. Whenever a connection between increased sugar consumption and increased rates of diabetes was reported, the sugar lobby waged an effective public relations campaign, to suppress information and to deny harmful effects from sugar.

Bolt from the blue, consuming high amounts of sugar is bad for your health! A landmark (2013) study carefully examined the link between higher amounts of sugar in the food supply and higher rates of diabetes across 175 countries, over the past decade.[7] An equivalent level of evidence that linked cigarettes to lung cancer in the 1960s has now been shown to link the consumption of sugar to increased rates of diabetes. The diabetes-sugar connection is statistically significant even after correcting for obesity, physical activity levels, age, and socioeconomic factors. Obesity was found to exacerbate, but not to explain, the connection between sugar availability and higher rates of diabetes.

Western-pattern Diet, Increases the Risk of Metabolic Dysfunction

A Western-pattern diet sometimes referred to as the "meat-sweet diet," is characterized by high intakes of red meat, high-fat dairy foods, refined carbohydrates, sugary drinks, and desserts. The typical American diet is about 50% carbohydrate, 15% protein, and 35% fat.[8] This exceeds the Surgeon General's (1988) dietary guidelines for the amount of fat (below 30%), surprisingly is below the guidelines for carbohydrate (above 55%), but is at the upper end of the guidelines for the amount of protein (below 15%).[9] The quality of carbohydrates (e.g., white bread versus broccoli), as well as quality of food preparation methods (e.g., fried versus raw) makes a fundamental and critical difference. Nonetheless, "Americans spend $33 billion annually on weight loss products and services, and a large portion of this money is spent on low-carbohydrate diets. . . . These diets induce greater weight loss at three and six months than traditional low-fat diets; however, after one year there typically is no significant difference in maintained weight loss."[10]

Medical analysis of tissues and organs exposed to a Western dietary pattern repeatedly demonstrate evidence for increased levels of oxidative damage to vital nucleic acids, proteins, and lipids.[11-13] Collectively, this data suggests that a wide range of illness related to the Western-pattern diet (e.g., diabetes, heart disease, autoimmune diseases) is likely due to the adverse oxidative stress effects. Increased oxidative stress levels are evident in experimental animal models of metabolic dysfunction, as well as humans linked to diet-induced metabolic dysfunction[14-15]

Figure 19. Graph of Increased Adult Diabetics in U.S.

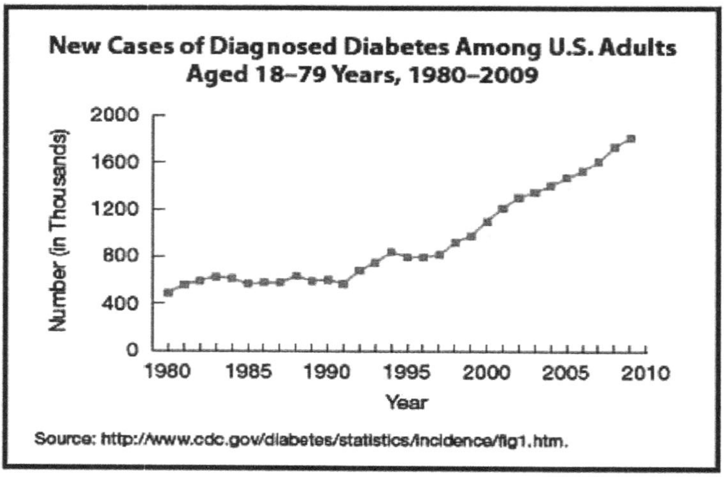

The effects of a Western-pattern diet, which increases the risk of metabolic dysfunction, are evident in the growing percentage of individuals who exhibit hyperglycemia, hyperinsulemia, insulin resistance, dyslipidemia, hyper-triglyceridemia, and free fatty acid disturbances. Most individuals do not quickly progress to diabetes, but exist in a prediabetic state (e.g., metabolic syndrome) for decades, where harmful metabolic complications slowly advance. Finding solutions to prevent or ameliorate prediabetic states, such as hypoglycemia and metabolic syndrome, is necessary to improve the quality of health for millions of people.

Why, such a dramatic increase in diabetes over a single generation? Researchers have now demonstrated that half of the known genetic risk variants for type II diabetes can be influenced by epigenetic changes (e.g., increased sugar consumption or lack of exercise), which in turn influences the function of insulin-producing cells.[16] The revolutionary field of epigenetics research studies mechanisms other than changes in DNA sequence, such as environment influences or behaviors that influence (switch on or off) genetic expression. For example, the cells of our body contain DNA, which contains genes. "We inherit our genes and they cannot be changed. The genes, however, have 'methyl groups' attached which affect what is known as 'gene expression' – whether the genes are activated or deactivated. The methyl groups can be influenced in various ways, through exercise, diet and lifestyle, in a process known as 'DNA methylation'."[17]

Scientists have discovered changes in thousands of genes, linked to type II diabetes, heart disease, and obesity ,all linked to altered DNA methylation, which responds to changes in diet and lifestyle. According to medical research (2013) at Lund University Diabetes Centre in Sweden, "19 of 40 known genetic risk variants for type II diabetes are affected by DNA methylation, which in turn changes the function of the insulin-producing cells."[18] Perhaps this explains why type II diabetes is exploding across our population and throughout the world, as more countries adopt a western-pattern diet along with decreased levels of physical activity.

According to Gicacco and Brownlee (2010) review article on oxidative stress and diabetic complications, "All forms of diabetes are characterized by hyperglycemia, a relative or absolute lack of insulin action, pathway selective insulin resistance, and the development of diabetes specific pathology in the retina, renal glomerulus [kidney], and peripheral nerve [outside brain and spinal cord]. Diabetes is also associated with accelerated atherosclerotic disease [artery wall thickens] affecting arteries that supply the heart, brain, and lower extremities."[19] In addition, diabetic cardiomyopathy, a heart muscle disorder commonly found in diabetics, frequently leads to heart failure, which is independent of most risk factors, such as hypertension and coronary artery disease.

Sugar is as old as Photosynthesis!

Microalgae occupy the bottom of the food chain. Ancient microalgae combine the properties of plant life (e.g., oxygenic photosynthesis) with the properties of single-celled microorganisms (i.e., no cell nucleus, fast growth rates, and the ability to accumulate or to secrete primary and secondary metabolites). All photosynthetic organisms, from colorful seaweeds to giant redwood trees, have inherited the ability to capture sunlight to create energy and food from cyanobacteria (blue-green algae). Their unique evolutionary history and remarkable nutritional properties have led scientists to explore the use of edible microalgae compounds to address fundamental health challenges, including metabolic disorders and complications arising from diabetes.

Sugar is chiefly produced by photosynthesis. Think again of cyanobacteria origins! Not surprising then that glucose (sugar) is the primary fuel for most life forms. One way or the other, directly (plants) or indirectly (animals), nutritional energy comes to us from the sun. Ultimately whether we eat steak or kale salad, eventually our body seeks out the sugars that food contains. Most of the glucose in our bodies comes from carbohydrates, yet when necessary even proteins and fats can be broken down to sugars. Sugar is a gift of the sun — our primary nutritional fuel. In a reversed genetic engineering triumph, we take in sugars derived from photosynthesis and use oxygen to help convert glucose into cellular energy (ATP), releasing water and carbon dioxide.

Figure 20. Cellular Respiration in Animals Offset by Photosynthesis in Plants

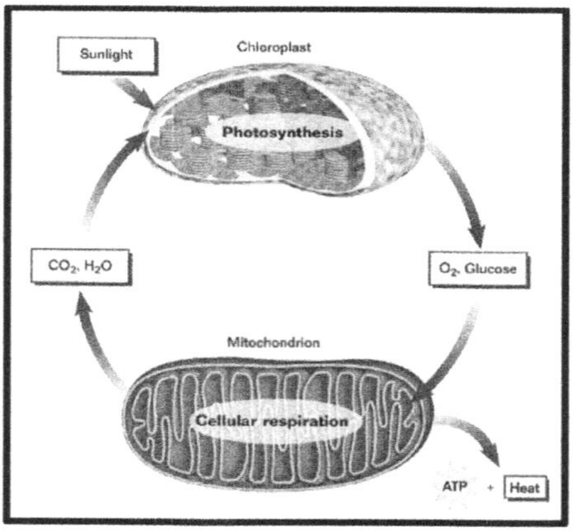

Although the steps involved are somewhat different, occurring in separate organelles (chloroplasts vs. mitochondria) — cellular respiration is functionally a reversal of photosynthesis.

The basic formula for *photosynthesis* is:

$6H_2O + 6CO_2$ + energy (photon) $\Rightarrow$ $C_6H_{12}O_6$ (glucose) + $6O_2$

The basic formula for *cellular respiration* is:

$C_6H_{12}O_6$ (glucose) + $6O_2$ $\Rightarrow$ $6H_2O + 6CO_2$ + energy (ATP)

While the basic formulas for photosynthesis and cellular respiration look similar, they are reversed. Aside from the reversal of beginning and end products, the major difference appears to be in the form of energy. Photosynthesis uses energy (sunlight), water, and carbon dioxide inside chloroplasts to create glucose and release oxygen. Cellular respiration uses glucose and oxygen inside mitochondria to produce energy (ATP) and to release water and carbon dioxide. Even today, "Depending on the availability of light and O_2, many purple photosynthetic bacteria can switch between photosynthesis and respiration, requiring only relatively minor reorganizations of their electron-transport chains."[20]

Elemental Redox Cycles of Life—Bottom of the Food Chain

When we take a big view of Earth's food cycle, we are all linked into elemental Redox cycles of life, along with nearly every other creature on Earth, from dandelions to whales. Redox (reduction-oxidation) reactions include all chemical reactions in which atoms have their oxidation state change. Biological energy is frequently stored and released by means of redox reactions. Cellular respiration, for instance, is the oxidation of glucose to CO^2 and the reduction of oxygen to water. Photosynthesis involves the reduction of carbon dioxide into sugars and the oxidation of water into molecular oxygen.

In a fundamental way, sugar is the worldwide currency of most life on Earth. How we manage sugar consumption can help us to maintain a balanced health account. However, when sugar consumption is imbalanced (excessive), sugar will contribute to heavy deposits (fat), or might burn a hole in our pockets (free radical damage). When chronic excessive amounts of sugar flood our cells or if our cellular mechanisms are unable to handle sugars, we experience the many health complications of diabetes. Diabetes is the new "poster-child" of modern illness. Once a medical rarity, an adaptive cellular survival strategy for times of famine; the underlying diabetic pathways are now evident across a myriad of so called chronic, degenerative, diseases of affluent, modern societies.

Almost 150 years ago, a paradigm shift in medicine occurred with Pasteur's discovery of the "Germ Theory of Disease." Now, every health professional knows how necessary good hygiene and sanitation are to maintain health. Yet, doctors are only just beginning to unravel the complex ways that our cells respond to sugar (glucose) and generate reactive oxygen species, which is crucial to understanding modern illnesses and the very process of aging itself. Dr. Cousens, M.D., (2008), describes diabetes as a form of "accelerated aging," a mitochondrial response to excess sugar, linked to the overproduction of reactive oxygen species, which then phenotypically damage various critical tissues and organs.[21]

The "evolution of antioxidant defenses must have been closely associated with the evolution of photosynthesis and of O_2-dependent electron transport mechanisms."

Our ancient evolutionary past links every cell in our body to require sugar (glucose) as our primary fuel—going back to the first photosynthetic organisms (cyanobacteria), whereby cells first learned to capture the sun's rays to power life activity. All photosynthetic organisms, from green seaweeds to giant redwood trees, have inherited the ability to capture sunlight to create energy and food from blue-green algae. Our basic cellular biology (cellular respiration) reveals how all human cells are directly or indirectly fueled by light. Yet in order to manage the mechanisms responsible for harnessing that sunlight – balancing the dynamic energy necessary to rip apart molecules of water and liberate oxygen, powerful redox antioxidant mechanisms were essential. According to

Dr. Halliwell (1999), the "evolution of antioxidant defenses must have been closely associated with the evolution of photosynthesis and of O_2-dependent electron transport mechanisms."[22]

Both plant and human cells harness energy by re-arranging atomic particles (e.g., electrons and protons). However, similar to nuclear reactors, our living cells are left with highly reactive, toxic substances from spent "fuel." Superoxide (O_2^-), hydrogen peroxide (H_2O_2), and OH radical/hydroxyl are deadly oxidative byproducts, which can damage unprotected nucleic acids, amino acids, and lipids.

Over billions of years of harnessing the powerful rays of light, microalgae relied upon and developed a number of powerful antioxidants to combat reactive oxygen species (free radicals), which are byproducts of photosynthesis and cellular respiration.[23] Microalgae contain some of the most powerful naturally occurring antioxidants, which happen to be of extremely low weight and size, capable of crossing the blood-brain-barrier and even entering into cellular mitochondria.

The Calvin Cycle—Making Food Out of Thin Air

Animals cannot take in CO_2 directly. Therefore, animals, including humans, must eat other organisms to get essential organic, carbon, compounds. Only autotrophs, like microalgae, can take inorganic molecules of CO_2 and turn them into high-energy organic molecules like glucose. Autotrophs are self-feeders, living primarily on light energy from the sun.

Microalgae's evolutionary ability to harness sunlight and transform carbon dioxide (CO_2) is called carbon fixation. The Calvin cycle is named after the American biochemist Melvin Calvin. In 1961, Melvin Calvin received the Nobel Prize in Chemistry for his research on the pathways of carbon dioxide assimilation in plants using *Chlorella*, a green edible microalga. The Calvin cycle results in carbon dioxide (CO_2) and water being transformed into oxygen and sugar (G3P) to produce glucose. The sugars produced during carbon metabolism yield carbon skeletons that can be used for other metabolic reactions like the production of amino acids and lipids.

Figure 21. The Calvin Cycle; Solar Food for Life

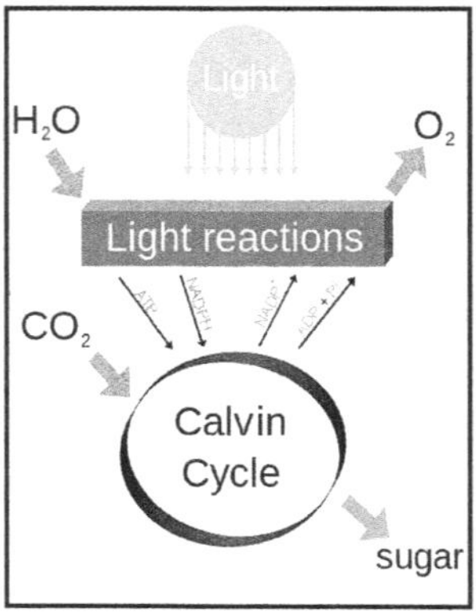

In plants, algae and cyanobacteria photosynthesis electrons are capture to eventually turn carbon dioxide into glucose (sugar), to help build carbohydrates, lipids and proteins

In a comparable way, we can get sun burned when exposed to excesses of sunlight; our cells can get "burned out" if we consume too many refined sugars. Sugar after all fuels our cells. Imagine putting rocket fuel into your car engine. The form of sugar and the speed at which sugar gets delivered into our cells is extremely important to our health.

Oxidative Stress: The Mitochondrial Dysfunction Model of Diabetes Complications

" What we discovered is that all of these different pathogenic mechanisms [related to diabetes] do reflect a single hyperglycemia-induced process and that this single unifying process is the overproduction of superoxide by the mitochondrial electron transport chain . . . a consistent differentiating feature common to all cell types that are damaged by hyperglycemia is an increased production of reactive oxygen species (ROS)." [24]

— Brownlee M, 2005

"Collectively, these observations suggest that mitochondrial dysfunction may be a central cause of insulin resistance and associated complications."[25]

—Kim J, Wei Y, Sowers J, 2013

According to Gicacco and Brownlee, in their 2010 review article on diabetic complications, "Oxidative stress plays a pivotal role in the development of diabetes complications, both microvascular and cardiovascular. The metabolic abnormalities of diabetes cause mitochondrial superoxide overproduction in endothelial cells [form inner layer of blood vessels] of both large and small vessels, as well as in the myocardium [muscular wall of the heart]. This increased superoxide production causes the activation of 5 major pathways involved in the pathogenesis of [diabetic] complications."[26]

A number of *in vitro* and *in vivo* studies have demonstrated that many protein, lipid, and DNA markers of oxidative stress are increased in vascular tissues from animals and patients with diabetes.[27-29] Several mechanisms for cellular oxidative stress have been postulated. The mitochondrial respiratory chain is composed of different cytochromes accounting for 85–90% of the oxygen consumed in the cell. Approximately 1–3% of this oxygen is incompletely metabolized and diverted into superoxide anion.[30] High glucose levels are reported to increase mitochondrial superoxide production in vascular endothelial cells.[31]

Growing evidence has indicated that NADPH oxidase may be the most important source for ROS production in diabetic vascular tissues. High glucose levels have been reported to stimulate ROS production in cultured vascular cells through a PKC-dependent activation of NADPH oxidase.[32] High glucose levels build up reactive oxygen species (ROS) through glucose auto oxidation, step up advanced glycation end products (AGEs) and increase xanthine oxidase. Xanthine oxidase is a superoxide-producing enzyme found in blood serum and the lungs; it can contribute to gout, liver, renal, and cardiovascular complications and oxidative eye injury. The regulation of this enzyme pathway represents an important metabolic contributor to diabetic symptoms.

NADPH oxidases are the major source of superoxide anions; the primary ROS produced by mitochondria, and are found in nearly every tissue of the body. "ROS produced by NADPH oxidases have been shown to modify virtually all other possible sources of ROS. Changes in the structure of several oxidases caused by the overproduction of ROS from NADPH oxidases lead to their dysfunction and to a further increase in generation ROS, forming a vicious cycle of oxidative stress. An example is endothelial nitric oxide synthase, which primarily produces nitric oxide, but generates ROS rather than nitric oxide when it is uncoupled under pathologic conditions. . . Similarly, production of superoxide anions by NADPH oxidase induces mitochondrial oxidative damage via structural changes to the inner mitochondrial membrane and disturbs flow in the electron transport chain, which enhances ROS production."[33]

It is well established that hyperglycemia is a major causative factor for microvascular disease. Several well-researched theories have been proposed to explain how chronic hyperglycemia causes the micro- and macro vascular derangements. Among various pathways, oxidative stress has been identified as a common mediator of these biochemical abnormalities. Oxidative stress linked to hyperglycemia-induced generation of reactive oxygen species (ROS), especially mitochondrial ROS, may substantially contribute to the development and progression of diabetes, including insulin resistance, and related vascular complications. Reactive oxygen species play key roles in regulating endothelial function and vascular tone, as well as dysfunction. NADPH oxidase has been described as the major source of ROS generation in the vasculature in response to high glucose and advanced glycation end-products.

Sustained activation, overproduction, of NADPH oxidase in diabetes can deplete intracellular levels of NADPH, an essential cofactor for endothelial NO synthase (eNOS) and several antioxidant systems, as well as led to mitochondrial dysfunction.[34] The characteristics and roles of NADPH oxidase make this enzyme an ideal candidate for a therapeutic target. Yet, ROS production via NADPH oxidase offers a double edged sword, potentially beneficial as well as damaging. For example, NADPH oxidase may provide both intra and inter-cellular feedback via redox signaling to help up-regulate antioxidant enzyme defenses. While sustained hyperglycemia likely only intensifies redox dysregulation, a positive feedback loops, leading to severe diabetic complications.[35] Japanese medical researchers, Sasaki and Inoguchi (2012) along with numerous others, have identified NADPH oxidase as an effective therapeutic target for the prevention of vascular complications in diabetes.[36]

Although the involvement of oxidative stress in diabetic vascular complications has been demonstrated *in vitro* and in animal studies, the effectiveness of antioxidant therapy is controversial because few human studies have demonstrated beneficial effects of antioxidants on diabetic vascular complications. Most dietary antioxidants, with the exception of some microalgae antioxidants, do not appear to work at the intra-cellular mitochondrial levels. Some medical researchers are working to develop newer, less toxic and more selective NADPH oxidase inhibitors. Still others are engaged in an ongoing search for naturally occurring compounds to prevent, reduce or delay the onset of diabetic complications. Because reactive oxygen species (ROS) not only have pathologic effects but, when under proper antioxidant control, participate in "redox signaling" pathways that are crucial for cellular functioning, selective and ideally reversible NADPH oxidase inhibitors are needed.

The Bilin-Diabetes Connection: Microalgae Bilin, Light Pigment Molecules, to the Rescue

Bilirubin once thought only a waste product of blood (heme) conversion, is now considered an important endogenous antioxidant[37] and reported to have a powerfully inhibitory effect on the activity of NADPH oxidase.[38] Bilirubin belongs to a phylogenetically old superfamily of tetrapyrrole compounds, which have multiple biological functions. Similar to bilirubin, microalgae light harvesting accessory pigments (e.g., like phycocyanin) exhibit powerfully inhibitory effects on the activity of NADPH oxidase.[39] Scientists have discovered that algal heme oxygenase is a component of the phycobilin biosynthetic pathway and algal biliverdin is a precursor to phycocyanobilin.[40]

From an evolutionary perspective, both blue green and unicellular red alga have analogous ancestral tetrapyrrole (bilin) biosynthetic pathways, leading to phycobilin photosynthetic accessory pigments, like phycocyanin (PC). On the other hand in animals, including humans, a similar tetrapyrrole (bilin) biosynthetic pathway leads to bilirubin. In humans, the conversion of heme requires oxygen and NADPH, and produces bilirubin and carbon monoxide in equimolar amounts. In blue green algae, the conversion of heme requires CO2 and NADPH and produces phycobilins and carbon monoxide in equimolar amounts.

> *In effect, microalgal phycobilins are molecularly created similarly to animal bilirubins. Overall, these findings indicate not only the protective effects of bilirubin per se; but additionally, other products of the heme catabolic pathway including biliverdin, phycobilins, and phycocyanin.*

In effect, micro algal phycobilins are molecularly created similarly to animal bilirubin. Overall, these findings indicate not only the protective effects of bilirubin *per se*; but additionally, other products of the heme catabolic pathway including biliverdin, phycobilins, phycocyanin, and heme oxygenase. Generally proteins are broken down into individual amino acids or small oligopeptides before absorption. However, the phycocyanobilin chromophore appears to enter the blood stream intact. Structurally phycocyanobilin resembles the bile pigment bilirubin, which can be absorbed from any part of the small or large intestine as long as it remains unconjugated.[41] A variety of biliproteins might be harnessed as powerful therapeutic agents for the prevention and treatment of a variety of metabolic illnesses.[42]

Medical scientists at Kyushu University in Japan further demonstrated that phycocyanin (blue-green pigment) derived from blue-green algae, and its light-harvesting complex, phycocyanobilin, with a chemical structure similar to that of biliverdin, protect against oxidative stress and adrenal dysfunction in type II diabetes. Phycocyanin decreased urinary and renal oxidative stress markers and normalized the expression of NADPH oxidase components. Similar antioxidant effects were observed following two weeks of oral administration of phycocyanobilin. Phycocyanobilin, bilirubin, and biliverdin all inhibited NADPH oxidase-dependent superoxide production in cultured renal mesangial [kidney] cells. In conclusion, the researchers proposed that, "oral administration of phycocyanin and phycocyanobilin may offer a novel and feasible therapeutic approach for preventing diabetic nephropathy [kidney disease]."[43]

A connection between bilirubin (a heme metabolic product) and diabetes was found when Japanese researchers studied diabetic patients with Gilbert syndrome, a genetic condition that elevates blood bilirubin concentrations. Japanese scientists at Kyushu University discovered a markedly lower prevalence of vascular complications, including kidney disease, in diabetic patients with Gilbert syndrome. Gilbert syndrome is a genetic disorder of bilirubin metabolism that can result in mild jaundice, found in about 5 % of the population.

Their study showed for the first time, in these patients a significantly lower prevalence of vascular complications, retinopathy, and coronary artery and cerebrovascular disease, as well as reduced markers of oxidative stress and inflammation. Furthermore, serum bilirubin concentrations were negatively associated with albuminuria [biomarker of kidney disease] in patients with type 2 diabetes.[44] Albuminuria is a pathological condition in which the kidney leaks protein (albumin) into the urine. Albumin is an indicator of kidney damage in patients with diabetes. The discovery of decreased levels of vascular complications in diabetic patients with Gilbert syndrome indicated that antioxidant therapy may be useful for preventing the development of vascular complications for most diabetic patients.

Other Japanese investigators (Yasuda et al., 2011) conducted a cross-sectional study involving 3,119 participants. After adjusting for factors known to be associated with diabetic retinopathy; people with the lowest bilirubin levels had far greater rates of diabetic retinopathy than those with the highest bilirubin levels. The researchers concluded that, "Elevated serum bilirubin levels may be protective against diabetic retinopathy among persons with either diabetes or impaired glucose metabolism, independent of known risk factors for diabetic retinopathy."[45]

Another Japanese cross-sectional study of 12,949 typical persons revealed a similar relationship between higher serum bilirubin levels and lower prevalence rates of type II diabetes. The researchers concluded that higher concentrations of serum bilirubin probably confer protection against the development of type II diabetes.[46]

Additional supportive evidence in a large Korean cross-sectional study involving almost 94,000 subjects, found that higher serum bilirubin levels was associated with a reduced risk of diabetes mellitus and diabetic nephropathy.[47] A subsequent Korean study involving more than12,000 participants with phenotypic Gilbert syndrome, exhibited a low prevalence of metabolic syndrome compared with the general population.[48]

Several researchers have reported that bilirubin and biliverdin (a precursor of bilirubin) have beneficial effects on diabetic vascular complications, including nephropathy, via antioxidant effects.[49] Japanese scientists at Kyushu University have demonstrated that bilirubin (BR) and biliverdin (BV) both offered beneficial treatments against the development of renal disease with a normalization of oxidative stress markers in diabetic mice and in rats. Both BIL and BVD significantly inhibited NADPH-dependent superoxide production. "Collectively, our findings suggest that BR and BV may protect against diabetic nephropathy and may lead to novel antioxidant therapies for diabetic nephropathy."[50]

Phycocyanobilin, Bilirubin, and Biliverdin all Inhibit NADPH Oxidase–Dependent Superoxide Production

Similar beneficial effects, comparable to those of BIL and BVD, were obtained with oral administration of phycocyanin derived from blue-green algae, and its chromophore phycocyanobilin, which has a chemical structure similar to that of biliverdin[51] Oral administration of phycocyanin for ten weeks protected against oxidative stress and kidney dysfunction in diabetic mice. The phycocyanin helped to normalize tumor growth factor-β and fibronectin expression.

Phycocyanin stabilized additional urinary and renal oxidative stress markers, along with a normalized expression of NADPH oxidase components. Similar therapeutic antioxidant effects were evident following oral administration of phycocyanobilin for two weeks. Phycocyanobilin, bilirubin, and biliverdin were all proven to inhibit NADPH oxidase–dependent superoxide production in cultured renal mesangial cells.[52]

Microalgae antioxidant therapies appear effective against diabetic vascular complications by targeting mitochondrial ROS dysfunctions. One astounding area where microalgae components appear effective is inhibiting the up-regulate of NADPH oxidase. Based on the Japanese research and other studies, the effectiveness of bilirubin proteins, including phycocyanobilin, can inhibit NADPH-dependent superoxide production. Antioxidant therapies that effectively target mitochondrial mechanism for NADPH oxidase activation are some of the most promising therapies for the prevention of diabetic vascular complications.[53] Microalgae-based antioxidant therapy may prove useful in preventing or delaying the onset of type 2 diabetes, as well as inhibiting diabetic-related complications.

Edible Microalgae and Diabetes Research

Despite significant progress that has been made in drug therapies and early-diagnosis tools, the vast majority of medical researchers and practitioners agree that the treatment of diabetes and cardiovascular diseases remains a critical challenge. At present medical complications and morbidity rates from cardiovascular and diabetic diseases still remains disturbingly high. According to Davis and Granner (1996) "currently available antidiabetic drug regimens too often fail to effectively address the medical complications associated with diabetes mellitus, and the drugs themselves are associated with side effects. Hence, the search is on for new antidiabetic compounds with multiple targets and with fewer unintended harms (e.g., less cardiovascular risks, less gastrointestinal problems, and less weight gain)."[54] Yet, despite better drug delivery methods and insulin monitoring systems; new antidiabetic compounds have been hard to find.

A new philosophy of medicine is emerging, one that embraces an environmental and evolutionary perspective on the types of diseases, like cancer, diabetes, heart disease and autoimmune conditions, increasingly found in modern, affluent and industrialized cultures. Diet plays a key role in the treatment of diabetes, and proper nutrition is vital to prevent diabetes complications. However, the diets prescribed for the treatment of diabetes are difficult to follow, and compliance is reported to be very poor.[55] In recent years functional and medicinal foods have helped patients with chronic diseases such as diabetes. Therefore, the search for microalgae solutions to act as an adjunctive therapy for optimal management of diabetes is well underway.

Diabetes a Disease Model for Illnesses associated with "Aging"

Diabetes has been described as a disease model for illnesses associated with "aging." An overlapping co-occurrence of cardiovascular diseases, depression, rheumatoid arthritis, diabetic disorders, and obesity encouraged researchers to investigate their common backgrounds. "Special attention is drawn to those factors which play a role in immunology and inflammatory processes . . . [revealing] evidence for significant disturbances in release or action of the same mediators in hypertension, heart failure, obesity, diabetes mellitus, metabolic syndrome, starvation, chronic stress, depression and other psychiatric disorders."[56] Cellular pathways that contribute to inflammation, oxidative damage, and an imbalanced immune response appear to stem from central redox mechanisms within the cell, which may express themselves phenotypically as different so-called illnesses. What is common across these various illnesses and perhaps central to the underlining process of aging, is the centralized role of mitochondria. Mitochondrial organelles energetic handling of oxygen has built-in vulnerabilities to oxidative risks.

> *The therapeutic benefits of microalgae compounds — including hyperglycerolemia, hypercholesterolemia, cardiovascular diseases, neurological and inflammatory diseases — symptoms all common in diabetes.*

Currently, increasing amounts of research are devoted to assessing the therapeutic benefits of microalgae compounds on an array of illness conditions — including hyperglycerolemia, hypercholesterolemia, cardiovascular diseases, neurological and inflammatory diseases — symptoms all common in diabetes.

Edible microalgae have received considerable attention due to their association with cholesterol-regulatory properties, their modulation of the host immune system, and their remarkable antioxidant and anti-inflammatory potentials.[57-58] Increasing scientific and medical evidence suggests that microalgae may offer important aid in helping to control chronic diseases such as diabetes.[59]

Diabetes is virtually nonexistent in sea creatures, the organisms closest to the microalgae food chain. Furthermore, researchers report lowered diabetes risks in human who eat fish which contain abundant microalgae nutrients and fatty acids. Algae-derived omega-3 fatty acids (EPA and DHA) may improve the sensitivity of insulin receptors. Most people have heard about the various benefits of fish oil. For example, fish oil for diabetes, with its omega-3 fatty acid content, can lower blood pressure, help improve lipid profiles (lowering both triglycerides and very low density lipoprotein (VLDL) cholesterol) and may support our blood vessels by supporting their elasticity. However, most people do not know that fish get their abundance of omega 3 fatty acids primarily by consuming microalgae (the original omega-3 fatty acid source).

A series of studies conducted on fish oil and diabetes prevention, observed that the most impressive results appear to be from consumption of the whole fish or microalgae, rather than fish oils. "Eating at least two servings of fish each week seems to protect people with diabetes who also have kidney disease, according to a long-range study of more than 22,000 adults in England. The study, published in the *American Journal of Kidney Diseases*, the official journal of the National Kidney Foundation, shows that fish consumption lowers abnormal levels of protein in the urine in people with diabetes. Abnormal amounts of protein appear in the urine when the kidneys are damaged; in fact it is a key indicator of kidney disease. Previous studies have shown that fish and fish oil consumption decreases protein in the urine, increases glucose tolerance, decreases fats in the blood, and lowers blood pressure — all benefits to people with diabetes."[60]

In countries with a high prevalence of obesity, the incidence of developing type 2 diabetes is significantly reduced with high fish and seafood consumption.[61] The widely acclaimed "Mediterranean diet" lowers the risk of type 2 diabetes, metabolic syndrome, and obesity. The Mediterranean diet is low in refined carbohydrates and red meat but high in fish, natural sources of antioxidants like red wine, garlic, red peppers, whole grain sources of carbohydrates, and rich in monounsaturated fats (olive oil).[62]

Ming Zhang and colleagues in Canada (2013) completed a systematic review and meta-analysis of the relationship between fish and omega- 3 consumption and the incidence of Type 2 diabetes (T2D). They "found a significant protective effect of oily fish intake on T2D risk . . . Dose-response analysis suggested that every 80 g per day intake of oily fish may reduce 20% risk of T2D." Yet, there was no clear relationship between EPA and DHA intake and risk of T2D.[63]

Fish oils, while providing some benefits, tend to lack a number of additional protective factors, such as amino acids or pigments. Over and over again, evidence suggests that microalgae hold potent natural therapeutic agents for diabetes and diabetes-related complication.[64] The nutritional factors and phytochemical compounds responsible for diabetic benefits appear multiple—including glycoproteins, essential fatty acids, GLA (the only natural whole food source of gamma linolenic acid in the plant kingdom), trace minerals like chromium, bile analogs like phycocyanobilin, lipophilic (fat-soluble) and hydrophilic (water-soluble) antioxidants, B-vitamins, essential amino acids, and insulin-like protein molecules. Sulfate polysaccharides in microalgae exhibit immune-modulatory, antitumor, antithrombotic, anticoagulant, antimutagenic, and anti-inflammatory effects.[65]

Macro- and Microalgae Algae Are Functional Foods for Managing Diabetic Challenges

Marine algae are a diet staple in Asian countries, where diabetes rates are traditionally quite low. Macro- and microalgae algae are both recognized for their health properties, with a variety of high biological value, and bioactive compounds. Studies completed in the 1980s revealed modulatory effects of blue-green algae on the metabolism of both lipids and carbohydrates.[66] Animal research has shown blue-green algae *consumption* helps to regulate cholesterol, to increase antioxidant capacity, to improve insulin resistance, and to improve the uptake of glucose.[67-68] Overall, blue-green algae is reported to be effective for improving blood lipid profiles,[69] enhancing immune capacity, and reducing oxidative stress.[70] Since dyslipidemia (abnormal amounts of cholesterol and/or fats in the blood) and oxidative/ inflammatory stress are considered to be contributing factors for diabetes, blue-green algae is one of the most promising functional foods for management of type II diabetes.

Both macro and microalgae are rich in minerals, antioxidants, and soluble dietary fibers. "The plentiful soluble dietary fibers in algae help avoid obesity and diabetes. The total fiber content of several algae species, (~6 g/100g), is greater than that of fruits and vegetables promoted for their fiber content."[71] One promising application of algae health science involves components (within both macro- and microalgae), which contribute to "satiety signaling mechanisms" as well as "fill-gut" benefits. For example, certain soluble fibers in algae like alginates can absorb up to 200 times their weight in water, which fill-up the gut and thereby suppress appetite. Furthermore, certain algae-based compounds help activate gut to brain satiety [hunger reduction] signaling mechanisms. These signals contribute to feelings of fullness and reduce hunger. Blue-green algae has been used in the treatment of protein malnutrition disorders, as well as reported to reduce body weight in obese humans.[72]

Insulin-like Proteins in some Algae, Cyanobacteria and Plants

Most people think of insulin only as a drug. Other people understand that insulin is a hormone produced by our pancreas to remove excess sugar (glucose) from our blood, which otherwise would be toxic. But it is not widely known that the presence of "insulin-like proteins" in plants has been investigated since the discovery of insulin. "Collip (1923) reported hypoglycemic activity of several plant extracts when administered to normal rabbits and depancreatized dogs. He named the protein detected in these extracts 'glucokinin' instead of insulin because it did not originate from the islets of Langerhans in the pancreas. Later, Silva et al. (2002) detected insulin-like antigens in a wide range of green plants, algae, fungi, and a cyanobacterium by an ELISA assay and western blotting . . . Recently, in our labs (Anwer et al. 2012), we have shown insulin-like protein from *Spirulina* [a blue green algae] that has the same molecular mass, immunoreactivity, and retention time by RP-HPLC as that of bovine insulin."[73]

Anwer (2012) published research describing the "Treatment of diabetic rats with crude, aqueous extract, ethanolic extract, and insulin-like protein of *Spirulina* successfully ameliorated diabetic complications by increasing body weight and significantly decreasing the levels of blood glucose, glycosylated hemoglobin, serum glutamic oxaloacetic transaminase, serum glutamic pyruvic transaminase, alkaline phosphatase, total bilirubin, serum creatinine, serum uric acid, and blood urea nitrogen."[74]

Blue-green Algae can Improve Blood Glucose Control and help to Reduce Insulin Resistance

Several studies conducted both with animals and with humans have demonstrated that ingesting blue-green algae can improve blood glucose control and help to reduce insulin resistance.[75-80] Certain algae compounds may improve the sensitivity of the cells to the action of insulin and improve glucose tolerance and help to normalize blood sugar. Moreover, algae polyphenol extracts have antidiabetic effects through the modulation of glucose-induced oxidative stress.

> *Certain algae compounds may improve the sensitivity of the cells to the action of insulin and improve glucose tolerance and help to normalize blood sugar.*

Mani et al. (1998) showed a significant decrease in patients' fasting blood sugar levels after 21 days of supplementation with *Spirulina* at 2 grams per day.[81] In another study, rats were given *Spirulina* for 45 days at a dosage of 15 mg per kg of body weight, and the researchers found a significant reduction in the animals' blood glucose level.[82] The researchers theorized this might be due to a high content of fiber that interfered with the absorption of glucose.[83] Another theory is that the peptides and polypeptides in blue-green algae might have an insulin-stimulating effect on the pancreas.[84]

Indian researchers, collaborating with the World Health Organization, undertook a series of studies on the use of blue-green algae to benefit diabetic patients. Their initial findings showed that two grams of blue-green algae a day for a period of two months resulted in a significant reduction in fasting and postprandial blood glucose in patients with type II diabetes.[85] Earlier master's thesis research, at the same Indian University (Shah, S., 1997), demonstrated clinically improved levels of lipid metabolism in patients with hyperlipidemia, when they took two grams a day of blue-green algae for two months.[86]

Impressively, these researchers further discovered that supplementing patients with two grams of blue-green algae a day for a period of four months resulted in a "steady and substantial improvement" in their glycemic status. They reported, "Significant reductions not only in the fasting but also in the 2-h postprandial blood glucose levels were noticed after two months as well as after four months of intervention." The blue-green algae supplementation resulted in a significant decrease in the glycated hemoglobin (HbA_{1c}) levels, a useful blood sugar level index, over two- to three-month intervals. Additionally, there was a significant reduction in patients' blood levels of triglycerides, total lipids, and total cholesterol, while the consumption of blue-green algae boosted the status of the "good" cholesterol (high-density lipoprotein cholesterol).[87]

The reduction of low density lipoproteins and the raising of high density lipoprotein (cholesterol ratios) have been shown in both epidemiologic and intervention studies to decrease mortality. Dyslipidemias, a pathological ratio of serum lipid levels, are among the major etiologic factors for arterial occlusive diseases (blindness). Also, dyslipidemias contribute to the most prevalent causes of death, stroke and coronary heart disease.

Blue-green Algae proves more effective than Metformin, a Leading Antidiabetic Drug

Metformin is now believed to be the most widely prescribed antidiabetic drug in the world. In the United States alone, more than 48 million prescriptions were filled in 2010 for its generic formulation.[88] Two independent experimental studies on rats compared the effectiveness of blue-green algae (*S. maxima)* solutions versus a standard antidiabetic drug, metformin. First, hyperglycemia and hyperlipidemia were caused in rats by feeding the animals high-fructose diets for one month. Then both studies compared the benefits of either blue-green algae or the drug metformin to counteract the negative diabetic effects associated with a high-fructose diet. Remarkably, the benefits of blue-green algae were found superior to those of metformin for significantly decreasing blood glucose levels.[89-90] Quite a dramatic finding, as metformin is a "first-line drug of choice for the treatment of type II diabetes" and blue green algae solutions outperformed this well-known oral antidiabetic drug.

Remarkably, the benefits of blue-green algae were found superior to those of metformin, for significantly decreasing blood glucose levels.

Both of the metformin comparison studies demonstrated that blue-green algae exhibited antihyperglycemic, antihyperlipidemia, and hepatoprotective activity in rats fed high-fructose diets. Significant (P<0.001) reductions in blood glucose, lipid profile (triglycerides, cholesterol and LDL, VLDL) and liver function markers were recorded along with elevated levels of HDL cholesterol after 30 days of administration of a liquid extract of blue-green algae liquid extract. Blue green algae extracts (either 5% or 10% solutions) were demonstrated to offer significant protection against the negative effects of excessive fructose, with superior liver and blood glucose markers in the algae group compared the metformin drug control group.[91]

Due to declared antidiabetic properties, along with blue-green algae's known remarkable nutritional profile, Iyer (2008) and colleagues explored how microalgae might be incorporated into several traditional Indian foods (e.g., rice dishes, chapatti, and snacks). Importantly, "on comparing the glycemic response of *Spirulina*-supplemented recipes with that of unsupplemented recipes carried out on T2DM [type 2 diabetes], a marked reduction in GI [glycemic index] was noticed. A similar response was observed with respect to the lipemic [lower amounts of fats in blood serum] response." "A favorable reduction in the glycemic response was observed with 2.5 g of *Spirulina* supplementation."

The medical researchers concluded that blue-green algae can be incorporated into various vegetable dishes, while meeting consumer taste and sensory acceptability and may "be effectively used as a supportive therapy in the management of hyperglycemia and hyperlipidemia."[92] Hyperlipemia is where excessive amounts of fat and fatty substances are in the blood. Hyperglycemia is where there is abnormally high level of glucose in the sugar. Both problems are commonly associated with type II diabetes and appear to respond favorably to an oral intake of a few grams of blue green algae.

Korean Researchers use Blue-green Algae with Human Diabetic Subjects

Kim and Kim (2005) reported that healthy elderly subjects receiving blue-green algae supplementation for eight weeks exhibited improved total antioxidant status and a decrease in markers of lipid oxidation (cell damage).[93] However, the effects of blue-green algae on healthy people and people with type II diabetes are sometimes mixed. For example, Lee (2008) reported that blue-green algae given to healthy elderly people with normal fasting blood glucose (FBG) brought a significant reduction in mean fasting blood glucose, from 105 mg/dl to 100 mg/dl. However, in type II diabetes patients consuming blue-green algae revealed no significant reductions—on fasting blood glucose, on glycated hemoglobin (a measure of average blood glucose control over the past two to three months), or on insulin measures. The researchers concluded that blue-green algae may not have beneficial effects on blood glucose concentration for all diabetic patients.[94]

Korean researchers found that giving blue-green algae supplementation to healthy elderly people resulted in significant reductions of plasma triglycerides, total and LDL cholesterol, and blood pressure along with improvement in antioxidant status and inflammatory status and concluded that it was suitable as a functional food for elderly people (Park & Kim, 2003; Park et al., 2008).[95-96] A study (Eun Hee Lee et al., 2008) of 37 Korean patients with type 2 diabetes was conducted at a diabetes clinic in Seoul.[97] Patients were randomly assigned to a blue-green algae supplemented (8 g/day) experimental group or a nonalgae control group. During the 12-week study, patients were asked to stay with their usual diet and told not to take any dietary supplements. Patient body size measures did not vary for either group. However, patients who ate blue-green algae showed significantly lowered plasma triglycerides levels. Also, blue-green algae supplementation resulted in a substantial reduction in free radical damage, along with an increase in plasma adiponectin levels. Adpionectin is a beneficial hormone secreted by adipocytes (fat cells) that helps regulate the metabolism of lipids and glucose and influences our body's response to insulin. Higher concentrations of adiponectin are associated with a reduced risk of heart attacks.

Yet, their results also showed that the lipid lowering effect of blue green algae supplementation differed according to initial patient serum lipid levels, which was assessed prior to the intervention. The patients with higher initial triglyceride levels showed higher reduction in plasma triglyceride and blood pressure. Likewise, the patients with higher initial total cholesterol and LDL-cholesterol levels showed higher reduction in plasma concentrations of total cholesterol, LDL-cholesterol, IL-6, and blood pressure. It appears that blue green algae supplementation is more effective in patients with dyslipidemia. Altogether, the physiological changes evident in patients eating blue green algae could play a role in the prevention of diabetic complication. The researchers concluded that blue green algae supplementation is likely useful in diabetes management; improving blood lipid profiles, inflammatory variables, and antioxidant capacity in Korean patients with type II diabetes.

Microalgae's Antioxidant Properties Useful for Diabetic Inflammatory Symptoms

Microalgae, especially blue-green algae, provide beneficial antioxidant effects. Numerous studies have carefully documented multiple and strong antioxidant (ROS scavenging) benefits of edible microalgae.[98-101] In a study of hamsters fed on diets designed to promote a chronic inflammatory response in their arterial walls (atherosclerosis), administration of an antioxidant from blue-green algae significantly reduced oxidative stress and NADPH oxidase expression.[102] Importantly, an excessive production of reactive oxygen species (ROS) via NADPH oxidase has been implicated in the pathogenesis of diabetic nephropathy, as well as diabetic retinopathy.

One of the primary benefits of blue green algae for diabetes treatment or prevention may well be cyanobacteria's NADPH oxidase inhibition ability. NADPH oxidases are a family of enzyme complexes whose primary function is to catalyse the transfer of electrons from NADPH to molecular oxygen resulting in the generation of superoxide. This enzyme function makes NADPH oxidases unique and distinguishes it from all other oxidases where superoxide production may occur as a by-product of another oxidative reaction (e.g. the mitochondrial electron transport chain). Superoxide is released by phagocytes, as part of the immune system, to kill invading microorganisms. However, most superoxide is released as a by-product of mitochondrial respiration, capable of both deleterious and beneficial roles. Excessive production of superoxide can contribute to the development of many diseases, including complications associated with diabetes.

One of the primary benefits of blue green algae for diabetes treatment or prevention may well be cyanobacteria's NADPH oxidase inhibition ability.

Edible blue-green algae can help to reduce the production of pro-inflammatory molecules (cytokines) by inhibiting the nuclear factor kappa B pathway.[103] Consumption of edible blue green algae may also reduce risks of cataracts and age-related macular degeneration.[104] As well, use of blue-green algae *(S. maxima)* was shown to improve vascular tone in aortic rings in a rat model.[105] An impairment of vascular dilation and constriction is a hallmark of type 2 diabetes mellitus.

Using an animal experimental model, microalgae helps to prevent or to reduce complications associated with steatohepatitis, a non–alcohol related liver condition, sometimes called fatty liver disease.[106] Microalgae are shown to be protective against liver damage caused by deadly aflatoxin (produced by fungus) and by cisplatin, a potent and widely used chemotherapy drug.[107-108]

There is an international effort to find for solutions to what is now a world-wide epidemic of diabetes. In addition to blue-green algae, brown and green algae have also been studied for their potential antidiabetic properties.[109-111] Research at the Department of Ophthalmology, Xinhua Hospital, School of Medicine in Shanghai in China, reported that astaxanthin, a carotenoid and powerful antioxidant found in microalgae, seafood, and various plants, can prevent the death of retinal ganglion cells (evident in diabetic retinopathy), significantly reduce oxidative stress markers (reduce superoxide anions), reduce lipid peroxidation (MDA status), decrease oxidative DNA damage, and improve antioxidant activity in renal tissue in experimental studies with diabetic mice.[112]

Scientists use animal models of diabetes to precisely measure key physiological changes in controlled experiments. Yet, unless a person is a medical investigator, their language can be rather overwhelming. For example, Korean researchers (2013) reviewing the various antidiabetic mechanisms associated with phlorotannins in brown algae stated: "[The] majority of the investigations on phlorotannins derived from brown algae have exhibited their various antidiabetic mechanisms such as α-glucosidase and α-amylase inhibitory effect, glucose uptake effect in skeletal muscle, protein tyrosine phosphatase 1B (PTP 1B) enzyme inhibition, improvement of insulin sensitivity in type 2 diabetic db/db mice, and protective effect against diabetes complication."[113] Serious medical researchers are carefully assessing the physiological efficacy of various macro- and microalgae components for treatment of diabetes-related complications.

To model and research diabetic complications in animals, scientists often rely on genetically engineered animals (e.g., KKAy mice) that rapidly develop diabetic conditions. Another experimental approach is to cause diabetes in animals by injecting them with toxic chemicals (e.g., alloxan) or feeding them high-fructose or high-cholesterol diets. For example, KKAY mice are genetically prone to over-eat, which leads to obesity, excessive insulin secretion, blood sugar dysregulation, and extreme levels of fats in the blood. As a result, KKAy mice develop early structural and functional defects in their pancreas and kidneys.

Chinese researchers studying the use of the blue-green pigment (phycocyanin [PC]) from microalgae reported, "Histopathological results showed that PC administration suppressed the abnormal enlargement of islets observed in the pancreas of KKAy mice . . . The antidiabetic effect of PC on KKAy mice is most likely due to its ability to enhance insulin sensitivity, amelioration of insulin resistance of peripheral target tissues and regulation of glycolipid metabolism. Therefore, PC may have a potential clinical utility in combating type-II diabetes."[114]

Diabetes mellitus is associated with metabolic dysfunctions, including alterations in circulating lipid levels and fat tissue accumulation. Microalgae can have hypoglycemic and hypolidemic properties, as well as anti-inflammatory action as, well demonstrated in diabetic animals.[115] Blue-green algae supplementation is shown to result in a significant decrease in plasma concentrations of triglyceride, total cholesterol, and LDL cholesterol levels in patients with type II diabetes.[116-117] Also, rats fed a 20% water-soluble fraction of blue-green algae showed a significant improvement in their total cholesterol scores (LDL/HDL ratio).[118]

> *Blue-green algae supplementation is shown to result in a significant decrease in plasma concentrations of triglyceride, total cholesterol, and LDL cholesterol levels in patients with type II diabetes.*

Reductions in blood triglyceride levels (high triglycerides raise the risk of heart disease) and blood pressure–lowering effects of blue-green algae have been reported. Iwata et al. (1990) proposed the triglyceride reducing effect of blue-green algae might be due to an effect on the metabolism of lipoproteins. Lipoprotein lipase is an important enzyme in the human body that metabolizes triglycerides. They reported that rats fed a diet containing blue-green algae showed a significant increase in the activity of lipoprotein lipase compared to rats fed a high-fructose diet.[119]

Blue-Green Algae can Lower Harmful Blood Fats

Many animal and human studies have repeatedly reported the lipid-lowering effects of blue-green algae. When *Aphanizomenon flos-aquae,* was fed to rats, for 43 days cholesterol levels were significantly decreased when compared to the control group. The researchers proposed that the hypocholesterolemic effect may be due to the algae's high chlorophyll content, which was said to stimulate liver function and to decrease blood cholesterol.[120] Park and Kim (2003) reported that the plasma levels of triglycerides, total cholesterol, and

LDL cholesterol in Korean elderly people were significantly decreased after blue-green algae supplementation for 24 weeks.[121] A correlation analysis of this study revealed that elderly patients with higher initial blood levels of triglycerides and cholesterol showed the largest reductions in lipid concentration, blood pressure, and inflammatory cytokine levels. Therefore, blue-green algae supplementation may be most effective for lowering total cholesterol in people who have moderately high blood cholesterol levels.

Early research by Kato et al. (1984) showed a reduction in the levels of serum cholesterol in animals supplemented with *Spirulina*.[122] Additional work by Blé-Castillo et al. (2002) found that blue-green algae supplementation decreased total hepatic (liver) lipids by 40% in mice administered drugs and a hypercholesterolemia diet.[123] Iwata et al. (1990) observed that *Spirulina* supplementation inhibited plasma increases in HDL cholesterol, triglycerides, and phospholipids.[124] On the other hand; there was no statistically significant difference observed between the control group and the groups supplemented with *Spirulina* when liver lipid levels (in contrast to plasma levels) were compared. Furthermore, the authors reported an increased level of lipoprotein lipase enzyme activity in the animals that received *Spirulina* supplementation.

Using an experimental model, Kato et al. (1984) fed rats a diet rich in cholesterol with and without *Spirulina* supplementation. In this study, the authors observed an increase in total cholesterol levels, LDL + VLDL cholesterol, and phospholipids in the serum of the group that did not ingest *Spirulina*. However, there was a significant reduction in the levels of these cholesterol fractions when the animals were supplemented with 16% *Spirulina*.[125]

Yet, the relationship between lipid profiles and microalgae intake requires more controlled studies. Recently Cheong et al. (2010) acknowledged that the cholesterol control mechanisms of blue-green algae are not well understood. These authors cite the hypothesis "that the addition of this alga into the diet diminishes the intestinal absorption of cholesterol as well as the re-absorption of bile acids in the ileum." Blue-green algae might therefore be "considered a functional food capable of reducing the levels of cholesterol and consequently preventing atherosclerosis."[126]

Recent research using a diabetic animal model (2011) reported that "Based on our results, we conclude that *Spirulina* intake can provide lower levels of circulating LDL cholesterol in comparison to aerobic training in diabetic rats. Furthermore, both *Spirulina* intake and physical exercise cause an improvement in hepatic steatosis [fatty liver] in these animals." Yet a comparison of blue-green algae alone vs. exercise alone vs. control (no exercise or blue-green algae) showed a striking difference in levels of total hepatic lipids (24% vs. 21% vs. 117% levels, respectively). In conclusion, treatment interventions of blue-green algae supplementation and/or exercise "seem to reduce the accumulation of total hepatic lipids in diabetic rats, thereby attenuating hepatic steatosis."[127]

While the mechanisms underlying the effect of blue-green algae on blood pressure are less well understood, Guan et al. (2007) reported that the high potassium content might have a beneficial effect with respect to reducing high blood pressure.[128]

Blue-Green Algae may Prevent Nonalcoholic Fatty Liver Disease

One diabetic complication, among other pathologies, is the development of nonalcoholic fatty liver disease (NAFLD). In experimental studies with rats, blue-green algae effectively prevented fatty liver disease in rats given a high fructose rich diets,[129] or prevented animals from developing fatty liver induced by carbon tetrachloride treatment.[130]

Moura et al. (2011) investigated the effects of physical exercise and *Spirulina* intake on the control of NAFLD in rats with diabetes.[131] First the rats were made diabetic by an injection of a chemical (alloxan). Then all diabetic animals were randomly divided into four groups: 1) rats fed with a control diet and allowed no physical exercise (control group), 2) rats given a control diet and exercised (CE group), 3) rats whose diets included blue-green algae (BG group), and 4) rats given blue-green in their diet and exercised (BG/E group).

Rats supplemented with blue-green algae and those provided exercise (the BG, BG/E, and CE groups) all presented with lower plasma concentrations of LDL cholesterol than the control rats (without exercise or blue-green algae). Additionally, lower levels of total liver lipids were evident in the BG, BG/E, and CE in comparison to levels in the control group. Thus, blue-green algae appears to be effective in reducing total circulating levels of LDL cholesterol and hepatic lipids, alone or combined with physical exercise in diabetic rats. According to the researchers, "The principal finding of our study was that *Spirulina* in the diet of these rats resulted in an attenuation of fatty deposits in the liver as well as lower plasma levels of LDL cholesterol in comparison to the diabetic control group that did not receive treatment."[132]

Simply adding a gram or two of edible microalgae to one's diet offers a low risk, high potential way to help balance blood glucose and to improve the glycemic index of foods.

All together, the remarkable extent and depth of microalgae research and unique algal properties for use of prevention, reduction and treatment of type 2 diabetes is most promising. While more experimental and clinical outcome trials are always needed for the advancement of medical science; our world-wide epidemic of type-2 diabetes requires immediate lifestyle changes and natural solutions. Simply adding a gram or two of edible microalgae to one's diet offers a low risk, high potential way to help balance blood glucose, improve the glycemic index of foods, provide antioxidant support, and likely improve one's overall lipid and cardiovascular health profile, for millions of people.

CHAPTER

11

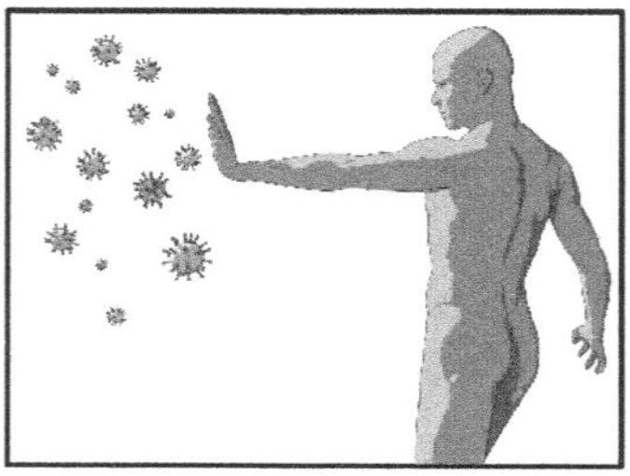

Give Your Immune System a Fighting Chance

"We may be different in gender, color of hair and skin, religion, and job. But we have a common bond—we are survivors. Our parents survived long enough to conceive us. Grandparents had the same claim for your parents. The thing that made this possible is that precious commodity—the immune system."

— Schmidt, Smith, Sehnert, 1993

Microalgae Hypothesis VI: Immune Support

Microalgae can improve overall immune function.

Microalgae can enhance innate, nonspecific immunity function.

Prebiotics can aid intestinal mucosal barrier.

Microalgae can support our specific, adaptive immunity.

Microalgae activate the circulation of macrophage and NK cells.

Phycocyanin powerfully inhibits influenza viral activity.

Microalgae aids production of hematopoietic stem cells (HSCs).

Microalgae may help autoimmune disorders through regulatory T cells.

Microalgae Offer Essential Nutrients that Support Our Immune System

"Deficiencies in any of a large number of single essential nutrients can produce dysfunctions in the immune system and other host defensive mechanisms . . . Single nutrients that impact importantly on protein synthesis [can] influence every aspect of immunity . . . Deficiencies of essential amino acids can also depress the synthesis of proteins, including those that contribute uniquely to host defenses . . . Trace [mineral] elements that function as the key components of metalloenzymes are also known to have some effects on the internal structure and function of lymphocytes."[1]

Nutritionally acquired immune deficiency syndromes (NAIDS) are primarily found in patients in Third World countries where malnourishment is the norm. Yet while "NAIDS is seen most frequently in children throughout the world, it is often seen in elderly individuals, and it is an all too common complication of severe medical or surgical diseases, malignancies, burns and other forms of trauma being treated in our most modern hospital centers . . . Children are highly susceptible to infection and each febrile infection causes them to lose vital body nutrients. Multiple, closely spaced or severe infections are the most frequent cause of childhood malnutrition."[2]

Chemistry professor Karl Abrams describes how the carotenoids in *AFA* enhance the immune system by protecting the thymus gland and offering antioxidant protection to immune cells (e.g., white blood cells). "Beta carotene and the other carotenoid compounds of AFA bio stimulate our immune system by increasing the number, activity, circulation of antiviral thymus helper cells.[3] And once beta carotene has been biochemically transformed into vitamin A, white blood cells may use it to increase their abilities to kill invading viruses. This vitamin A precursor is also known to increase B-cell activity, thus increasing antibody production, especially IgA, when necessary."[4] However, beta-carotene's immune-enhancing effects are maximized only in cases when it occurs in combination with a wider family of carotenoids, such as with microalgae.[5]

In a study of 100 children diagnosed with a zinc deficiency and given either zinc sulfate or blue-green algae tablets, those given blue-green algae demonstrated a superior immune response. The zinc found in blue-green algae may be about three times more effective than zinc from mineral sources.

From a clinical point of view, important individual nutrients that support immunological functions include vitamin A, beta carotene, zinc, iron, B-vitamins, amino acids (especially arginine and glutamine), polyunsaturated fatty acids, and nucleotides—that are found abundantly in microalgae. Micro algal carotenoids such as beta carotene, canthaxanthin, and astaxanthin have demonstrated immune-modulating activities using *in vitro* ("test tube") cell culture experiments.[6]

Glutathione produced in blue green algae is also an important nutrient to the immune system. Glutathione is essential to the replication of white blood cells known as lymphocytes, as well as vital to the activity of natural killer (NK) cells.

The immunological effects of zinc are many, are well studied and have a variety of possible mechanisms. Zinc is required for nucleic acid metabolism and is essential to over 100 metalloenzyme activities. In addition, zinc is a fundamental component of thymic hormones, which help to stimulate the activities of T-lymphocytes throughout the body and to increase antibody production.

In an experimental study of 100 children diagnosed with a zinc deficiency with half given either a dose of zinc sulfate or blue-green algae tablets, those given blue-green algae demonstrated a superior immune response. The zinc found in blue-green algae was found to be twice as effective in curing zinc deficiency in children, compared to zinc from mineral sources.[7] The medical researchers concluded that naturally chelated zinc in blue-green algae improved mineral absorption, general health and the children's immune systems.

In two separate studies which reviewed school records of children consuming *AFA* blue-green algae individually revealed a dramatic improvement in class attendance. Both research teams, along with school personnel reports, suggested that the increased attendance of students who ate blue-green algae was directly related to a decrease in sick days.[8-9] Likewise; animal feeding studies found that blue green algae given to fish and poultry resulted in better disease resistance, improved survival and growth rates, as attributed to enhanced immune function.[10-11]

Our Complementary Specific and Nonspecific Immune Systems

The principle purpose of our immune system is to detect and eliminate invading pathogens, such as bacteria, viruses, fungi, and parasites. Our immune system also assists in the elimination of damaged and abnormal cells. We are continually exposed to infectious agents and the normal breakdown of cells. Most of the time our immune system provides the means to overcome these daily challenges.

There are two primary types of immune responses: a) nonspecific, innate, reactions and b) specific, adaptive, reactions. Both immune responses utilize different types of white blood cells as part of a defense strategy. Our innate or host immune system acts as the first line of defense, providing physical, anatomical, barriers (e.g., intestinal flora, mucus membranes or saliva), along with quick acting reactions that involve inflammation and phagocytes.

Phagocytes are 'white blood cells' that protect the body by ingesting harmful foreign particles, bacteria, and dead or dying cells. These phagocytes roam throughout our bloodstream awaiting chemical signals of foreign invasion and inflammation. When triggered, the white blood cells rush off to the site of infection. However scientists are discovering that certain innate immune cells (macrophages) do more than "act as antimicrobial warriors, they also play critical roles in immune regulation [balance] and wound-healing."[12]

Alternatively, our specific, adaptive, immune system takes more time to react. Still, our adaptive response has a detailed "immune memory," which remembers the invading organisms, and is antigen specific. For example, vaccinations are given to assist our specific immune system to recognize potential invaders, or infectious agents. Specific immunity has mainly to do with how our lymphocytes (e.g., T and B and Natural Killer (NK) cells) learn to identify specific bacteria or viruses and become better prepared to launch an attack. For most of the lifetime of the animal, including humans, these memory cells will "remember" each specific pathogen encountered, and are ready to launch a strong and rapid response if the pathogen is detected again. All parts of our immune system actually work together in this orchestrated manner.

For example, natural killer (NK) cells were originally defined as part of innate immunity system. "More recently, a more nuanced view of NK cells has emerged. NK cells are now recognized to express a repertoire of activating and inhibitory receptors that is calibrated to ensure self-tolerance while allowing efficacy against assaults such as viral infection and tumor development. Moreover, NK cells do not react in an invariant manner but rather adapt to their environment. Finally, recent studies have unveiled that NK cells can also mount a form of antigen-specific immunologic memory. NK cells thus exert sophisticated biological functions that are attributes of both innate and adaptive immunity, blurring the functional borders between these two arms of the immune response."[13]

Microalgae Enhances Our Innate, Nonspecific, Immunity

Microalgae support our entire immune system, but chiefly acts to strengthen our adaptive, nonspecific, immune system. Several studies, both animal and human, have demonstrated the ability of microalgae to increase macrophage movement.[14-15] The dietary use of blue-green algae is reported to enhance the immune response in laboratory mice, by stimulating macrophage functions, phagocytosis [ingestion other cells and foreign particles] and enhanced interleukin-1 [helps to regulate inflammation] production.[16-17]

For example, oral "ingestion of 1.5 g of dried whole *A. flos-aquae* (*AFA*) resulted in a transient reduction in peripheral blood NK cells in 21 healthy human volunteers, suggesting increased NK cell homing into tissue. We have now identified an extract from *A. flos-aquae* that directly activates NK cells in vitro and modulates the chemokine receptor profile . . . The low-molecular-weight fraction <5,000 of *AFA* was responsible for the most robust NK cell activation, suggesting novel compounds different from previously reported macrophage-activating large polysaccharides."[18]

Two independent research teams (Manoukian R, et al., 1998; Jensen, et al., 2000)[19-20] demonstrated that consumption of *AFA* algae rapidly activated blood lymphocytes, natural killer (NK) cells. NK cells seek out and destroy virus-infected cells or tumor cells without activation of the specific immune system, antibodies. The usual way of measuring NK activity is by measuring levels in the blood or in vitro (e.g. test tubes). However, because NK cells are believed to work at the tissue level, indirect assessments are involved.

Human participants were given 1.5 grams of *AFA* algae after their immune function, blood parameters, were assessed. Ingestion of *AFA* algae had an immediate effect on blood lymphocytes, natural killer (NK) cells. NK cells were rapidly reduced from the blood by 60%. While the mechanism for this finding is not known, the researchers hypothesized that *AFA* algae induced the migration of NK cells into the tissues, by activating their adhesion mechanisms. Thus, the *AFA* algae increasing the surveillance by NK cells after *AFA* consumption would strengthen the volunteers' innate immune system response. Because the activation of NK cells was so rapid, the *AFA* algae may have stimulated nerves in lymphoid tissues (tissues associated with immune function). Neuropeptides might have drawn the recruitment of immune cells from blood circulation.

*AFA A*lgae Increases Circulation of White Blood Cells

Gitte Jensen, Ph.D., an immunologist at McGill University, working with a team of researchers at the Royal Victoria Hospital in Montreal, demonstrated improved trafficking of immune cells to be among the effects of *AFA* algae on the human immune system. Many immune cells (e.g., natural killer [NK] cells) do their primary work outside of the bloodstream in the tissues. *AFA* algae increased the number of white blood cells that moved from the bloodstream into the tissues to do their search-and-destroy mission.[21]

In a follow-up double-blind study, Jensen's team replicated the initial results and also found that long-term consumers of *AFA* demonstrated greater benefits than those taking algae for the first time. Yet even short-term consumers showed some benefits. Dr. Jensen's team found that within two hours of eating *AFA* there was a significant migration of natural killer cells from the blood into the surrounding tissues. Those whose who consumed the *AFA* also had more adhesion molecules which aid the immune cells to attach to tissue cells. Natural killer cells play a key role in our defense system as they "patrol for invading microbes and infected or transformed precancerous cells." This gentle immune boost was rapid, short-term, and cell-type specific.[22] It is also noteworthy that high doses of *AFA* are not required to realize these benefits, in Dr. Jensen's studies. The positive immune effects were seen with low amounts of *AFA* algae (1.5 grams), available in food supplementation.

According to Dr. Jensen, these milder, killer-cell specific, episodic effects are preferable to a stronger, prolonged, and more global immune activation. For example, it is possible to put something into the blood-stream that will provoke a hyper activation and a more global immune response. There are, however, substantial drawbacks to over-stimulation, including undesirable levels of inflammation. For example macrophages are the main source of several cytokines that promote an inflammatory response and are often referred to as "proinflammatory cytokines." A more moderate immune response is preferable.

Research on blue green algae has demonstrated features that can stimulate macrophage functions, such as phagocytosis and production of chemokines. This effect may be attributable at least partially to the polysaccharide fraction. In contrast the same data suggests that C-phycocyanin from *Spirulina* may down-regulate these macrophage activities.[23] The fact that blue green algae has been shown to both decrease inflammatory responses and increase immune responses, suggests a more balanced or gentle immune support. Many studies have demonstrated the anti-inflammatory benefits of microalgae.

Over a two month trial period, over fifty percent of the subjects (ages 40 to 65 years) who drank a hot-water extract of blue green algae, daily – showed increased natural killer cell mediated destruction of harmful or infectious cells related to improved interferon-*y* (IFN*y*) production. Furthermore, IFNy has been shown to play a major role in promoting specific mechanisms of innate, host, defense.[24]

Lipopolysaccharides and C-phycocyanin in blue-green algae have been shown to stimulate macrophage activation and stem cell differentiation potential.[25-26] Also, cell wall fragments of lipopolysaccharide, lipid A, and glycolipoproteins are known to have immune-enhancing effects that strengthen the entire immune system.[27-28] When blue-green algae, such as *AFA*, is freeze-dried and encapsulated, such complex polysaccharide compounds of the cell wall are broken down into easily absorbable fragments. *AFA* microalgae, like human breast milk, contains a peptide known as substance P, which also acts as an immune booster.[29]

Unique high molecular weight polysaccharides have been isolated from food-grade microalgae, which are potent activators of human monocytes/macrophages. They are termed, "Immulina" from *Spirulina platensis* , "Immunon" from *Aphanizomenon flos-aquae*, and

"Immurella" from *Chlorella pyrenoidosa*. These three polysaccharides are all structurally complex and water soluble; making-up between 0.5 to 2.0% of the micro algal dry weight. These micro algal "polysaccharides are between one hundred and one thousand times more active for in vitro monocyte activation than polysaccharide preparations that are currently used clinically for cancer immunotherapy."[30] For instance, it was found that a smaller dose of Immunon was needed to create the same immunostimulatory response as bacterial lipopolysaccharide (LPS). LPS is found in certain bacteria and triggers a strong immunostimulatory response in animals, including humans. Surprisingly, the researchers also discovered that mRNA (that helps to duplicate DNA strands) within those substances (e.g., interleukin-1*B*) that regulate the immune system were increased, too.

Microalgae 'Prebiotics' Aid Intestinal Protective Barrier

From an evolutionary perspective, bacteria and microalgae have been interacting over billions of years. According to various scientists, interactions between algae and bacteria are not random, but highly specific. "There are numerous data which report that the extracellular products from algae are capable to stimulate the growth of bacterial strains through the excretion of carbohydrates, organic acids, nitrogenous substances and vitamins. Some studies in natural ecosystems have determined that organic substances derived from phytoplankton are used by bacteria as a substrate for growing. However, microalgae also inhibit bacterial growth by production of organic exudates or toxic metabolites. There are several reports suggesting a synergistic action between microalgae and its bacterial flora associated."[31]

The oral consumption of certain microalgae can help to support the integrity and health of intestinal micro flora. Andrew Valencia reported that *AFA* significantly reduced "leaky gut syndrome" in patients, based on functional clinical test measures.[32] Compromised intestinal permeability increases the uptake of toxic compounds and macromolecules that can exhaust the detoxification capability of the liver. A combination of leaky gut and dysfunctional liver detoxification may lead to increased tissue stores of toxic compounds and depressed immune status.

Microalgae Supports Specific, Adaptive, Immunity

Evidence exists that suggests certain edible microalgae may aid the function of our specific immunity, as well. Researchers have found both increased antibody production and enhanced immune function in animals supplemented with blue-green algae.[33-34]

For example, in a study of 463 calves, kept under the same conditions of tending and feeding, the experimental groups fed supplemental algae were the least likely to contract disease. The experimental conditions were as follows: Group 1 received *Chlorella* microalgae and a hyper immune antibovine rabbit serum. Group 2 received either a) microalgae or b) hyper immune antibovine rabbit serum. Group 3 received vaccines for both mucosal disease (MD) and infectious bovine rhinotracheitis (IBR). Results showed that Group 1, treated with the microalgae and the hyper immune serum contracted the least diseases (20% infection rate); Group 2-a and 2-b treated with either algae or hyper immune serum had similar results (27% infection rate, each); Group 3 with vaccines only, had the worst results, (42% infection rate) of the calves contracted a disease.[35]

Heat Shock Proteins can Increase Antibodies in Mammals and Aquatic Animals to Fight Diseases

Scientists at the Marine Biotechnology Institute in Malaysia announced they were developing research to produce a type of algal protein, known as Heat Shock Protein (HSP 70), "that can increase antibodies in mammals and aquatic animals to fight diseases. . . When it is completed, this product will be the first of its kind in the world and it could be commercialized for aquaculture breeders. . . He added that the technique developed will be a safe way to increase protein levels in animals as well as help reduce stress in livestock."[36]

Japanese researchers report that oral consumption of a hot water extract of *Spirulina* helped to activate microbial-sensing proteins, called Toll-like Receptors in blood cells. Toll-like Receptors can detect a broad range of human pathogens, as well as other molecules that signal tissue damage. These Toll receptors can activate both sides of the immune system – both the innate and adaptive responses which can work together to ward of infectious or other attacks. The researchers discovered cells from both innate (e.g., macrophages) as well as specific (e.g., natural killer cells) were up-regulated by ingestion of the blue green algae extract.

Boosts Production of Hematopoietic Stem Cells (HSCs)

Almost all immune cells ultimately come from the same progenitor, hematopoietic stem cells located within our bone marrow. Hematopoietic stem cells (HSCs) are blood-forming cells produced primarily within the bone marrow. HSCs have a greater ability than other immature blood cells to exit the bone marrow barrier. HSCs shift back and forth though the bloodstream, usually between bone marrow sites, and can move virtually anywhere in the body. If they settle in the thymus, they may develop into T cells. HSCs have remarkable ability to transform into all the different blood cell types, which make up our immune system.

Figure 22. Key Immune Cells from Hematopoiesis

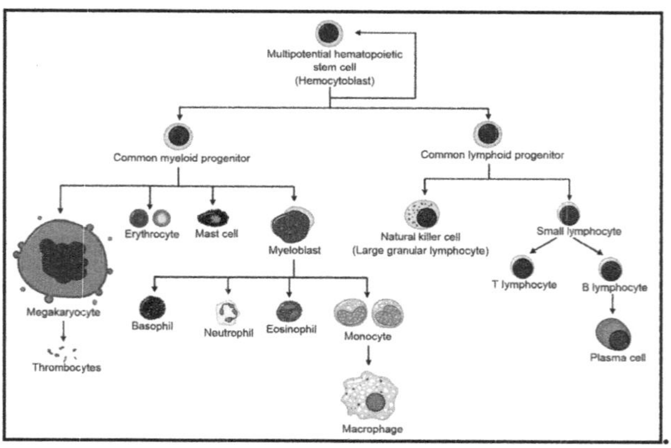

Simplified chart by A. Rad (July 21, 2009)
http://en.wikipedia.org/wiki/File:Hematopoiesis_simple.svg

Macrophages, natural killer cells, T- and B lymphocytes are just a few examples of immune cells. Yet progenitor, hematopoietic stem cells are relatively rare. Only an estimated 1 in 10,000 to 15,000 cells in the bone marrow is a hematopoietic (blood-forming) stem cell (HSC). Any nutrition that ultimately supports bone marrow stem cells, like blue-green algae, would support immune function as well.

Researchers have studied the effectiveness of oral microalgae consumption to support immune function and report that components of blue green algae, phycocyanin, will increase the proliferation of bone marrow cells.[37] For example, mice orally treated with the blue-green algae extracts experienced immune competent-cell proliferation and differentiation; increased neutrophils and lymphocytes in the peripheral blood and bone marrow.[38]

Phycocyanin Helps Autoimmune Disorders via Induction of Regulatory T Cells

Regulatory T cells (Tregs) are a subpopulation of T cells which help maintain balance in the immune system. Tregs are an important regulatory component of the immune system that suppresses the immune response of other cells. This is an important "self-check" built into the immune system to avoid excessive reactions, like in autoimmune diseases. Mouse models have suggested that modulation of Tregs can help treat autoimmune diseases and cancer and be used to prevent the rejection of transplanted organs.

Mark McCarty's (2007) article on 'Clinical Potential of Phycocyanobilin for Induction of T Regulatory Cells in the Management of Inflammatory Disorders," posits that since oral phycocyanin is beneficial in various rodent models of autoimmune disorders, it is reasonable to suspect that blue-green algae extracts of C-phycocyanin, may have clinical potential for boosting Treg activity in human autoimmune or allergic syndromes. "Exposure of human mononuclear cells to phycocyanin in vitro is reported to promote generation of Treg cells. Induction of heme oxygnase-1 (HO-1) in lymphocytes has a similar effect, and it is not likely an accidental that a key product of HO-1 activity, biliverdin, is homologous to the structure of phycocyanin's chromophore phycocyanobilin (PhyCB)."[39]

Dr. McCarty establishes a connection between Treg induction exhibited in mice following injection with bilirubin, biliverdin's chief metabolite. He states that "These considerations suggest that bilirubin, generated within lymphocytes by HO-1 activation, may play a physiological role in the promotion of Treg immunomodulation." If so, than the ingestion of C-phycocyanin and phycocyanobilins are likely to increase Treg induction as well. Since there are "no rich natural sources of biliverdin or bilirubin – which are expensive to synthesize – and *Spirulina*, on the other hand is a safe natural food that can contain 20% or more phycocyanin by dry weight, it should be an urgent priority to study the impact of *Spirulina* or of PhyCB-enriched *Spirulina* extracts on autoimmune disorders.

An early European study (1996) examined the effect that blue-green microalgae on patients with multiple sclerosis, found that "intake of *Spirulina* makes for lengthening of remission in those patients with disseminated sclerosis (Buletsa et al, 1996).[40] Later empirical observations by Penton-Rol and colleagues (2010) suggested that interferon-treated multiple sclerosis patients tend to have a better clinical course when they concurrently ingest blue-green algae.[41]

Multiple animal studies have proven orally administered of a whole blue-green algae or C-phycocyanin has exerted profound anti-inflammatory effects. Therefore, orally administered C-phycocyanin might duplicate the antioxidant and anti-inflammatory benefits achieved physiologically by biliverdin/bilirubin.[42-43] A safe and nutritional strategy for boosting Treg induction might be beneficial in a wide range of autoimmune, allergic, septic shock, and various inflammatory disorders (e.g. in the lungs), and as well as potentially useful for preventing graft or organ rejection.

Microalgae as an Immune-Enhancer and Bio modulator

Edible microalgae offer a large variety of different components that synergistically drive the therapeutic effects of the whole compound. In Gitte Jensen et al., (2001) review article on "Blue-Green Algae as an

Immuno-Enhancer and Biomodulator," she suggests that, "Research results based on the numerous isolated compounds from blue-green algae warrant the exploration of using whole algae as conjunctive therapy due to the possible synergistic effects of many phytochemicals within the whole algae . . . A significant body of data suggests that blue-green algae immunoenhancing properties could be useful in the adjunct treatment of various disease involving 1) suppressed or exhausted immune system, and 2) inappropriate immune response including allergies, autoimmune disease, and chronic inflammatory conditions."[44]

In a retrospective review of medical cases, researchers found positive evidence that *AFA* blue-green algae might be useful in the treatment of chronic fatigue, Epstein Barr infection, fibromyalgia, and AIDS. These diseases all involve significant immune system, and sometimes viral, components. Such anecdotal evidence suggests that at least some autoimmune diseases may respond favorably to blue-green algae. Some species of microalgae, *AFA* in particular, have been shown to be rich sources of polyunsaturated fatty acids (PUFAs), which increase cell membrane fluidity and enhance immune function. Our immune system depends on optimal cell membrane fluidity. For example, when white blood cells in our immune system develop more rigid cell membranes, it becomes more difficult for them to maneuver through tissues to attack viruses, bacteria, and unhealthy cells. Additionally essential fatty acid (EFA) deficiency is known to substantially lower the number of thymus suppressor cells and thus impair overall immune function.[45]

Nourishment is an essential factor in maintaining immune health. For example, according to Kenneth Bock, M.D., beta carotene affects the immune function by: "Enhancing lymphocyte production; increasing macrophage cytotoxicity and cytotoxic T cell activity; helping detoxify pollutants; enabling the growth and development of cells; maintaining the membrane receptors that are essential for immune function; and modulating the release of prostaglandins and leukotrienes."[46] Microalgae provide the richest source of the best quality and most easily absorbed beta carotene. Also microalgae contain potent immune enhancing polysaccharides. Algal polysaccharides extracted from *Spirulina platensis* , *Aphanizomenon flos-aquae* and *Chlorella pyrenoidosa* have been documented to be 1,000 times more active than other polysaccharides that are currently used for cancer immunotherapy![47]

Yet, microalgae provide not single nutrients, but a wide spectrum of compounds that together support and feed our immune system. This is the key to understanding why consuming blue-green algae appears to have benefits for human aging. While some people may be rapidly struck dead by a lethal virus or flesh-eating bacteria; most people die in a far slower manner. Over decades, most elderly people experience a gradual weakening of their immune system, along with a slow decline of cellular integrity, kindled by inflammation and oxidative stress.

Researchers report a vital benefit of blue green alga is to activate the innate immune system.[48] Our innate immune system is the body's initial line of defense, which usually brings on an increase in inflammation. Thus, the finding that blue green algae assists our innate immunity and decreases issues associated with inflammation is excellent news; a sustainable immune-support approach for the myriad of challenges associated with aging.

CHAPTER

12

Cancer Protection for Cells Gone Wild

"At present we have overwhelming evidence . . . (that) none of the risk factors for cancer is . . . more significant than diet and nutrition."

— *Reddy B, 1992*

"Traditionally, cancer was thought to be caused by changes in DNA sequence, or mutations, that allowed for uncontrolled cell growth. That's still true. However, there's also increasing interest in the role played by epigenetics, in which such factors as diet, environmental toxins, and lifestyle affect the expression of genes — not just in cancer, but also cardiovascular disease, diabetes, and neurological disorders."

— Science Daily, *2012*

"Today, treating cancer is a huge business. Every 30 seconds another American is diagnosed as having the disease. Typical cancer patients spend over $25,000.00 to try to treat their condition . . . Every 55 seconds, another American dies of cancer."

— *Robbins J, 1987*

Microalgae Hypothesis VII: Cancer-Protection

Microalgae can improve cell signaling to detect, repair or eliminate malignant cells.

Polysaccharides can stimulate immune-function and increase circulation of T cells and macrophages.

Blue-green algae can slow growth of certain cancer cells by inhibiting NADPH oxidase.

Microalgae can inhibit tumors found in heart, stomach, liver and other forms of cancers.

Microalgae can decrease the side effects of cancer treatments.

Microalgae can produce therapeutic drugs to treat cancer.

Microalgae Provide Protective Factors, which Lower the Risk of Cancer

Cancer is defined as cellular replication that has gone awry. The same factors within microalgae that optimize cellular regeneration, defending against genetic errors, and protecting mitochondrial function, may offer significant cancer-preventative effects. Despite the "war on cancer," conducted over the last 25 years and expending hundreds of billions of dollars, cancer still stands as the second leading cause of mortality in developed nations, trumped only by cardiovascular disease. Environmental factors and diet play a large role in the development of certain forms of cancer. An illuminating water-shed was an article by Doll and Peto (1981) that estimated 35% of cancer incidence was related to nutritional factors.[1]

Notwithstanding the ongoing debate—with an increasing recognition of the multi-causal factors involved with the etiology of cancer—the role of diet remains a pivotal one. In the last few decades researchers have increasingly focused on foods, vitamins, and plant chemicals to provide protective factors that lower the risk of cancer.[2-3] For example, in 2013 a study of more than a half a million Americans over the age of 50, found that people with the healthiest eating habits are 15 percent less likely to develop pancreatic cancer than those with the poorest diets.[4]

Weight appears to be a Significant Risk Factor for Cancer

The Centers for Disease Control and Prevention (CDC) stated there is "convincing evidence of an association between excess weight and increased risk of several cancers, including adenocarcinoma of the esophagus, colon and rectum cancer, kidney cancer and endometrial cancer."[5] The National Cancer Institute (NCI) adds thyroid and gallbladder cancer to this weight-associated list and predicts that other cancers will be added soon.[6] Weight appears to be a significant risk factor for cancer.

According to Jane Hart, M.D. (2013), "a number of mechanisms linking obesity and cancer have been purported . . . ranging from the presence of excess estrogen produced by fat tissue; increased levels of insulin and insulin-like growth factor-1 (IGF-1); fat-cell production of hormones that stimulate or inhibit cell growth; fat cells' effect on tumor growth regulators; and chronic inflammation associated with increased cancer risk."[7] Because edible microalgae appears to offer benefits in the prevention of complications associated with diabetes, as well as potentially reducing obesity, microalgae beneficial effects on inflammation and balancing insulin factors may also prove effective in cancer prevention. Furthermore, some macro- and microalgae components have been shown to inhibit tumors found in heart, stomach, liver and other cancers.

Microalgae Use Popular Among South East Asian Breast Cancer Survivors

Use of algae, both macro- and microalgae products have a broader cultural acceptance in Asia than in Europe or in America. For example, when breast cancer survivors were surveyed in two Malaysia hospitals, *Spirulina* was identified as the most common nutritional supplement taken by these women, outside of multivitamins.[8]

Spirulina, a more widely studied blue green algae, has been reported to inhibit the formation of tumors in experimental animals and to induce the regression of existing tumors.[9-10] In another experimental animal study, *Spirulina* pre-treatment prior to the carcinogen application demonstrated a significant inhibition of tumor burden as well as tumor incidence in both of the tumor models (skin and stomach tumors) studied. There was a significant activation of phase II enzyme activities associated mainly with carcinogen detoxification. Moreover, *Spirulina* dramatically increases a variety of antioxidant enzymes viz., superoxide dismutase, catalase, glutathione reductase, glutathione peroxidase and reduced glutathione.[11]

Several components of chlorophyll have shown tumor-suppressive activities.[12-13] A study published in *Cancer Research* reported that chlorophyll, abundant in microalgae, is a "potent inhibitor of hepato-carcinogenesis (the development of liver cancer) ...and may have important implications in intervention and dietary management of human cancer risks."[14]

Chlorophyll Rich Foods Partially Offset the Effects of Carcinogens

Scientists at Oregon State University Linus Pauling Institute (2012) found that consumption of spinach can partially offset the damaging effects of the carcinogen. In tests with laboratory animals, it cut the incidence of colon tumors almost in half, from 58 percent to 32 percent. The researchers credited the high amount of chlorophyll present in spinach as one of the active biomolecules responsible. Notably, microalgae have a far higher concentration of chlorophyll than spinach.[15]

The possibility that simple chlorophyll derivatives might aid in cancer prevention arose in the 1980s, when researchers in Italy, Japan, and the U.S. discovered that chlorophyllin could reduce the ability of certain mutagenic chemicals to damage the genes of bacteria and fruit flies. This was particularly interesting because chlorophyllin had several human applications without any evidence of toxicity. It has been used, for instance, as a green food dye in some countries, for accelerated wound healing, and for odor control in geriatric patients."[16]

Researchers at Oregon State University have postulated that the "ability of chlorophyllin to counteract malodorous chemicals may be a clue to its antimutagenic action." Japanese (Hikoya Hayatsu) and American researchers (Roderick Dashwood and Vibeke Breinholt), were able to show in the 1980s and 1990s that chlorophyllin has the ability to bind or "sandwich" certain classes of chemical mutagens and carcinogens.

Exposure to Aflatoxin, a Fungal Toxin in Foods (e.g., corn, peanuts), Related to Increased Cancer Risk

Exposure to aflatoxin, a fungal toxins found in certain foods (e.g., rice, corn, peanuts) has been related to increased cancer risk.[17] Chronic aflatoxin exposure has an additive effect; increasing the risk of liver cancer in all persons who ingest contaminated food and an increased the risk of lung cancer in workers who handle the grain. Approximately 4.5 billion persons living in developing countries are chronically exposed to uncontrolled amounts of aflatoxin.[18] Aflatoxin can be especially prevalent in grain stores in tropical and subtropical regions because moist grain is a very good environment for the fungi to thrive.[19] A person's risk of developing liver cancer may increase by as much as 30 fold, when someone is infected with either the Hepatitis B or C viruses combined with an exposure to aflatoxin.[20]

In some cases the survival rates of algae-treated mice increased nearly 80% over control groups. Such findings suggest that presurgical treatment with extracts of microalgae might decrease or prevent metastasis or tumor progression.

In the early 2000s, Dr. Kensler of Johns Hopkins University reported on a double-blinded, placebo-controlled, biomarker intervention trial with 180 Chinese volunteers. Two group of 90 people each were randomly assigned to receive either a chlorophyllin tablet or a placebo tablet with each meal for four months. Result indicated that those people consuming the chlorphyllin tablets had a 55% reduction in the urinary aflatoxin exposure biomarker compared with those receiving the placebo. These results provided "evidence that, for pennies per day, chlorophyllin supplements may cut the liver cancer death rate in aflatoxin-exposed populations in China and elsewhere at least by half . . . Chlorophyllin shows great promise for protection against major human cancers." Further longer-term outcome studies are needed to translate this research into a clinical treatment.[21]

Gastrointestinal Cancers Rank Second in Overall Cancer-Related Deaths

"Gastrointestinal cancers rank second in overall cancer-related deaths. Carotenoids, sulfated polysaccharides, and polyunsaturated fatty acids (PUFAs) from microalgae exhibit cancer chemo preventive features at different stages of carcinogenesis. For instance, sulfated polysaccharides bear a prophylactic potential via blocking adhesion of pathogens to the gastric surface, whereas carotenoids are effective against Helicobacter pylori infection. This effect is notable because *H. pylori* has been targeted as the primary cause of gastric cancer."[22]

Additionally, microalgae products have been suggested for the treatment of patients suffering from chronic inflammation, inflammatory bowel disease (IBD), and colon cancer. Researchers from the University of Seville in Spain have reported that numerous studies established the potential colon cancer preventive properties of phytochemicals derived from both plants and from marine environment, including microalgae species and their products. They advance that "these naturally occurring compounds may mediate chemo preventive effects on cancer. These actions include induction of cell cycle arrest and apoptosis [death of cancer cells], inhibition of cell proliferation, stimulation of antimetastatic [limits the spread of cancer cells] and antiangiogenic [blocks blood vessels into tumors] responses and increased antioxidant and anti-inflammatory activity."[23] All of these actions help to prevent the occurrence or spread of cancer cells.

Beta carotene is one of the Best-Known Anticancer Substances Found in Food

Many microalgae extracts are known to be rich in carotenoids. Extending "studies on human cells has demonstrated that many pigments beyond their ecological function as light harvesting molecules, also act as potent bioactive compounds on cancer cells and may have great potential in the prevention and treatment of cancers. . . . A large number of studies have demonstrated that purified carotenoids (carotenoids, lutein, zeaxanthin, lycopene, fucoxanthin, astaxanthin, neoxanthin) exert a direct antiproliferative activity on cancer cells grown *in vitro* and induce their apoptosis [cell death]."[24-30]

> *Studies on human cells has demonstrated that many pigments beyond their ecological function as light harvesting molecules, also act as potent bioactive compounds on cancer cells and may have great potential in the prevention and treatment of cancers.*

For example, a violaxanthin rich extract from a *Chlorella* species of microalgae, was demonstrated to reduce the spread, as well as initiating the death of cancer cells on human colon cancer cells.[31] Likewise, a study by French scientists (2011) , "unequivocally demonstrates the strong antiproliferative activity of violaxanthin on MCF-7 human mammary cancer cells grown *in vitro* and suggests that violaxanthin and derivatives obtained by pharmacomodulation should be studied as possible new drugs to cure breast cancer."[32]

Moreover, dozens of large-scale studies have disclosed evidence that eating fruits and vegetables rich in beta carotene reduces the risks of cancer.[33-36] However, it is important to note that isolated beta carotene (sold as a supplement on its own or in multivitamin formulas) does not provide the same benefits. In fact, the large-scale study referred to as "CARET" (Carotenoid and Retinol Efficacy Trial) found that synthetic beta carotene supplements were correlated with increased—not decreased—morbidity and mortality from cancer.[37]

Because microalgae are the foods richest in natural beta carotene, notably *Dunaliella, Chlorella,* and *Spirulina,* have been extensively tested for anticancer effects and these effects have been well documented.[38-47] *Chlorella vulgaris,* a unicellular green microalga, has also been shown to have immune-modulating and anticancer properties, with very promising results for its use in human cancer therapy.[48-52]

Microalgae Offer a Significant Contribution to the Prevention of Oral Cancer in Humans

Researchers at the Harvard University School of Dental Medicine demonstrated that algal extracts rich in beta carotene applied to cancerous tumors in the mouths of hamsters reduced the number and size of tumors or caused them to disappear.[53] In a further study, when an algal extract was administered to 20 hamsters pre-treated to develop mouth cancer, none of them developed the disease. By comparison, two pretreated control groups that did not receive any algal extract (40 animals) all developed mouth cancer. Interestingly, when synthetic beta carotene alone was given (provided by Sigma Chemical Company) fully half the animals developed cancer.[54] This research team has continued to replicate these effects, repeatedly demonstrating the ability of blue-green algal extracts to inhibit and prevent tumor growth and cancer.[55-57]

Additionally, there is evidence that blue-green algae can make a significant contribution to the prevention of oral cancer in humans. Researchers at Maryland's Human Nutrition Center administered a single gram of algae a day to tobacco chewers in Kerala, India, who had pre-cancerous mouth lesions (oral leukoplakia). After one year, the sores had vanished or shrunk significantly in more than 50% of the individuals, with complete regression in 45% of the human study subjects (N = 44). Lesions disappeared in only 7% of the people in the placebo group (N = 43). Within one year of discontinuing algae supplements, 45 percent of the experimental subjects developed recurrent lesions. There was no experimental intervention to change the carcinogenic behavior of their tobacco chewing.[58]

Novel Algal Protein Compounds may be useful in the Treatment of Difficult, Drug-Resistant Tumors

Beta carotene is not the only cancer-protective substance to be found in microalgae. Cancer researchers at the University of Hawaii isolated a blue-green algal pigment, called cryptophycin, which demonstrates powerful anticancer properties—especially useful in the chemotherapy of drug-resistant tumors.[59-60] Several metabolites from cyanobacteria (e.g. cryptophycins, dolastatins, curacin A) have, for example, been investigated with respect to anticancer activities (i.e. inhibition of cell-division).[61-63] Other new algal protein compounds have also exhibited "multidrug-resistance reversing activities" that may be useful in the treatment of difficult, drug-resistant tumors.[64-66]

Unicellular algae have some compounds in their cell walls that are similar to those found in bacteria and yeast. Scientists discovered that these algae protected mice against experimentally grafted sarcoma in much the same way as the other microorganisms.[67] Bacteria and yeasts are able to induce, when inoculated into laboratory rodents, a general stimulation of defenses through an immune process. Glycoproteins extracted from algae demonstrate potent antitumor properties in animal studies.[68-72] Initial human trials with these antitumor peptidoglycans from microalgae have shown promising effects.[73]

Microalgae in Combination with Adjuvant-Based Antitumor Immunotherapy

An "adjuvant" is a substance that increases the body's immune reaction and can assist in the prevention, amelioration or cure of disease. Nonspecific adjuvant immune therapy using bacteria-derived "BCG cell-wall skeleton" treatments has been shown to prolong survival time in lung cancer patients[74] and effective in the treatments of other types of cancers.[75]

Prior findings that "oral administration of hot-water extract of *Spirulina*, cyanobacterium *Spirulina platensis* , leads to augmentation of NK cytotoxicity in humans;" encouraged a team of Japanese researchers to investigate the possible mechanisms behind blue-green algae's "antitumor NK activation" effects. The team discovered that blue green algae "enhances tumoricidal NK activation through the MyD88 pathway"; which are pathways involved in "adjuvant-based antitumor immunotherapy. The team next found that "*Spirulina* and BCG-cell wall skeleton synergistically augmented IFN-gamma production and antitumor potential" immune responses. The researchers suggest that blue-green algae might be usefully combined in the development of "adjuvant-based antitumor immunotherapy."[76]

Marine unicellular algae, diatoms, have displayed antiproliferative effects against solid human tumor cell cultures—lung carcinoma, kidney carcinoma and melanoma—and have been shown to inhibit nonsmall-cell bronchopulmonary carcinoma in *in vitro* cell studies.[77-78] As yet, however clinical research on human cancer patients with microalgal diatom-derived therapies remains to be initiated.

Microalgae Phycocyanin, Blue-Green Pigment, may be a Potent Antitumor Agent

"In research in Japan, phycocyanin (the blue pigment of blue-green algae) was extracted and . . . [given] orally . . . [to] mice with liver cancer. The survival rate of the treatment group was significantly higher than the control group not given phycocyanin. After five weeks, 90% of the phycocyanin group survived, but only 25% of the control group were still alive. After eight weeks, 25% of the phycocyanin group still survived, yet none of the control group was alive. This suggests eating phycocyanin may increase the survival rate of cancer stricken organisms."[79]

Researchers at Dainippon Ink and Chemicals Company have a patent pending for using phycocyanin as an "anti-tumor agent." They base their claim on a study demonstrating that after two weeks the levels of white blood cells of a cancer group treated with phycocyanin were significantly higher than those of the untreated cancer group and equal to or higher than those of the normal group without cancer. This suggests phycocyanin may raise lymphocyte activity.[80]

Another patent claim filed by Dainippon Ink and Chemicals and Tokyo Kenkyukai is for "Anti-tumor agents containing phycobilin—also used to treat ulcers and hemorrhoid bleeding."[81] This patent application documents that taking a small dosage of algal extract daily maintains or accelerates normal cellular control functions that prevent generation of malignancy or inhibit its growth or recurrence. "Radiation and chemotherapeutic treatments of cancer often result in undesirable reduction of defensive white cells. Studies published in *Medication and New Drugs* in 1966 reveal that 2 to 3 grams of *Spirulina* [blue-green algae] will slow this loss of white blood cells. In addition, patients reported less nausea and lassitude from cancer treatment when *Spirulina* was added to their regimens."[82]

Blue-Green Algae may slow Spread of Cancer Cells and Impede Blood Flow into Developing Tumors

Numerous studies with unicellular green algae have demonstrated antitumor and cancer inhibitory effects in laboratory rodents.[83-89] For the most part the antitumor effects appear to be mediated by an increased host immune response. In some cases the survival rates of algae-treated mice increased nearly 80% over control groups.[90] Such findings suggest that pre-surgical treatment with extracts of microalgae might decrease or prevent metastasis or tumor progression.[91]

Oxidative stress produced either by tumors or by certain types of immune cells, is extremely toxic to NK cells and cytotoxic lymphocytes, impairing their tumor-killing capacities and even killing them altogether. The NADPH oxidase enzyme complex is associated with much of this cellular oxidative stress. 1). A key phytonutrient found in blue-green algae, phycocyanobilin, can function as a potent inhibitor of the NADPH enzyme complex, addressing the root of oxidative stress. 2). Moreover, like probiotic bacteria, blue-green algae cell walls contain polysaccharides, which stimulate immune-function. 3). Also, by inhibiting NADPH oxidase, blue-green algae can act directly on some cancer cells to slow their growth, and 4). Microalgae compounds might be used to impede blood flow into developing tumors (angiogenesis). In sum, blue green algae appears to work in at least four complementary ways to slow tumor growth and support immune defense against tumor formations.[92]

Cancer Cells Exhibit a Deviant Energy Metabolism, and a Higher Rate of Glycolysis

Energy-modulating vitamins, especially those with antioxidant and mitochondrial supportive properties, have been shown to prevent breast cancer in rats.[93] Cancer cells exhibit "a deviant energy metabolism, and a high rate of glycolysis," when cancer cells are rapidly growing. The overall production of ATP by the intracellular organelles, mitochondria, are greatly diminished in cancer; leading to the wasting syndrome with loss of weight, muscle atrophy, and fatigue seen in late-stage cancer patients.

Indian researchers demonstrated improvements in mitochondrial energy production using vitamins (B vitamins and Coenzyme Q10) in rats with mammary carcinoma. "These [mitochondrial] activities were restored to a greater extent in animals treated with energy-modulating vitamins. From these experimental results, one may hypothesize that the combination therapy of energy-modulating vitamins could be of major therapeutic value in breast cancer."[94] Moreover, one of the pathogenic mechanisms of mitochondrial disorders is the accumulation of reactive oxygen species, which may be benefited by micro algal antioxidants. Particularly as microalgae deliver good sources of B vitamins, along with a "mitochondrial cocktail" of beneficial nutrients and antioxidants, clinical trials on the complementary and combined effectiveness of microalgae with various cancer treatments is indicated.

Microalgae may Decrease the Unwanted 'Side-Effects' of Various Cancer Treatments

> *"Whole body irradiation" animal studies suggest there may be a potential benefit for cancer patients given algal beta-carotene before and after radiation treatments to protect against cellular damage caused by free radicals induced from irradiation.*

Blue-green and other microalgae's may find a role in enhancing the effectiveness or decreasing the side-effects of various cancer treatments. For example, pretreatment supplementation with blue-green algae has been found to effectively increase phase II detoxification enzyme pathways associated with carcinogen detoxification in all the organs studied in animal induced-cancer models: liver, lung, kidney, and stomach.[95]

"Whole body irradiation" animal studies suggest there may be a potential benefit for cancer patients given algal beta carotene before and after radiation treatments to protect against cellular damage caused by free radicals induced from irradiation.[96] Additionally, Japanese researchers using an animal model found that components of unicellular algae may be beneficial in the alleviation of cancer chemotherapy side effects (e.g., immune suppression) while supporting the antitumor activity of the chemotherapeutic agents.[97]

Green Microalgae Produces a Human Therapeutic Drug used to Treat Cancer

Biologists at UC San Diego (2012) genetically engineered a species of green microalgae in order to produce a very complex and expensive human therapeutic drug used to treat cancer. The breakthrough technology is "the culmination of seven years of work in Mayfield's laboratory to demonstrate that *Chlamydomonas reinhardtii*, a green alga used widely in biology laboratories as a genetic model organism, can produce a wide range of human therapeutic proteins in greater quantity and more cheaply than bacteria or mammalian cells."

> *What the development of the malarial vaccine showed us was that algae could produce proteins that were really complex structures. . . once we made that, we were convinced we could make just about anything in algae."*

The UC team reports that "You can't make these drugs in bacteria, because bacteria are incapable of folding these proteins into these complex, three-dimensional shapes," said Mayfield. "And you can't make these proteins in mammalian cells because the toxin would kill them." Mayfield's team has produced over 7 therapeutic proteins for humans, including the development of a malarial vaccine. "What the development of the malarial vaccine showed us was that algae could produce proteins that were really complex structures. . . once we made that, we were convinced we could make just about anything in algae."[98]

DHA Could Become a New Agent for Treating Neuroblastoma and Other Cancers

A 2010 Swedish study stated that "The next treatment for cancer might come from fish [or microalgae] . . . scientists show that the omega-3 fatty acid, "docosahexaenoicacid" or" DHA," and its derivatives in the body kill neuroblastoma cancer cells. This discovery could lead to new treatments for a wider range of cancers, including neuroblastoma, medulloblastoma, colon, breast, and prostate cancers, among others. . . . Scientists administered DHA to neuroblastoma cells from the nervous system and analyzed the cells for by products as the DHA was metabolized into the cells. Researchers then examined the effect of both DHA and its derivatives on the growth of cancer cells. Results showed that DHA killed the cancer cells, but that the toxic derivatives produced by DHA were even more effective at killing the cancer cells. This suggests that DHA could become a new agent for treating neuroblastoma and possibly many other cancers."[99]

Again it is crucial to emphasize that fish cannot synthesize the omega-3s, including DHA. Microalgae have the capacity to make omega-3 fats. Smaller fish eat microalgae and accumulate omega-3s in their tissues. Next, bigger fish prey on the smaller fish and amass even higher concentrations of omega-3s in their flesh, which can be extracted for omega-3 and DHA fish oils. Pure sourced microalgae derived DHA are readily available as supplements. While, the Swedish scientists were not using oral doses of DHA, other cancer researchers have demonstrated that oral use of DHA in experimental animals reduced the size of tumors and enhanced the positive effects of the chemotherapy drug cisplatin, while limiting its harmful side effects.[100]

Conflicting and widely publicized data from the Fred Hutchinson Cancer Research Center reported a link between high blood concentrations of omega-3 fatty acids and an increased risk for prostate cancer. Their study found an overall 43 percent increase in risk for all prostate cancers with statistically higher intakes of fatty acid concentrations of fish oils.[101]

Yet, Dr. Barry Sears noted that while the Hutchinson Center researchers found statistical significance for omega-3 fatty acid levels; from a clinical standpoint the group differences of Omega-3 levels were meaningless. According to Dr. Sears, all of the EPA (omega-3) amounts reported across all of the Hutchinson Center subjects were far below optimal recommended levels.[102] In contrast, Dr. Sears refers to the Japanese who consume a great deal more omega 3s in their diets than other nations, including the U.S. Yet Japan has one of lowest rates of prostate cancer rates in the world. Remarkably, Japanese rate of prostate cancer incidence is 10 times lower than the United States![103] What's more, the Japanese mortality rates of dying from prostate cancer are about 5 times less than prostate-related deaths in the United States.[104]

CHAPTER

13

Detoxify For Life

"Toxic substances are everywhere – in the air we breathe, the food we eat, and the water we drink. Even our bodies and the bacteria in the intestines produce toxic substances. It can be strongly said that the health of an individual is largely determined by the ability of the body to detoxify."

— Murray, Pizzorno, 1991

"Since 1950, at least 70,000 new chemical compounds have been invented and dispersed into our environment through new consumer commodities, industrial products, and food. We are by default conducting a massive clinical toxicological trial. And our children and their children are the experimental animals."

— Needleman H, Landrigan P, 1994

Microalgae Hypothesis VIII: Detoxification Benefits

Microalgae can help eliminate xenbiotics.

Microalgae may protect endocrine function (e.g., thymus and thyroid).

Microalgae can protect liver cells from damage.

C-phycocyanin can prevent formation of kidney stones and protect kidney function.

Microalgae can help eliminate harmful heavy metals.

Prebiotics support healthy intestinal mucosal barrier.

Our Ability to Detoxify and Eliminate Toxins Shapes our State of Health

Detoxification of harmful substances from the body is a life-long process. Our ability to effectively detoxify and eliminate toxins such as heavy metals, solvents, pesticides and microbial toxins shapes and determines our state of health. Two major organ systems, the liver and kidneys, are primarily involved in body detoxification. The liver is the largest solid organ in the human body. At any one time our liver hold's over one tenth of the body's total blood supplies. The liver is so essential to life that liver cells have the ability to regenerate quickly; if cells are lost in a healthy liver, they can be replaced within a few hours. Not to be overlooked, our kidney's process an even greater blood flow than the brain, liver or heart. Additionally, healthy bacteria and a healthy mucosal lining are essential to reduce the absorption of toxins into the blood through the intestines. Microalgae can support, directly and indirectly, these major detoxification pathways to better eliminate toxins. Microalgae also offer detoxifying compounds, like the amino acid methionine, to help eliminate toxic build-up.

"Methionine was probably one of the first amino acids available in Earth's ancient primordial seas billions of years ago. This amino acid was (and is still) used by primitive bacteria and blue-green algae to biosynthesize glutathione, possibly Earth's first antioxidant (protection) tripeptide molecule. Methionine in this form has been shown to help humans detoxify lead and copper contamination in the blood."[1]

Xenobiotic are Toxic Chemical Compounds

In the late 18th century, a xenobiotic-based detoxification hypothesis was first proposed. Xenobiotic are chemical toxic compounds that are foreign to a living organism. Typically ingested toxic compounds are first transformed into water-soluble substances and then excreted out of the body by urine. For more than 100 years, scientists collected urine from a variety of animals, purifying and then chemically analyzing the compounds to better understand how the body managed to remove various xenobiotics. In the process, scientists discovered how "conjugating substances," can convert a nonwater soluble compound into substances passed-out of the body by urine.

R. T. Williams in his landmark 1947 monograph on "Detoxification Mechanisms," proposed that xenobiotics are biotranformed through two phases: 1) *functionalization* which uses oxygen to form a reactive site and 2) *conjugation*, which results by addition of a water-soluble group to the reactive site. These two steps, functionalization and conjugation, are termed Phase I and Phase II detoxification pathways, respectively.[2] Detoxification is not a single process but a process that involves multiple reactions, REDOX pathways, enzyme, cellular and organ systems.

When our bodies are challenged by high xenobiotic toxic loads, greater and more rapid numbers of enzymes are made to ensure a faster rate of detoxification. Both Phase I and Phase II detoxification pathways must work in concert. When these pathways are out-of-balance, an increase in reactive compounds occurs, which may damage essential DNA, RNA, and proteins. "Induction" is a detoxification process whereby increased enzyme production occurs as a response to higher levels of toxic exposures. Bio active molecules such as flavonoids that are found in fruits, vegetables, and microalgae have been demonstrated to support greater enzyme "induction" via detoxification pathways. In addition, they support redox balance between the different pathways. The primary health benefits of eating fruits, vegetables, and microalgae may be explained, at least in part, by their enhancement of detoxification pathways which lowers our risk for acquiring various cancers, heart disease, or metabolic disorders.

Blue-Green Algae are one of the Richest Food Sources of Detoxifying Polypeptides

Blue-green algae are one of the richest food sources of detoxifying polypeptides, including methionine and glutathione, along with B-vitamin precursors. Once ingested, these molecules are modified as needed. Such polypeptides are essential in the protection of DNA and are essential in the chemistry of detoxification. Also glutathione, along with ascorbate, may help to protect against polyunsaturated fatty acid (PUFA) oxidation.[3]

> *Blue-green algae have multiple liver-protective factors, including amino acids (e.g., methionine, arginine, and isoleucine), chelating trace minerals, and potent antioxidants, such as phycocyanins and superoxide dismutase (SOD).*

Methionine's most useful metabolite, glutathione, is a peptide made of three amino acids—glycine, cysteine, and glutamic acid—all present in microalgae. Moreover, methionine can be restored from homocysteine—its alter-ego, sister form—with the help of B vitamins and especially folic acid in conjunction with B-6 and B-12. An elevated blood homocysteine level is a risk factor for heart attack, the number one killer of adults in America.[4]

Glutathione is the most vital component in the body's detoxification arsenal. In fact, detoxification is the most crucial activity in the body's entire biochemistry and the biggest consumer of energy for making new molecules. Glutathione is the body's most important antioxidant and is especially critical within the organs most responsible for cleaning our body: the kidneys and liver. Biochemically, the glutathione reductase enzyme present in blue-green algae shows amino acid sequence similarities to human reductases.[5] Only two groups of prokaryotes, the purple bacteria and the cyanobacteria (blue-green algae) produce glutathione.[6]

Prescription Drugs' "Unsought" Effects are a Prominent Cause of Death

Large numbers of our elder population depend upon multiple pharmaceutical drugs to maintain health. However, the long-term use of many of these pharmaceuticals typically places an undesirable burden upon our detoxification pathways, rather than supporting them. Drugs and their reactive metabolites, spin-off molecules, typically have some unwanted toxic or "side" effects. Depending upon the kind of drug and its metabolites, toxicity may affect one or more organs that are not targeted for treatment. Thus, a number of medicinal agents, when taken in excess or even at times when taken at the prescribed level, may subsequently injure internal organs. Organ-specific drug toxicity affecting the liver, kidney, heart, lungs, stomach, or intestines occurs too frequently. Anyone watching pharmaceutical 'drug' commercials on television hears a litany of potential serious "side-effects." In fact, unexpected death due to prescription medication usage and misusage are in the top five causes of death in the United States. This alarming statistic does not include fatalities related to illegal drug usage. Yet, the beneficial use of nutraceuticals like nutrition or herbs as protective agents to guard against the unwanted toxicity or damage caused by drugs, has been overlooked by most doctors.

Our Liver Acts as the Major Organ for Detoxification

Our liver acts as the major organ for detoxification. Liver cells are both highly specialized and highly susceptible to damage. After our food is digested and absorbed through the intestine, nutrients are transported to the liver for further processing. All blood returning from the stomach, intestines, spleen, and pancreas is "detoxified" by the liver, reducing and helping to eliminate many toxins. More than 900 drugs have been implicated in liver injury. Drug induced liver injury is responsible for 5% of all hospital admissions and 50% of all acute liver failures.[7] Blue-green algae have multiple liver-protective factors, including amino acids (e.g., methionine, arginine, and isoleucine), chelating trace minerals, and potent antioxidants, such as phycocyanin and superoxide dismutase (SOD).[8-10]

One notorious liver toxin is carbon tetrachloride (CC14). Before 1970s, CC14 was widely used as a cleaning fluid (e.g., spot remover) and a common chemical inside of fire extinguishers. Additionally, until it was banned for food use in 1986, the chemical was commonly applied as a grain fumigant. Even today CC14 is utilized in the manufacturing of various propellants and other industrial chemicals. Therefore contamination levels may still be found in some drinking water. Because CC14 quickly injures the liver, the substance has been widely used as a model of liver toxicity in animal studies.

An Indian study that compared carotenoid-producing micro-alga, *Dunaliella bardawil* and *Spirulina*, demonstrated superior liver protective effects with *Dunaliella* in CC14-liver injured animals. *"Spirulina* contains only beta carotene and *Dunaliella* contains other carotenoids and xanthophyll. The increase in protection with *Dunaliella* indicates that mixed carotenoids exhibit better biological activity than beta carotene alone. The results of this study indicate that carotenoids obtained from an algal source have a higher antihepatotoxic effect, compared with synthetic beta carotene and with beta carotene alone from a natural [micro algal] source."[11]

Another study by the same Indian team, studied rats treated with *Dunaliella*, a microalgae extract, for 14 days to assess its affect against carbon tetrachloride induced liver toxicity. A battery of biochemical marker enzymes for liver and blood demonstrated better serum enzyme values for *D. bardawil* treatment, with lowered lipid peroxidation measures. Furthermore, the Indian scientists "demonstrated protection against DNA strand breaks in hepatocytes [liver cells]" Also, the microscopic analysis of the rats' liver tissue indicated "less severe [damage] with *D. bardawil* treatment, supporting the apparent protective action of 14-day treatment on hepatic oxidative injury."[12]

An extract of microalgae (*Aquilegia vulgaris*) inhibited in vitro microsomal lipid peroxidation, moderately abolished superoxide radicals and eliminated harmful iron ions in experimental rats exposed to carbon tetrachloride-induced oxidative stress. The micro-algal extract effectively reduced microsomal lipid peroxidation in the liver of toxic exposed rats by 40 percent. The researchers concluded that "the protective effect of the *A. vulgaris* extract in CCl4-induced liver injury is mediated by inhibition of microsomal lipid peroxidation and restoring activity of some antioxidant and related enzymes."[13]

When micro algal supplementation was given to rats consuming a high-fructose (60%) diet, a preventive effect on the liver triglyceride level was observed, along with lowered plasma cholesterol. The researchers reported that the microalgae helped reduce liver fats that were elevated by the excessively fructose-rich diet. [14]

Chlorophyll, Essential Fatty Acids and Carotenoids Pigments, Abundant in Microalgae, can Protect Liver Cells

Chlorophyll, which is found in abundance in microalgae, helps to stimulate liver function, increase bile secretion, and protect cellular functions.[15] Also, "chlorophyll appears to promote regeneration of damaged liver cells."[16]

Researchers from the National Institute of Oceanography in Haifa, Israel, reported that increased absorption of essential fatty acids (EFAs) which occurs with consumption of certain microalgae supports healthier liver function.[17] Unfortunately exposure to air oxidizes many oils in food, including EFAs; light and heat both accelerate this process. Eating oxidized oils places a burden on our liver.

Experimental animals fed oxidized (i.e., rancid) oil combined with a *Dunaliella* algal extract were better able to maintain their hepatic stores of beta carotene and vitamin A than control animals given a synthetic all-trans beta carotene. The control group showed significant liver (pro) vitamin losses, likely due to a greater utilization of their antioxidant vitamin reserves. Beta carotene assists in reducing hepatic and erythrocyte peroxidation associated with the consumption of oxidized oils, as well as protects liver cells from damage. [18-19]

In Japan Algae has a Long Tradition as a Master Cleanser

In the Orient algae has a long tradition as a master cleanser. Macrobiotic counselor, Steve Gagné reports, "When ingested, algae begin to go to work like a janitor, cleaning, purifying and strengthening the internal environment." [20] Japan has been subject to several tragic incidents of severe chemical poisoning. The use of algae has stimulated the excretion of some contaminants, notably cadmium, at an accelerated rate in patients. Lead and mercury are also excreted, without the detrimental effects associated with conventional chelation therapy.[21]

The study presents positive pharmacological implications of phycocyanin against oxalate mediated nephronal impairment and warrants further work to tap this potential aquatic resource for its medicinal application.

Our kidneys are major sites of activity in the removal of toxic drugs and pollution from our blood stream and elimination through urine excretion. Because our kidneys act as a major filtration system, and are rich in vasculature, they tend to accumulate higher loads of toxic metabolites than other body organs. Several researchers have tested the effectiveness of microalgae on supporting kidneys exposed to toxic damage. Collectively, a number of studies have confirmed that blue-green algae offers protection against drug- and chemical-induced renal (kidney) toxicity.[22] C-phycocyanin has been shown to protect against oxalate-mediated kidney injury in rats.[23]

C-phycocyanin Might Prevent the Formation of Kidney Stones

Kidney membrane injury is a prime spot for the deposit of crystals. Especially "oxalate-induced" oxidative stress contributes to cell injury and promotes kidney deposits of calcium (oxalate) crystals. These damaging crystals impact the formation of "kidney stones." NADPH oxidase has been identified as "one of the mechanisms responsible for oxalate-induced oxidative injury in renal tubular epithelial cells."[24] Because NADPH oxidase is inhibited by the blue-green pigment in microalgae; researchers tested whether pretreatment with C-phycocyanin, might act as a protective agent against oxalate toxicity in the kidneys.

Experimental rats were injected with sodium oxalate (two experimental groups with or without pretreatment of C-phycocyanin). "Assessing calcium oxalate crystal retention in renal tissue using polarization microscopy and renal ultrastructure by electron microscopy reveals normal features in phycocyanin – pretreated groups." This contrasted with the kidney damage evident in the control group, without C-phycocyanin. Also, C-phycocyanin use significantly enhanced antioxidants levels in both kidney tissues and in red blood cells, along with markers of less free radical damage. All together these results show that microalgae derived C-phycocyanin has powerful antioxidant effects on oxalate-mediated oxidative stress, in the kidneys. The authors concluded, "Thus the study presents positive pharmacological implications of phycocyanin against oxalate mediated nephronal impairment and warrants further work to tap this potential aquatic resource for its medicinal application."[25]

Microalgae Reduced Kidney Toxicity from High Doses of Mercury and Pharmaceutical Drugs

Japanese researchers from Chiba University found that blue-green algae effectively reduces kidney toxicity from high doses of mercury and pharmaceutical drugs administered to laboratory rats.[26] These scientists used two indicators for kidney toxicity: blood urea nitrogen (BUN) and serum creatinine. They checked levels of both indicators before and after administering various toxic doses of mercury, para-aminophenol (a common pain-killer), gentamicin (an antibiotic), and cis-dichloro-diamino-platinum (an anticancer drug). In all cases, rats fed blue-green algae (up to 30% of their diet) exhibited significantly reduced BUN and serum creatinine levels which reverted back to near normal levels after the experimental poisoning. This was not the case with the poisoned, untreated, control animals.

Medical researchers have demonstrated that green microalgae increase the detoxification of harmful chemicals like chlordecone (a highly toxic pesticide), dioxin, and PCBs.[27] In a study of chlordecone-poisoned rats, ingested algae decreased the half-life of the chemical toxins from 40 to19 days.[28]

Microalgae Treatments may Reduce Harmful Side-Effects from Anticancer Chemotherapy Agents

After reviewing numerous studies, medical researchers concluded that blue-green algae "offers a protective antioxidant effect against the nephrotoxicity caused by various toxicants, heavy metals, and drugs including the anticancer chemotherapeutic agents"[29] For example, cisplatin (CP) is an effective chemotherapeutic agent used in the treatment of a wide array of both pediatric and adult cancers. Kidney damage is a major toxic "side-effect" of cisplatin treatment, which often results in patients having to halt or reduce their treatment regimes.

Blue-green alga "offers a promising pharmacological intervention strategy in the humans encountering the clinical drug-induced nephrotoxicity caused by therapeutics during transplantation and cancer treatment."

Growing evidence has revealed the critical role that reactive oxygen species contribute to renal dysfunction and point to the therapeutic potential of blue green algae in cisplatin-induced nephrotoxicity. A single dose of cisplatin in experimental animals produced marked renal oxidative and nitrosative stress and significantly worsened renal (kidney) functions.[30]

In a significant and dose-dependent manner, blue green algae treatments have restored kidney function, reduced lipid peroxidation, and enhanced reduced glutathione levels and superoxide dismutase and catalase activity.[31] Further studies demonstrated that "Pretreatment with *Spirulina* protected the rats from CP-induced nephrotoxicity. The rise in plasma urea, creatinine, urinary beta-NAG, plasma and kidney tissue MDA and histomorphological changes were significantly attenuated by *Spirulina*. In vitro studies using human ovarian cancer cells revealed that *Spirulina* did not interfere with the cytotoxic effects of CP on tumor cells."[32] Thus, a potent chemical cancer therapy (cisplatin) might be administered in combination with microalgae treatments, to help mitigate the harmful side-effects while maintaining the benefit of the treatment.

Blue-Green Algae Lowers Incidence of Acute Kidney Failure Related to Gentamicin, an Antibiotic Drug

Gentamicin (GM) is a drug widely used in clinical practice for the treatment of serious Gram-negative bacterial infections. The clinical utility of GM is limited by the frequent incidence of acute kidney failure. Experimental evidences suggest that oxidative stress is a major factor in GM induced kidney injury (nephrotoxicity). Recent studies have demonstrated that blue-green algae offered significantly and dose dependently, protection against gentamicin-induced oxidative stress, kidney cell alterations, and kidney dysfunction in rats.[33-34] Furthermore, the damaging effect of heavy metals has been implicated in a variety of pathological conditions, including hypertension and kidney, brain and liver disorders. Lipid peroxidation of cell membranes is often evident in these conditions. Blue-green algae has been reported to effectively decrease lipid peroxidation and to restore the levels of antioxidants in several organs such as the kidney, liver, heart, lung, and brain of experimental animals.[35]

No Safe Lead Exposure Levels are Known

Lead poisoning, sometimes referred to as "painter's colic" or "plumbism," is a medical condition caused by elevated levels of heavy metal in the body. Lead interferes with the development of the nervous system, harms a variety of organs (e.g., heart, kidneys, intestines, bones, brain) and is particularly toxic and damaging to children. Numerous studies have demonstrated a strong relationship between childhood learning disabilities and body stores of heavy metals, particularly lead.[36-37] No safe threshold of lead exposure has been discovered. Even the very smallest amounts of lead cause bodily harm. Mice exposed to lead toxicity were randomly assigned to two groups: a) control (lead exposure only) and b) two treatment groups, pre- and post-treated *Spirulina* groups (plus lead exposure). Significant improvements were seen in animals' overall weight, tubular diameter, and testes weight in animals given *Spirulina* prior to lead exposure compared to the other groups.[38]

Blue-Green Algae Protective against Lead

Consumption of *Spirulina* in lead exposed animals exhibited a significant decrease in the deposition of lead in their brain tissues, inhibited lipid peroxidation and restored the levels of endogenous antioxidants to normal (pre-lead exposure) levels.[39] Oxidative damage has been proposed as a possible mechanism involved in lead toxicity, especially affecting the liver and kidney. Another study conducted in Mexico evaluated the antioxidant activity of *Spirulina* against lead acetate-induced hyperlipidemia and oxidative damage in the liver and kidney of male rats. The blue-green algae fed rats showed significant protection of their liver and kidney functions, along with enhanced antioxidant status, compared to lead-only comparison animals.[40]

An Indian research team (2001) found that ". . . *Spirulina* had significant ($p < 0.001$) antioxidant activity thereby protecting the animals from lead induced toxicity," in an experimental study of lead treated [including control group] animals.[41] The same Indian team did a follow up study (2003) and found that "administration of *Spirulina* to lead exposed animals significantly ($p < 0.001$) inhibited lipid peroxidation and restored the levels of endogenous antioxidants to normal. To conclude, *Spirulina* had a significant effect on scavenging free radicals, thereby protecting the organs from damage caused by the exposure to lead. Furthermore, *Spirulina* showed a significant ($p < 0.05$) decrease in the deposition of lead in the brain."[42] The other organs examined (liver, lung, heart and kidney) did not show a change in lead levels, but did have lower levels of lipid oxidation. Blue-green algae treatment was especially effective at crossing the blood-brain barrier and helping to eliminate lead from the brain.

Other studies on blue-green algae have demonstrated an ability to reduce lead toxicity, as well. The beneficial heavy metal removing effects of blue-green algae may be due to the presence of the abundance of antioxidants, including chlorophyll, C-phycocyanin, beta carotene and SOD enzymes.

Microalgae Solutions Appear Promising for Prevention of Chronic Arsenic Poisoning

Tens of millions of people, worldwide, especially in India, Bangladesh, Taiwan and China are routinely exposed to high concentrations of arsenic through contaminated drinking water, inhalation, or skin absorption. There is no specific treatment. After entering the body, arsenic poisoning can affect multiple body systems and it is very difficult to diagnosis in its early stages. Malignancy generally requires at least 10 years of arsenic exposure, resulting in cancers and other health issues. Once high levels of arsenic poisoning are reached in the body it results in rapid mortality. Medical researchers in Bangladesh randomly assigned forty-one patients diagnosed with chronic arsenic poisoning to either a) blue-green algae (*Spirulina*) plus 2 mg of zinc supplement treatment group or b) a placebo control group. The results showed that the *Spirulina* (250 mg) plus zinc (2 mg) twice daily for 16 weeks provided a useful treatment for chronic arsenic poisoning. In the treatment group there was a significant elimination of arsenic through urine and hair scalp measures; with a significant reduce of signs of melanosis (changes in skin pigment) and keratosis (rough skin nodules).[43]

In a poor country like Bangladesh, where expensive chelation therapies might not be either readily available or affordable, microalgae solutions appear especially promising. In a follow-up study, rat liver tissues were first incubated with arsenic; then blue-green algae extracts was employed to remove arsenic from arsenic loaded liver tissues. Depending upon the type of extract (alcohol or hexane), the blue green algae effectively removed between 33.8% and 83.0% of the arsenic from the liver tissues, respectively.[44]

Several Grams of Microalgae Eliminated Excessive Aluminum in Children

Studies have found that heavy metals such as mercury, cadmium, lead, aluminum, and tin affect chemical synaptic transmission in the brain and the peripheral and central nervous system. Blue-green algae may be especially helpful for dialysis patients, who have a greater risk for aluminum accumulation and an increased risk of neurotoxicity.
In a three-month study Bruno J, et al. (2001) demonstrated that daily consumption of several grams of *AFA* blue-green algae eliminated excessive aluminum from the systems of children. Both low amounts (1.5 grams) and high amounts (3.0 grams) of daily AFA over a three month period of time resulted in significant reductions of aluminum as determined by hair analysis. Also, the parents reported significant decreases in negative health symptoms suggestive of improved detoxification pathways.[45]

Figure 23. Pre- and Post *AFA* Hair Analysis for Aluminum

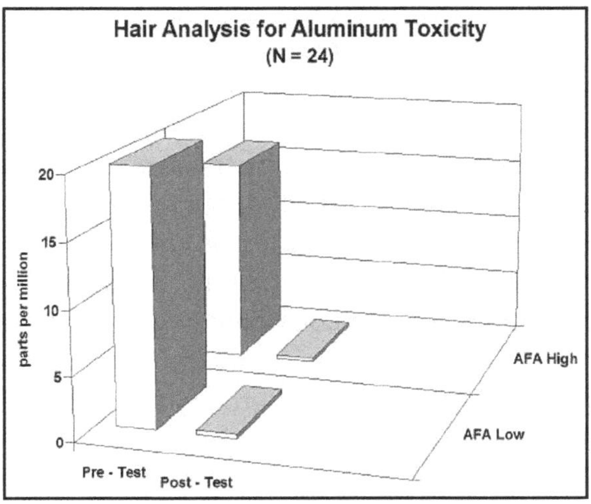

Aluminum exposure in humans is unavoidable. Common exposures to aluminum include aluminum cookware, antiperspirants, antacids, processed cheese and other processed food, lipstick, medications and drugs (antidiarrheal agents, hemorrhoid medications), "softened" water, and tap water.[46]

Increasingly studies suggest that aluminum contributes to neurological disorders such as Alzheimer's disease, Parkinson's disease, dementia, depression, clumsiness of movements, learning challenges and speech impairments.

Our Body's Initial Contact with Most Toxic Molecules is in our Gastrointestinal Tract

Oxidative stress can impair digestive enzymes, which are essential for optimal digestion. Fortunately "free-form" amino acids, within blue-green algae, need no further digestion or conversion. They are already completely broken down to the form needed for efficient absorption. Most literature on detoxification refers to the important of liver enzymes, as the liver is a principle organ of detoxification. However, the first contact our body makes with most toxic molecules is inside our gastrointestinal tract. Over the course of an average adult lifetime, we digest more than 25 tons of food. This represents the largest load of xenobiotic (toxins), including most drugs, and antigens processed by our body. Several important factors of gut integrity influence how much toxic chemicals or antigens will be absorbed by our blood stream, requiring later detoxification by the liver. In most adults, the gastrointestinal tract if it were stretched out flat; the resulting surface area would be about the length of a tennis court.

Therefore, protecting the integrity of this physical "gut" barrier, which is our chief physical, molecular contact, with the outer world is essential to shut out unwanted environmental toxic compounds. Additionally, high concentrations of detoxification enzymes are found at the tip of villi (tiny, mobile, hair-like projections) in the intestines. A healthy balance of probiotics and gut mucosa provides a "first pass metabolism" of xenobiotics. A compromised gut mucosa/probiotic barrier will result in accessibility of xenobiotic, drugs, and antigens to transit into blood circulation, without an initial opportunity for detoxification.

Therefore, supporting healthy gut mucosa is essential to decreasing toxic load. Moreover, gut micro flora can produce compounds that either support or inhibit detoxification activities. For example, pathogenic bacteria can produce toxins that enter our blood stream and increase our toxic load. Whereas, the use of beneficial probiotics may assist in the breakdown of xenobiotics, as well as support protective gut mucosal barriers. The term "probiotics" refers to "live microorganisms," taken as food or food supplements, which have a benefic effect upon our intestinal 'gut' microbiological balance. Andrew Valencia reported that *AFA* significantly reduced "leaky gut syndrome" in patients, based on functional test measures.[47]

Compromised intestinal permeability increases the uptake of toxic compounds and macromolecules that can exhaust the detoxification capability of the liver. A combination of leaky gut and dysfunctional liver detoxification may lead to increased tissue stores of toxic compounds and a depressed immune status.

Treatment with both reduced glutathione and selenium-enriched *Spirulina* algae helped to normalize the intestinal permeability of rats exposed to an experimental anaphylaxis reaction. Reduced glutathione administered on its own did not help normalize the intestinal permeability of the sensitized rats.[48]

Microalgae Supports Healthy Intestinal Function

Colonic Anastomosis is a surgical procedure that allows people to have normal bowel function even after their colon and rectal lining are removed. At a Department of General Surgery in Ankara, Turkey "Seventy-two rats were randomized to three groups. Group 1 was fed with standard diet for 15 days, before and after the colonic anastomosis. Groups 2 and 3 were fed with a malnutrition diet for 15 days prior to colonic anastomosis and then with the basic diet for 15 days thereafter. Group 3 also received an extract of algae derived from *Chlorella* . . . . postoperatively, in addition to the basic diet." Detailed microscopic observations of the colon tissues "revealed significantly better healing patterns for group 3 than for groups 1 and 2 ($P < 0.05$)." [49] Thus, an extract from the green microalgae, *Chlorella* was demonstrated to have "favorable effects" on healing from colon surgery in animals.

Next the Turkish medical team studied the effects of two microalgae extracts, *Chlorella* and *Spirulina*, on intestinal barrier function and oxidative stress in experimentally jaundiced rats. They found that "*Chlorella* sp. microalgae supplemented enteral diet has significant protective effects on intestinal mucosa barrier in obstructive jaundice, and reduces intestinal translocation of bacteria and endotoxin." *Chlorella* protected the jaundiced rats "against oxidative stress, as demonstrated by reduction of intestinal lipid peroxidation, increase of the antioxidant reduced glutathione (GSH), and decrease of the oxidized glutathione (GSSG)."[50] Surprisingly these results were not reported to reach significant levels in the *Spirulina* treated animals.

Chlorophyll, the green pigment in algae, was found to be helpful in controlling body and fecal odors in 62 geriatric patients.[51] Chlorophyll help to restore intestinal balance by reducing the quantity of toxic bacteria and fungi in the walls of the intestines. Chlorophyll is one of the most important chelates found in nature. Chelates are "claw-like" molecules that excel at drawing out other elements. In natural therapies, chlorophyll is considered an exceptional "detoxifying agent."

Chlorophyll content is higher in microalgae than in most plants. For example, wheat grass is often believed to be the premier "green food." However, AFA microalgae has almost six times more chlorophyll content, per gram (*AFA* has 30.0 mg; Wheat Grass has 5.5 mg.). Chlorophyll stimulates the intestinal lining, increase peristaltic action, reduces inflammation, balances excessive pepsin levels and helps to relieve constipation.

CHAPTER

14

A Healthy Heart Matters

"The human heart so far surpasses all known motors in functional capacity that we can hardly hope to improve on it, even with the most ingenious machine produced by man . . . It beats 100,000 times per day, approximately 40 million times in a year . . . It pumps two gallons of blood per minute and 100 gallons per hour, through a vascular system of about 60,000 miles in length — 2 1/2 times the circumference of the earth."[1]

— Bircher-Benner, 1973

Microalgae Hypothesis IX: Cardiovascular Benefits

Heart attacks are the leading cause of death in the United States.

High-energy tissues (e.g., heart) are vulnerable to oxidative stress.

Phycocyanin pigments inhibit clotting of blood platelets.

Microalgae can lower blood triglyceride and lessen LDL oxidation.

Algae-derived omega-3 fatty acids support heart function, reduce blood viscosity, decrease arteriosclerosis and lower high blood pressure.

Microalgae B12 and detoxifying polypeptides improve homocysteine status.

Microalgae support healthy sexual function (e.g., arginine).

Alginate sustains heart repair mechanisms after injury.

Heart Attacks are the Leading Cause of Death in the Western World

Heart failure is one of the most frequent medical issues faced by people over the age of 50 and is the leading cause of death in the Western world.[2] Substantial evidence links oxidative stress with atherosclerosis and thrombosis (blood clot formation). When a blood clot blocks one of the main arteries feeding the heart, a person suffers a heart attack. Also, whenever a region of the heart is starved of oxygen, some of the muscle tissue dies. As a result, in place of healthy heart muscle a section of scar tissue, called an infarct, develops. If there is excessive scar tissue, over-time the heart muscle fails to pump sufficient blood, which may result in congestive heart failure. This can set off a vicious cascade of increased oxidative stress which occurs due to reperfusion injury following hypoxia (lack of oxygen to the tissue). Reperfusion injury owing to oxygen-derived free radicals is a type of tissue damage, caused when blood rapidly floods back to the tissue following an oxygen deficiency. Most patients do not survive more than five years after a heart attack (75% mortality in men).[3]

Many High School Students Have Heart Unhealthy Lifestyles

Albert Sanchez, Ph.D., a public health expert who examined obesity and heart disease risk factors in 249 high school students, reported, "We are appalled at the horrendous diet that 80 to 90% of our children are eating," to physicians at the 49th annual conference of the American College of Cardiology in Anaheim, California, in March, 2000.[4] More than 80% of the students in Sanchez's study consumed diets that exceeded the recommended levels of total fat and saturated fat, with more than a third having elevated levels of LDL ("bad") cholesterol. Furthermore, the presence of an early buildup of fatty deposits in the carotid artery, as revealed by ultrasound images, was not uncommon. Cumulatively, these risk factors are predictive of potential heart disease and other serious health problems for these students only a few decades down the road.

Medical concerns about rising cholesterol levels, along with effective pharmaceutical marketing efforts, have resulted in a dramatic rise in prescription rates for cholesterol-lowering statins, which can have some dangerous side-effects. Statin drugs are one of the most commonly prescribed medications for adults in the United States. However, according to British cardiologist, Dr. Malhortra, moderating cholesterol through a change in diet is a healthier way to go. Specifically he reports that, "adopting a Mediterranean diet after a heart attack is almost three times as powerful in reducing mortality as taking a statin."[5]

Many Heart-Related Conditions have benefited from Microalgae Interventions

Micronutrient deficiencies are associated with the heart failure syndrome itself. Fortunately, microalgae nutritional factors may improve cardiovascular benefits including a) easily absorbable dietary proteins, including all of the essential amino acids; b) unique fatty acid profiles, high in Omega 3s, DHA and GLA; c) rich sources of B-vitamins; d) powerful antioxidant pigments; and e) one of the widest ranges of trace minerals. The beneficial effect of adding microalgae to one's diet for reducing cardiovascular disease and arteriosclerosis may depend more on the synergistic effects of all the different "whole food" nutrients than any one nutrient factor. For example, edible microalgae offer far more than just omega 3 fatty acids or simply antioxidants, being a food source with numerous beneficial micronutrients.

Microalgae's potent range of antioxidants, in addition to its healthy balance of EFAs, offers top-quality cardiovascular support. "A high consumption of fruit and vegetables, which are good sources of antioxidants, is associated with a lower coronary risk."[6] Yet, while microalgae are excellent, documented, sources of antioxidants, certain microalgae (e.g., blue-green alga) offer the only natural whole food source of gamma linolenic acid (GLA) in the plant kingdom. "GLA is a precursor for the body's prostaglandins (PG). The prostaglandin PGE1 is essential for regulating a variety of basic biochemical functions in the body including the regulation of blood pressure, cholesterol synthesis, inflammation, and cell proliferation."[7]

There are numerous cardiovascular benefits of adding microalgae to one's daily diet. The impacts of microalgae interventions on a wide array of heart-related conditions have been studied including: high cholesterol, blood sugar imbalances, inflammation, high blood pressure, blood lipid imbalances, oxidative stress, and recovery from experimental heart attacks and strokes. The principle cardiovascular benefits of microalgae consumption primarily stem from its antioxidant, anti-inflammatory and hypolipidemic actions.[8]

High Energy Tissues, like Heart Muscle, are Especially Vulnerable to Oxidative Stress

NADPH oxidase activity is significantly elevated in atherosclerotic lesions, leading to increased superoxide production. Superoxide is the chief source of reactive oxygen species (ROS), free radicals, which contribute to oxidative stress, within our bodies. Oxidative stress is a pivotal mechanism involved in cell injury and is associated with a number of disorders, including atherosclerosis, ischemia/reperfusion, arthritis, stroke, and neurodegenerative diseases. In addition, reactive oxygen species have detrimental, oxidizing, effects on proteins, lipids, and DNA and promote severe tissue damage and cell death especially in vulnerable, high energy, tissue like heart muscle.[9] Several studies have confirmed that Blue-green algae components inhibit the production of superoxide anions and NADPD oxidase levels.

Hamsters are commonly used in animal research. The hamster model of atrial thrombosis is the sole animal model of spontaneous thrombosis. The frequency of atrial thrombosis increases with age in hamsters, and both sexes are equally affected. The effects of blue-green algae, *Spirulina platensis*, were studied on plasma cholesterol, early atherosclerosis, cardiac production of superoxide anions and NADPD oxidase expression in hamsters. A reduction of all these heart risk factors was reported. "A sparing effect in liver glutathione peroxidase (87%) on average and superoxide dismutase (56% on average) activity was observed. . . Aortic fatty streak area was significantly reduced in the experimental groups. . . Cardiac production of superoxide anion significantly decreased." The researchers concluded that especially selenium-enriched "*spirulina* phycocyanin powerfully prevents the development of atherosclerosis. The underlying mechanism is related mainly to inhibiting pro-oxidant factors and to a lesser extent improving the serum lipid profile."[10]

Deng and Chow (2010) reported on the hypolipidemic, antioxidant, and anti-inflammatory activities of *Spirulina*, that "great attention and extensive studies have been devoted to evaluate blue-green algae's therapeutic benefits on an array of diseased conditions including hypercholesterolemia, hyperglycerolemia, cardiovascular diseases, inflammatory diseases, cancer, and viral infections." The researchers concluded the cardiovascular benefits of this blue-green alga were mainly due to its hypolipidemic, antioxidant, and anti-inflammatory effects.[11]

Phycocyanin, the Blue-Green Pigment, can Inhibit Blood Platelet Clotting

Researchers at the University of California - San Francisco found that children with an inherited tendency for high cholesterol levels benefited from antioxidant vitamins. Antioxidants seem to improve the condition of their blood vessels by helping to neutralize the free radical molecules or "reactive oxygen species" (ROS) that may prevent endothelial cells lining the blood vessels from releasing nitric oxide. Nitric oxide is responsible for the dilation of blood vessels.[12]

Two separate independent animal studies found that a blue-green alga, "*Spirulina* maxima, may decrease vascular tone by increasing the synthesis and release of both a vasodilating cyclooxygenase-dependent product of arachidonic acid and nitric oxide, as well as by decreasing the synthesis and release of a vasoconstricting eicosanoid from the endothelial cells."[13-14] When blood vessels become too narrow, an excessive "vascular tone," it increases the risks for high blood pressure and clotting.

Researchers in Kaohsiung Medical University in Taiwan studied the mechanisms involved in the antiplatelet effect of phycocyanin and reported that it inhibits platelet aggregation by means of several different mechanisms, including suppression of the activity of thromboxane synthase, assisting mobilization of intracellular calcium, decreasing the expression of glycoprotein on the surface of the platelets, and increasing membrane fluidity and cyclic AMP levels through inhibition of phosphodiesterase activity.[15] Furthermore, algal polysaccharides "are known to be potent anticoagulant agents of mammalian blood and represent a potential source of compounds for antithrombotic therapies."[16] Thrombosis is the formation of blood clots that can lead to heart attacks or brain strokes, or swollen legs depending upon where the clots occur in the body.

Further, microalgae are the foundational source of Essential Fatty Acids (EFAs). "EFAs have lubricating qualities and increase cell membrane flexibility. They are known to reduce blood cholesterol and thus help to prevent cardiovascular disease...EFAs are especially useful because of the efficiency with which they increase the solubility of cholesterol deposits and wash these deposits away from our artery walls...As the consumption of fish oils or essential fatty acids found in *AFA* increases, the tendency for blood platelets to aggregate decreases and blood pressure goes down."[17]

Microalgae may help Lower Blood Triglyceride Levels and Normalize Elevated LDL Oxidation

Hyperlipidemia includes three main types: a) elevation of lipoproteins, b) high triglycerides and c) elevated cholesterol. Hyperlipidemia is caused by an elevated level of any kind of lipid (fat) in the blood stream, while hypercholesterolemia is only caused by high cholesterol in the blood stream. All forms of hyperlipidemia, which can include high cholesterol, may cause cardiovascular disease and are associated with other health problems, including diabetes, liver and kidney disease or thyroid disease.

Dietary supplementation with algal beta carotene may normalize the elevated LDL oxidation in patients with diabetes, and thus delay the onset and further development of atherosclerosis in these patients. Twenty patients with long-standing noninsulin-dependent diabetes mellitus were studied in comparison with age- and gender-matched control subjects. Diabetic patients showed overall greater LDL oxidative effects compared to the controls. An algae-derived beta carotene supplement (60 mg daily dose) was given for 3 weeks. Upon supplementation, a marked reduction in oxidative effects was seen in the patients. Supplementation with algae naturally rich in beta carotene appears to normalize the diabetic-enhanced LDL oxidation levels and consequently may be of importance in delaying the accelerated development of atherosclerosis in patients.[18]

Fibrates have been found to be effective in lowering blood triglyceride levels (high cholesterol) and are often combined with statins to prevent heart disease. Fibrates reduce the number of nonfatal heart attacks, but not fatal heart attacks, and therefore are used by themselves only in those not tolerant to statins. Israeli researchers at the Institute of Lipid and Atherosclerosis Research conducted a double-blind clinical study measuring patients' high density lipoprotein (HDL)-cholesterol levels. They discovered that a combination treatment of fibrate, plus beta carotene-rich *Dunaliella* microalgae powder, was superior to the use of fibrate drug alone, to improve good HDL-cholesterol levels.[19]

Japanese white rabbits fed a ten-week series of a high-cholesterol diet and powdered *Chlorella,* showed a significant suppression of total and beta-lipoprotein cholesterol levels, along with reduced aortic atheromatous lesions. However, rabbits in the control group, with no algae in their diet, showed a dramatic increase in serum total cholesterol and beta-lipoprotein cholesterol levels, with resulting symptoms of atherosclerosis.[20]

In this study, *Chlorella,* both in powder form and in glycolipid (GL) and phospholipid (PL) extract form, suppressed the increase of serum cholesterol levels. There was no significant difference between using the algal powder form or the extracts in terms of effect on the serum lipid. But the algal extracts were more effective in increasing the fecal excretion of steroids (mostly cholesterol, deoxycholic, and lithocholic acid). It was concluded that algal extracts inhibited the absorption of exogenous steroids and promoted the turnover of bile acids in the liver, which suppressed the increase of serum cholesterol levels caused by the high-cholesterol diet.[21]

Scenedesmus (a green microalga) powder fed to rats on a cholesterol-enriched diet prevented the excessive deposition of cholesterol in their livers. The algae-supplemented rat livers weighed about one third less than the control rats. Also the levels of plasma cholesterol and triglycerides were significantly improved in the algae-supplemented rats.[22]

The docosahexaenoic acid (DHA) and alpha-linolenic acid (LNA) in certain microalgae support heart function and may help prevent heart attacks.[23] Also, the carotenoids that are so plentiful in microalgae may reduce the risk of heart disease.[24]

Dietary hyperlipidemia caused by a high-fructose diet in rats was improved by blue-green algae supplementation. Male rats fed on a high-fructose diet (68%) showed heightened hyperlipidemia, while experimental, algae-supplemented rats had increased lipoprotein lipase enzyme activity, along with suppressed levels of hyperlipidemia.

[25] Furthermore, according to Japanese research, a water extract of blue-green algae reportedly "may inhibit the intestinal absorption of dietary fat by inhibiting lipase activity"[26] and contribute to the treatment effects.

Vitamin B12, Hemoglobin and Chlorophyll all Share a Remarkable Molecular Similarity

Vitamin B12 and its derivatives are the most complex tetrapyrroles synthesized in nature, and are involved in various cellular functions such as DNA repair and methionine synthesis.[27] B12, hemoglobin and chlorophyll all share common biosynthetic pathways in certain prokaryotes, along with a structural (molecular) similarity. Like hemoglobin, vitamin B12 is bright red. B12 is essential for digestion (metabolism); DNA production and repair; neural (brain) communication; and even assists in the manufacture of hemoglobin.

All three molecules exhibit a twelve-around-one geometric arrangement, called a porphyrin ring. The main differentiation between the three molecules is in the central metal element: chlorophyll has a magnesium center, hemoglobin an iron center, and vitamin B-12 with a cobalt center. The value of metal-bound porphyrin rings for the transfer of electrons was established early in evolution, as evidenced by the multitude of cytochromes that pass electrons along various respiratory sequences. Thus, essential porphyrin rings are common to algae, plants, and animals. They assist plants in oxygen production and help mitochondria with cellular respiration.

B vitamins are vital catalysts, which accelerate biochemical reactions. Within the family of B-vitamins, B12 (cobalamin) was identified nearly 85 years previous as the antipernicious anemia factor in liver. Vitamin B12 is vital for cell division, the synthesis of red blood cells and nervous system function. Correspondingly B12 plays a central role in supporting the peak energy organs of the body, namely the brain and heart. Chronic B12 deficiency, if not corrected, can lead to critical illnesses, such as cardiovascular disorders, depression, Alzheimer's disease, anemia, and even certain types of cancers. Research supported by the National Institute on Aging (2011) documented that elderly people with lower levels of B12 exhibited brain shrinkage, smaller total brain volumes, and associated cognitive problems.[28]

Homocysteine is a Biomarker for Risk of Heart Attack, Stroke and Neurological Problems

Over the past four decades, an association between elevated plasma homocysteine levels and oxidant stress-induced cardiovascular pathology has been well demonstrated.[29] Elevated blood levels of homocysteine are a significant predictor for risk of heart attack, stroke, renal failure, cancer and neurological problems. High homocysteine levels in the blood can damage the lining of the arteries and increase the risk of blood clots, which can eventually lead to blood vessel blockages and coronary artery disease. Research has shown that adverse changes in heart structure, such as abnormal thickening of the heart valves, is related to high levels of homocysteine even in the absence of hypertension.

Certainly, there is strong evidence of a correlation between elevated plasma homocysteine levels with heart failure and cardiovascular disease. Furthermore, homocysteine can be changed into the amino acid, methionine, a beneficial 'alter-ego' form with the aid of folic acid, B-6 and B- 12. Yet, clinical trials of B-vitamin supplementation have not demonstrated any consistent benefits in preventing or treating cardiovascular disease. These findings suggest that B-vitamins may only be one portion of a greater issue. Recently, several scientists have proposed that "an abnormal homocysteine levels seen in pathologic states may reflect a graver disturbance of methoxistasis," or an abnormal redox-methylation balance.[30]

The "methoxistasis" theory describes how the methionine-homocysteine cycle and folate metabolism are part of a larger feedback-loop system that controls, and is influenced by redox balance, sulfur amino-acid metabolism, and methylation reactions. Even the oxidation of blood heme can increase enzyme (CBS) activity, which acts as "a major node in the methionine-homcysteine cycle that regulates the partitioning of homocysteine generated from methylation reactions into the transsulfuration pathway or remethylation back to methionine. . . Generally, the transsulfuration pathway appears to be activated by an increase in oxidant molecules, with the reverse occurring in the case of enzymes involved in methylation of homocysteine and other methylation reactions."[31] Not surprisingly then total homocysteine levels are influenced by a number of environmental factors, as well as can be a useful predictor for risk of cardiovascular disease (CVD), renal failure and even cancer. The study of epigenetics, how individual conditions and environmental influences can shape one another is certainly related to homocysteine levels.

Figure 24. Total Homocysteine, Status, and Illnesses

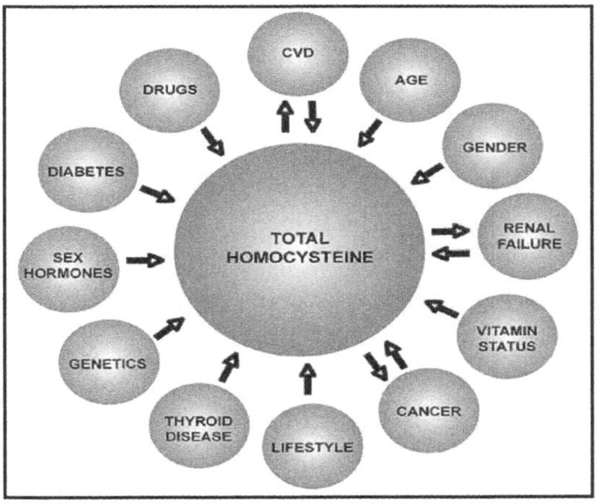

A Debate about Microalgae as a Useful Form of B12

The great debate about whether microalgae offer a useful form of B12, especially for vegetarians, is further complicated by the challenge of assessing B12 deficiencies. "Part of the difficulty derives from the lack of a diagnostic gold standard among the many available [B12] tests, part from the overwhelming numerical preponderance of patients with subclinical deficiency (in which isolated biochemical findings exist without clinical signs or symptoms) among the cobalamin deficiency states, and part from the decreased availability of reliable tests to identify the causes of a patient's cobalamin deficiency and thus a growing de-emphasis of that important part of the diagnostic process."[32]

Italian researchers from a Venice hospital's department of Neurorehabilitation decided to examine whether B12 from *AFA* blue green algae was bioactive. They compared the vitamin B12 status of 15 vegan study participants after 3 months on a "Klamath algae product" (*AFA* along with a proprietary combination of enzymes) to the B12 status of the same individuals at the end of a 3-month control period, during which they had not taken any B12 supplements. (All participants had stopped taking their usual B12 supplement at the beginning of the study.) At the end of the 3-month supplementation period, the researchers found that participants B12 status was improved, with a significant reduction in the blood concentration of homocysteine ($p = 0.003$). (A high homocysteine level is an indicator of B12 deficiency, and reduction in homocysteine levels indicates effective absorption.) The authors concluded, "The Klamath algae supplement appears to be, in a preliminary study, an adequate and reliable source of vitamin B12 in humans."[33]

Microalgae are an Important Source of B-vitamins and Methionine

Microalgae are an important source of B-vitamins and methionine. Prokaryotes, cyanobacteria and bacteria, are the ultimate source of B12. Meat products are often considered critical B12 suppliers. Yet, where do herbivorous such as cows, obtain this vitamin? Plants do not offer it because they have no cobalamin-dependent enzymes. Vitamin B12 is actually manufactured by the bacteria located in the cattle's intestinal tract. Seafood is an excellent source because fish consume algae, which is rich in vitamin B12. Some algae, such as Nori, contain as much cobalamin as liver. "A survey of 326 algal species revealed that 171 species require exogenous vitamin B12 for growth, implying that more than half of the algal kingdoms are cobalamin auxotrophs."[34] Nearly half of the algae species sampled require an environmental source of B12. Their source of cobalamin (B12) appears to be bacteria cobalamin an essential cofactor for methionine synthase in algal metabolism. Importantly, B12 plays a crucial role in human health assisting the conversion of harmful homocysteine back into beneficial methionine.

British scientists discovered that most algae do not possess the necessary genes to facilitate vitamin B12 production. All B12 is made by bacteria and "marine cyanobacteria, where all sequenced genomes appear to contain the B12 biosynthetic pathway."[35] All land plants, fungi, and many algae have an alternative metabolic process which does not require the vitamin. However, within the algal kingdom, about half of all micro-algal species need vitamin B12 as a growth supplement and must therefore have a source. "The scientists realized that the amount of vitamin B12 required to grow the types of algae that do need the vitamin in the laboratory is much higher than natural levels in the seas and rivers. They discovered that in the natural environment, symbiotic bacteria supplied the necessary vitamin B12 the algae needed. However, the relationship between the bacteria and algae was not one-way. The scientists found that the algae supported the bacteria by providing them with carbon from their own photosynthesis."[36] A symbiotic, mutually beneficial, relationship exists between numerous algae and B12 producing bacteria and cyanobacteria. Coastal nutrient-rich waters appear to have highest concentrations of B vitamin producing cyanobacteria and bacteria, while open, deep, ocean habitats contain the least.

In 2012, a team of researchers from Woods Hole Oceanographic Institution and the J. Craig Venter Institute discovered "a previously unknown protein in algae that grabs an essential but scarce nutrient out of seawater, vitamin B12 . . . a protein they described as 'the B12 claw.' Stationed at the algae's cell walls, the protein appears to operate by binding B12 from seawater and draws it into the cell. When B12 supplies are scarce, algae compensate by producing more of the protein, officially known as cobalamin acquisition protein 1, or CBA1."[37] B12 producing bacteria can even attach to marine algae (e.g., diatoms), using a sort of "glue," (extracellular polymeric substances), to create an intimate, symbiotic, association.[38]

Various species of algae contain large amounts of cobalt, essential for B12, and often substantial amounts of B12, itself. B12 concentrations vary widely, however, ranging from 0.004 micro-grams per gram for *kelp* to 0.6 micrograms per gram for *calothrex parientina*. Two ounces of the latter would provide an adult with an annual supply of B12. By comparison, it would require about a third of an ounce of *kelp* per day to obtain all one's B12 needs. A critical issue facing bioavailability of B12 from algae sources is the effect of processing, especially use of heat, on B12. For example, raw nori has been demonstrated to contain measurable amounts of genuine cobalamin (B12).[39] Yet heat dried nori does not offer usable forms of B12, but holds unhelpful, altered-forms (pseudo-B12). Corresponding research proved that actual B12 in raw nori is altered into harmful B12 analogues by the drying process.[40] Therefore, the bioavailability of B12 may be adversely affected by heat processing of microalgae. Differences in heat processing might contribute to the discrepant findings of why some blue-green algae appear to only offer pseudo-B12 analogues, whereas others appear to offer clinically useful forms of B12.[41-43]

Omega-3 Health Benefits

AFA alga has high concentrations of polyunsaturated fatty acids (PUFAs) which account for almost 10% of its dry weight. Even more significant, it contains a high percentage of the omega-3 fatty acids, a much higher percentage than is found in most plants, seeds, nuts, and other microalgae. "These long chain omega-3 fatty acids provide significant health benefits to the human population, particularly in reducing cardiac diseases such as arrhythmia, stroke and high blood pressure. Additionally, they have been seen to offer beneficial effects to depression, rheumatoid arthritis and asthma."[44]

The reason *AFA* has more concentrated amounts of valuable PUFAs than tropical algae like *Spirulina* is that it grows in a much colder environment. *AFA* compensates for the cold by manufacturing more of the flexible omega-3 and cis-unsaturated fatty acids. (Cis-unsaturated forms of fatty acids are healthier than the trans-fatty acid forms, because their curved shape further contributes to cell membrane flexibility.)

While both omega-3 and omega-6 forms of EFAs are important, it is omega-3s which are critically lacking in the standard Western diet. *AFA* algae contain an ideal ratio of essential omega-3 to omega-6 fatty acids and are especially high in the essential omega-3 fatty acid, alpha linolenic acid (3-5% by weight). Omega-3 and omega-6 fats are absolutely vital to our health, especially when optimized to the correct balance. These two families of essential nutrients have been shown to strongly influence genetic "switches", and hormone-like agents called prostaglandins. In turn, the gene switches and prostaglandins activated by omega-3 and omega-6 fatty acids control inflammation and key aspects of our cardiovascular, neurological, immune and metabolic health.

The benefits of omega-3 fatty acids to cardiovascular health have been reported for some time. People indigenous to Greenland, whose traditional diet was high in cold-water fish that eat blue-green algae, exhibit little cardiovascular disease.[45] A recent study by the Harvard School of Public Health (HSPH) and the University of Washington suggests that elderly people could reduce their overall risk of mortality by 27% and their risk of mortality due to heart disease by 35%, by increasing their blood levels of omega-3 fatty acids, the fatty acids found in microalgae and concentrated in fish oils. The study examined 2,692 adults with an average age of 74 years and no prevalent heart disease, cerebrovascular accident or underlying heart failure. Those with the highest bloods levels of omega-3s survived an average of 2.2 years longer than their counterparts with lower levels and faced a 27 % lower risk of mortality.

The research team identified three fatty acids that appeared to be responsible for observed reductions in mortality. Each was linked to a different ailment: coronary heart disease, nonfatal heart attack, or stroke.[46] These and other research findings confirm that adequate levels of omega-3 fatty acids are essential to cardiovascular health. The long-chain omega-3s (EPA and DHA) found only in certain microalgae and concentrated within seafood (especially fatty fish), support optimal human health in vital organs, body systems, and are especially necessary during critical developmental life stages.

EPA and DHA May Benefit the Following Conditions:

- Aging, dementia prevention
- Vision
- Metabolism, diabetes prevention
- Inflammation
- Cardiovascular, heart disease prevention
- Fetal development, associated with improved IQ
- Skin, hair, and nails
- Joints and movement
- Cognition, attention, mood, and memory
- Possibly reduce risk for certain cancers

Algae-derived omega-3 fatty acids may support heart function, reduce blood viscosity, decrease arteriosclerosis (a disease of hardened arterial walls) and lower high blood pressure, according to research of Zvi Cohen at the Laboratory for Micro-algal Biotechnology in Israel and Helen Norman at the United States Department of Agriculture.[47] The flexibility of any cell membrane is directly proportional to the amount and type of polyunsaturated fatty acids (PUFAs) it contains. Research reveals that algae supplementation can significantly reduce high levels of arachidonic acid (AA) in the blood and liver lipids and cause a significant increase in the percentages of the omega-3 polyunsaturated fatty acids (PUFAs).[48]

> *Dr. Krylov (University of Illinois) and colleagues concluded, after examining case histories of patients with hypertension who had consumed AFA blue-green algae, this particular algae appears promising for the treatment of hypertension.*

Dr. Rafail Kushak and colleagues demonstrated that *AFA* essential fatty acids are more easily assimilated than those of soybean oil and offer superior cardiovascular benefits. While both soybean oil and blue-green algae contain the essential omega-3 fatty acid, alpha-linolenic acid (LNA), the scientists found that rats required triple the amount of soybean oil in their diets to achieve the same level of circulating LNA as rats fed algae. Also, *AFA* significantly increased both eicosapentaenoic acid (EPA) and docosahexaenoic acid (DHA) blood plasma levels far more effectively than did soybean oil.[49] Both EPA and DHA are essential for optimal cardiovascular and brain function and can be synthesized in the body from LNA.

Dr. Krylov (University of Illinois) and colleagues concluded, after examining case histories of patients with hypertension who had consumed *AFA* blue-green algae, that this particular algae appears promising for the treatment of hypertension.[50]

Other studies suggest that in addition to blue-green algae providing an excellent PUFA source, algae oils might help the body better utilize other vegetarian sources of PUFAs. For example, while flaxseed oil offers high levels of omega-3s, many people have difficulty assimilating them from this source. Yet when microalgae is combined or taken with vegetable oils, like flaxseed oil, blood levels of omega-3s are significantly increased. Thus, microalgae's ability to raise PUFA levels might be a two-fold action: acting both as a superior source of PUFAs and helping to aid the assimilation of other dietary sources of PUFAs.

Microalgae-derived DHA supplements markedly enhanced the DHA status levels in serum and platelets of healthy vegetarian subjects. Researchers also found a substantial increase of EPA and a lowering of total and LDL-cholesterol: HDL-cholesterol ratios, suggestive of a decreased risk factor for heart disease in the DHA algal supplemented group.[51] Microalgae supplementation may be especially important for vegetarians who have a limited intake of fish and eggs.

Vegetarians, especially vegans who eat no animal products (i.e., no fish, eggs, or dairy), tend to have lower serum and platelet phospholipid levels of DHA and EPA than omnivores.[52] DHA is found in high levels in the brain and retina, where it functions in mental performance and visual acuity, respectively.[53] EPA and DHA may help to reduce the risk of developing cardiovascular diseases. Although vegetarians tend to have a lower risk for cardiovascular disease, due partly to their lower serum cholesterol levels, their thrombogenic (stroke) risk factors may still be significant.[54]

The ability of *AFA* algae to lower blood cholesterol in animals was also reported by Kushak's Massachusetts General Hospital research group. Rats fed diets supplemented "with 10% and 15% algae decreased their blood cholesterol level to 54% and 25% of ... [the levels of the] rats fed the standard diet." The researchers concluded that the cholesterol-reducing effect of *AFA* is probably influenced by something other than just its fatty acid content, because both experimental and control diets were rich in PUFAs. They propose that micro algal pigments, such as chlorophyll, might contribute to this cholesterol-reducing effect.[55] Prior studies have demonstrated reduced blood cholesterol levels with the dietary intake of microalgae.[56-58]

Astaxanthin may help the Prevention and Treatment of Thrombotic Cardiovascular Complications

Hawaiian and Japanese scientists have developed a way to grow and extract heart-healthy astaxanthin, derived from *Heamatacoccus,* a green microalga. Astaxanthin is a microalgae carotenoid pigment that provides the red color to many organisms including salmon, flamingos, lobsters and shrimp; as well as offering strong antioxidant benefits. Visually, astaxanthin acts like an evolutionary "make-up" drawing potential mates to individuals who display the brightest and healthiest range of colors. Astaxanthin is reported to have an antioxidant action up to 500 times that of vitamin E.

A randomized, double-blind trial in association with Michigan State University, evaluated whether astaxanthin, a natural antioxidant from microalgae, decreased DNA oxidative damage and reduced blood serum levels of C-reactive protein (CRP). CRP is an indicator of low-grade arterial inflammation and one of the single strongest predictors of future heart problem risk in apparently healthy men and women.[59] Overall, this study shows that dietary astaxanthin enhanced immune response, and decreased a DNA oxidative damage biomarker and inflammation in young healthy females. Medium but not high levels of astaxanthin significantly lowered CRP in young adult female subjects.[60]

Natural algae-rich beta-carotene supplementation appears to normalize the diabetic- enhanced LDL oxidation levels and consequently may be of importance in delaying the accelerated development of atherosclerosis in these patients.

In 2010 a published study on the effect of astaxanthin on thrombus formation used both a mouse model of arterial thrombosis, along with human endothelial cells to investigate potential mechanisms of action. When compared to control mice, the astaxanthin fed group exhibited significant increases in basal arterial blood flow and significant delays in occlusive thrombus formation following the onset of vascular endothelial injury. Human endothelial (umbilical vein) cells and platelets isolated from rats treated with astaxanthin demonstrated significantly increased levels of released nitric oxide (NO) and significantly decreased peroxynitrite (ONOO-) levels. These studies were reported to support the potential of astaxanthin in the treatment or prevention of thrombotic cardiovascular complications.[61]

A research paper published in 2011 on the effects of astaxanthin at different doses found that lower doses only decreased elevated blood pressure, but higher dose levels decreased elevated blood pressure as well as helped normalize glucose-insulin levels produced by heavy sucrose ingestion in rats. Increasing the amount of astaxanthin resulted in both a lowering of elevated systolic blood pressure and enhanced insulin sensitivity. Also astaxanthin at medium and higher doses lessened restraint stress in captive animals. The rodents were confined in "tightly fitting, plastic tubes normally used to measure tail blood pressure. The values for the Medium and High astaxanthin (Asta) groups were significantly lower at the 15 and 30 minutes marks," based on blood pressure measures compared to the control groups. Previous research demonstrated that astaxanthin ameliorated the harmful effects of heat stress[62] and helped overcome stress-induced gastric ulcers in rats fed astaxanthin compared to controls.[63] Two United States patents discussed the ability of astaxanthin to be an "antistress agent."[64-65]

AstaReal Technologies, a subsidiary of Fuji Chemical Industry Company of Japan is building a thirty million dollar astaxanthin facility in Moses Lake, Washington. Mitsunori Nishida, President and CEO of Fuji Chemical Industry said, "We see the 21st century as the era of antiaging," he said. "We believe that preventive medicine will play an even greater role in the future of healthcare as advances in antiaging, lifestyle and disease research drive improvements both in the treatment and prevention of disease and illness. To that end, we have succeeded in realizing the world's first industrial production of natural astaxanthin, which is both safe and high in antioxidant potency. . . . AstaReal Technologies will produce natural astaxanthin using state of the art photo-bioreactors to cultivate microalgae."[66]

Alginate Supports the Heart's Regeneration Process

In the November 2004 issue of Scientific America, Israeli doctors reported on discovering how an algae derivative, alginate, could be used to regenerate damaged heart tissue. While researching the use of fetal heart muscle cells to repair damaged heart tissue, the Israeli doctors made a fortuitous and unplanned discovery. Initially the researchers had little success seeding fetal heart muscle cells over infarcts in animals with experimentally induced heart attacks. Finally they tried an alginate bio-scaffold to lay the fetal 'stem' cells upon, next to the damage heart tissue. For the first time they saw significant improvements in the growth of various heart repair tissue indexes. "When we looked closer at what happened to the cells, however, we found something strange," continued Cohen. "Only some of the cells that we implanted differentiated into adult heart cells, and even these did not populate the whole implant, so we could not explain the results. Later on, when we examined the control animals, in which we had only implanted scaffolding - the results were similar. These results recurred in over 200 animals. We realized that we had a winning biological substance, and that perhaps it was doing everything, and there was no need to implant cells."[67]

Alginate is a complex sugar molecule, or polysaccharide, found in a wide variety of brown seaweeds. Medically alginate is used for wound dressings and one of the most frequently used dental materials, to capture teeth impressions. Alginate is biochemically similar to heparin sulfate, a linear polysaccharide found in all animal tissues. Heparin sulfate supports cellular growth factors and has been considered an agent for wound healing. Alginate has an advantage over heparin sulfate in that it is less rapidly broken down by enzymes found within injured tissue.

Dr. Cohen and Dr. Leor conducted research on 10 pigs suffering from infarcts. Pig hearts are more similar to humans and have been used for human transplant operations. Five of the pigs received implants of alginate scaffold on their infarcts and five were left untreated. All the untreated pigs eventually suffered severe heart failure, and two of them died, whereas all five pigs that received alginate implants survived and none displayed the tell-tale signs of heart failure.

Israeli doctors were able to show that alginate increased the secretion of this powerful stem-cell attractant protein for several months after the initial heart damage.

Further research demonstrated that alginate increased a secretion of a protein (SDF-1), which causes the movement of stem cells to the damaged area of the heart. Typically this protein is secreted by heart cells after a myocardial infarction, remaining at elevated levels for only a few weeks. However, the Israeli doctors were able to show that alginate increased the secretion of this powerful stem-cell attractant protein for several months after the initial heart damage.

According to Cohen, "The alginate may be acting as a stimulant that supports the heart's regeneration process. The heart tests on the pigs also showed an increase in the local contraction of the infarct and the general contraction of the heart, which indicates an increase in the total mass that is contracting. The positive results led us to develop a bio-substance that can be administered by injection [via a catheter]. In the future, perhaps there will be no need for open heart surgery, and it will be possible to inject the substance directly into the heart."[69] Leor J. et al and Cohen (2009) successfully verified that intracoronary injections of an alginate hydrogel could repair myocardial infarctions in swine.[70] Swine hearts are very similar to human hearts and human trials are likely. Now that is algae to the rescue!

CHAPTER

15

Antiviral, Antibacterial, and Antifungal Effects

"It says something profound about the human condition that history takes much greater account of wars and military mayhem than of the enormous losses caused by communicable diseases."

— Doherty P, 2012

"We have given too much attention to the enemy and have to some extent overlooked our defenses."

— Behar M, 1974

Microalgae Hypothesis X: Antiviral, Antibacterial, Antifungal Effects

Microalgae offer antimicrobial agents against bacteria, fungi and viruses.

Microalgae extracts may impede the AIDS virus.

CV-N is the first molecule known to reduce Ebola infection.

Allophycocyanin is a potential antienterovirus agent.

Microalgae can inhibit the growth of Salmonella bacteria.

Chlorophyll lessens infections and accelerates wound healing.

Microalgae support growth of friendly intestinal flora, which produce natural antibiotic effects.

Antibacterial, Antiviral, and Antifungal Properties in Microalgae Species

From an evolutionary perspective, bacteria and microalgae have been interacting over billions of years. According to various scientists, interactions between algae and bacteria are not random, but highly specific. "There are numerous data which report that the extracellular products from algae are capable to stimulate the growth of bacterial strains through the excretion of carbohydrates, organic acids, nitrogenous substances and vitamins. Some studies in natural ecosystems have determined that organic substances derived from phytoplankton are used by bacteria as a substrate for growing. However, microalgae also inhibit bacterial growth by production of organic exudates or toxic metabolites. There are several reports suggesting a synergistic action between microalgae and its bacterial flora associated."[1]

Human utilization of microalgae dates back 2,000 years as recorded by the Chinese, who used *Nostoc* to survive during times of famine. In the early 1950's, a growing recognition of the exponential increase in the world's population and predictions of an insufficient protein supply led food scientists to search for unconventional protein sources. Microalgae appeared to be a good candidate for meeting protein needs. Moreover, the systematic analysis of microalgae for biologically active substances, rapidly led to interest in microalgae's antibiotic properties.[2]

Microorganisms, bacteria, and fungi have been exploited for almost a century to provide useful antibiotics and other pharmacologically active compounds.[3] Antimicrobial agents, active against bacteria, fungi, and even viruses, have been isolated from marine algae, especially macroalgae.[4] For instance, sulfated polysaccharides from marine microalgae, with a low molecular weight, can inhibit viral and bacterial infections. Micro-algal polysaccharides are used as wound healing agents, blood plasma expanders and vaccines. Alginate has shown an unusual combination of biological activities and physicochemical properties, which has contributed to the development of novel or improved pharmaceuticals from microalgae.[5]

Certain algae mediums and bacteria have been utilized conjointly in nearly every hospital for the past century. Since its first clinical applications by Koch in 1881, agar medium has been used in nearly every hospital and bacterial research laboratory for the culture and isolation of pathogenic microorganisms. Agar from algae has the desirable capacity of readily absorbing water and swelling. Agar is an excellent culture medium because it has relatively inert properties, which are not digested by most bacteria.

Beneficial Effects of Microalgae in Treating Leprosy

Beneficial effects of microalgae in treating leprosy were first observed in the 1940s. "Jorgensen and Convit fed a soup made from concentrated *Chlorella* to eighty patients at a treatment colony in Venezuela. The improvement in those patients' physical condition was the first documented evidence of the potential of microalgae as a health supplement."[6] These microalgae pioneers "grew freshwater algae in unglazed baked red clay bowls exposed to the open air and hot sun. The nutriment consisted of a commercial fertilizer suspended in water. Evaporation from the porous sides of the bowls kept the temperature at 26C, even in direct tropical sunlight."[7]

Subsequent positive results in the treatment of leprosy were documented in India with the use of *AFA* blue-green algae.[8] Leprosy is a bacterial infection; in some as yet unknown way microalgae appears to improve that condition.

Antibacterial, Antiviral, and Antifungal Properties in Microalgae Species

Antibacterial, antiviral, and antifungal properties have been discovered in dozens of microalgae species.[9-13] A complex array of unique and diverse biochemical properties may be involved. The antibacterial agents of blue-green algae often appear dissimilar from known cyanotoxins as well as known antibacterial substances. While some micro algal species extracts show specific and targeted antiviral or antibacterial effects, extracts of other species demonstrate an inhibitory effect across a wider range of viruses, fungi, and bacteria. A few of the species used in research are known to be toxic, while other species are edible.

Certain macro- and microalgae produce a wide variety of pharmacologically useful active compounds. The overutilization of antibiotics, in medical and in agricultural applications, has led the emergence of antibiotic-resistant bacteria. Yet fewer new antibiotics are being developed, so the options are diminishing for treating antibiotic-resistant microbes, sometimes called "super bugs." Microalga antibiotic compounds offer a number of potential and tangible alternatives to the widespread use of traditional microbe-derived antibiotics.

Because cyanobacteria are largely unexplored, including some estimated 2,000 strains worldwide, they represent a rich opportunity for discovery. Especially as the rate of discovery of new compounds from traditional microorganism sources is diminishing, the promise of turning to cyanobacteria to locate new bioactive molecules is capturing interest. "Cyanobacteria lipopeptides include compounds that may be categorized as cytotoxic (41%), antitumor (13%), antiviral

(4%), antibiotics (12%), and the remaining 18% activities include antimalarial, antimycotics, multidrug resistance reversers, antifeedant, herbicides, and immunesuppressive agents."[14] Cyanobacteria species can produce a wide variety of secondary metabolites.

Blue-Green Algae Extracts Remarkably Active Against the AIDS Virus

In 1989, the National Cancer Institute (NCI) announced that a number of extracts from blue-green algae were found to be "remarkably active against the AIDS virus."[15] NCI has searched the world for natural plants and organisms that display active anticancer, antiviral, and immune-enhancing effects. A relatively high percentage (about 15%) of aqueous extracts from freshwater and marine algae, cyanobacteria, marine invertebrates, and terrestrial plants have exhibited activity in the National Cancer Institute's primary AIDS-antiviral screening program.[16]

Numerous studies indicate that cultured or harvested blue-green algae may represent a novel source of compounds that inhibit reverse transcriptase viral activity, including that of HIV-1.[17-18] For instance, naturally occurring sulfolipid portions of glycolipids in blue-green algae have demonstrated powerful antiviral properties. Through *in vitro* and *in vivo* experiments, a sulfated polysaccharide derived from blue-green algae was demonstrated to be superior to dextran sulfate (DS) against human immunodeficiency virus type 1 (HIV-1).[19-21] Extracts of micro algal polysulfide's appear to: (1) block HIV replication in cell cultures at low concentrations without toxicity to the host cells; (2) inhibit cell to cell adhesion of HIV at its primary binding sites; (3) possibly act synergistically with anti-HIV drugs; (4) be very slow to cause virus-drug resistance and show activity against HIV mutants that become resistant to reverse transcriptase inhibitors, such as AZT and others; and (5) have the potential to be effective in a vaginal formulation to protect against HIV infection.[22] The efficacy of micro algal polysulfates in therapy or prevention of retroviral and opportunistic infections needs to be further demonstrated in animal models and humans.

Compounds and extracts from blue-green algae, as well as other microalgae, showing HIV inhibitory activity are often active against other retroviruses such as *Herpes simplex* virus types 1 & 2, simian immunodeficiency virus (SIV), cytomegalovirus, measles virus, mumps virus, and influenza A virus.[23-27] Because of the affinity of some bacterial species for sulphated glycoconjugates exposed on the epithelial cells of susceptible host animals, researchers have hypothesized that sulphated exopolysaccharides of microalgae can be used in antiadhesive therapies against certain bacterial infections, both in cold- and warm-blooded animals.[28]

CV-N is the First Molecule Known to Inhibit Ebola Infection

Ebola virus is one of the deadliest viruses found on the planet, with no known cure or vaccine, killing up to 90 percent of infected humans. A collaborative study (2003), of researchers from the U.S. Army Medical Research Institute of Infectious Diseases, the Centers for Disease Control and Prevention, and the National Institute of Diabetes and Digestive and Kidney Diseases demonstrated that an antiviral protein from blue-green algae, known as cyanovirin-N (CV-N) can extend the survival time of Ebola-infected mice. "Its antiviral properties were originally discovered through NCI's Laboratory of Drug Discovery Research and Development in a screening process designed to identify natural materials that act against HIV. The study also found that CV-N effectively inhibits HIV infection of cells grown in the laboratory. . . Furthermore, when researchers injected CV-N into mice prior to infecting the animals with Ebola, then continued to inject CV-N once a day, the onset of visible illness was delayed and the animals survived longer than those not treated with CV-N. CV-N is the first molecule known to inhibit Ebola infection by interfering with the virus's ability to enter cells,"[29]

Micro algal Pigment Allophycocyanin: Antienterovirus 71 Agent

In 1998 a massive viral outbreak of hand, foot, and mouth disease (Enterovirus 71) occurred in Taiwan causing over 120,000 children to fall ill and nearly 80 died. Enterovirus 71 is part of a family of viruses which include poliovirus and coxsackie viruses A and B. No effective treatments were available for enterovirus at this time. Subsequently, Taiwanese medical and university researchers were able to purify allophycocyanin (a red colored pigment) from blue-green algae and demonstrate antienterovirus 71 activities. The researchers concluded that "Allophycocyanin neutralized the enterovirus 71-induced cytopathic [damage to host cells from a virus invasion] effect in both human rhabdomyosarcoma cells and African green monkey kidney cells." They also demonstrated that allophycocyanin possess antiviral activity and has a potential for development as an antienterovirus 71 agent."[30] The blue-green algae pigment, allophycocyanin, had the best antiviral activity effect when added to the host cells before the virus entered. This finding suggests that the mode of action of this algal protein occurs in the early stages of viral absorption or penetration.

AFA Blue-Green Algae has an Inhibitory Effect on the Growth of Salmonella Bacteria

AFA blue-green algae shows an inhibitory effect on the growth of *Salmonella* bacterial strains, in amounts greater than 2 mg.[31] A hot water extract of the green alga, *Chlorella,* given to mice infected with *Listeria monocytogenes,* significantly increased the survival rates of mice, as well as demonstrating an increased immune cellular response.[32-33] Unicellular green algae have also been shown to increase resistance against *E. coli* and cytomegalovirus infections.[34-37]

Marine microalgae were screened for *in vitro* inhibition of viral replication for African swine fever virus and viral hemorrhagic septicemia virus of salmonid fish. Two out of ten species of marine microalgae tested produced a significant inhibition of both viruses in a dose-dependent manner. These two viruses were used because of their major economic importance. Thus, extracts from marine microalgae may have prophylactic utility against fish and mammalian viral disease.[38]

A team of medical researchers reviewed over 200 documented medical cases and concluded that *AFA* blue-green algae may be

helpful in the treatment of chronic fatigue, Epstein Barr infection, chronic ear infections, AIDS, and other conditions involving viral infections. An earlier retrospective study (1996) documented orally ingested *AFA* to improve numerous case outcomes involving viral mononucleosis and *Candida albicans* (yeast) infection.[39-40]
The studies of Harvard Medical School, University of Illinois, Toyama Medical and Pharmaceutical University and in Germany, Russia and Mexico respectively demonstrate that "blue-green algae extract has significant inhibitory effect against Picomaviridae as well as Family Paramyxoviridae, including measles virus, mumps virus, herpes virus and HIV. Moreover the master's dissertation of Tsai Kun-nan from Chang Gung University Basic Medicine Research Institute [demonstrated that] a group of water soluble proteins – phycocyanin uniquely found in algae are capable of inhibiting the reproduction of enterovirus and influenza virus which lead to cytophathic effect through two mechanisms: 1. preventing infection; and 2. delaying virus reproduction in infected cells."[41]

Phycocyanin Powerfully Inhibits Influenza Viral Activity

Various types of influenza may be classified dependent upon specific immune responses. Yet most people know influenza simply as the "flu." This is a family of viruses that typically occur in the winter or early spring, and which primarily target the lungs or sinuses. The influenza or "flu" virus can be particularly lethal for the very young and the very old, contributing to tens of millions of human deaths over the last century. According to historical records, the influenza virus is said to cause a worldwide flu epidemic about every ten years and feverish local outbreaks, much more frequently.

According to McCarty et al. (2010) "The most foolproof way to promote survival in epidemics of potentially lethal influenza is to target, not highly mutable viral proteins, but rather intracellular signaling pathways which promote viral propagation or lung inflammation." He proposes using microalgae sourced phycocyanobilin to target NADPH oxidase, "as this may be a major source of influenza-evoked oxidant stress in lung epithelial cells as well as in phagocytes attracted to lung parenchyma. A number of studies demonstrate that oxidant stress contributes to over exuberant lung inflammation and lethality in influenza-infected mice . . . The lethality of influenza is also reduced in mice pretreated with adenovirus carrying the gene for heme oxygenase-1; this benefit may be mediated, at least in part, by the ability of bilirubin to inhibit NADPH oxidase. It may be feasible to replicate this benefit clinically by administering biliverdin or its homolog phycocyanobilin," which is found within blue green algae.[42]

Medical researchers and drug companies have spent billions of dollars trying to find effective treatments to combat influenza outbreaks, such as vaccines or the drug Tamiflu. Tamiflu is a leading oral drug that inhibits the spread of influenza viruses. The antiviral drug Tamiflu was thought to be the most effective treatment option for reducing deadly influenza flu symptoms. As a result, governments stockpiled Tamiflu most notably during the Avian flu fears of 2006. Tamiflu is sold by Roche Pharmaceuticals. Sales in the first quarter 2009 were $347 million worldwide, of which Roche noted that "$264 million of these profits were the result of pandemic stockpiling and $84 million were from seasonal use."[43]

> *Phycocyanin, the blue pigment in blue-green algae, was 12 times more effective than Tamiflu at inhibiting damaging influenza viral activity. . . phycocyanin was shown to be more than three times safer for healthy cells, with far less negative side-effects.*

A study at the School of Medical Technology of Chang Gung University in Taiwan, found that phycocyanin, the blue pigment in blue-green algae, was 12 times more effective than Tamiflu at inhibiting damaging influenza viral activity. In addition to being more effective than Tamiflu; phycocyanin was shown to be more than three times safer for healthy cells, with far less negative side-effects, than Tamiflu. The Taiwanese study tested four types of virus, including type influenza, the broad class of flu virus that contains the avian flu (H5N1) strain.[44] Perhaps governments and citizens would be better served by stockpiling blue-green algae rich in phycocyanin to prepare for the next influenza pandemic.

Russian medical researchers reported that selenium enriched blue-green algae was "effective as part of therapy for acute pneumonia in children." Forty-five children, ages 3 to 15 years old `who had acute pneumonia and were being treated in the intensive care unit of a Moscow children's hospital, were given the microalgae as part of their treatment regime.[45]

Beta carotene, one of the highest pro-vitamin sources found in microalgae, may also decrease susceptibility to respiratory infections. As beta carotene is transformed into vitamin A, deficiencies associated with vitamin A—such as increased risk of respiratory disease—might be reduced.[46-47]

Chlorophyll used to Prevent Infections and Accelerate Wound Healing

Chlorophyll, for which microalgae constitutes a rich source, appears to have antibacterial components.[48] Prior to the advent of modern antibiotics, chlorophyll was often used to prevent infections and accelerate wound healing.[49] Remarkably the ability of green, chlorophyll-rich, algae to destroy bacteria developed a billion years ago. In 2013, a team of scientists captured images of green alga consuming bacteria.[50] This discovery both demonstrated that microalgae acquired the ability to neutralize and destroy bacteria, as well as provided a glimpse of how early organisms may have acquired free-living photosynthetic cells. The acquisition of light-harvesting pigments, chromatophora, which then became dependent upon their hosts, was "a critical first step in the evolution of photosynthetic algae and land plants." Researchers using an electron microscope saw how "algae draws bacterial cells into a tubular duct through a mouth-like opening and then transports these food particles into a large, acid vacuole where digestion takes place." Over time, in the process of consuming bacteria, microalga likely developed unique antibacterial defenses.

Microalgae can Support Growth of Probiotics

According to doctoral research by Bob Kay, the addition of *AFA* microalgae significantly increased the growth of beneficial intestinal microflora.[51] Antibiotics often disrupt the microbial environment of the colon by wiping out colonies of normal and beneficial bacteria. Without the presence of healthy colonies, infectious bacteria have a greater chance to take hold. Increasingly health care providers are recommending a course of probiotics (beneficial bacteria like bifidum bacteria or acidophilus) after taking antibiotics. Probiotics may lower the risk of diaharea, help to normalize pH levels and reduce the conditions that allow the growth of pathogenic bacteria.

CHAPTER

16

Allergy and Asthma Relief

"If you are sitting on a tack, it takes a lot of aspirin to make it feel good. If you are sitting on two tacks, removing just one does not result in a 50 percent improvement."

— Baker S, 1997

" The lung, as the interface between the atmosphere and the rest of the body, is particularly vulnerable to oxidative injury. In addition, the lung shares with other organs the risk of injury due to infections, inflammation, ischemia-reperfusion, and other insults. Lungs therefore require potent defense mechanisms, and in fact have higher levels of antioxidant enzymes than almost any other organ."

— Morse D, Choi A, 2005

Microalgae Hypothesis XI: Allergy and Asthma Benefits

Microalgae may reduce anaphylactic and immune-type allergic reactions.

Microalgae can help patients with allergic rhinitis (hay fever).

Microalgae can inhibit histamine production.

Algal oils can reduce blood levels of arachidonic acid.

Extracts of beta carotene offer protective effect against exercise-induced asthma.

Phycocyanin has potent anti-inflammatory properties.

Microalgae may be used to clean air and for life-support systems.

According to James Breneman, former chairman of the Food Allergy Committee of the American College of Allergists, "The incidence of food allergy is greater than the incidence of any other type of illness affecting mankind. By some estimates, 60 percent of the population have unknown food intolerances or allergies."[1]

Blue-green Algae may Help Reduction Anaphylactic and Immune-Type Allergic Reactions

Pediatric studies published in the Japanese journal *Pediatric Clinics* in 1962 reported the use of decolorized green algae in the diets of infants sensitive to milk and soybean formulas. The algae reduced allergic sensitivity and provided a range of nutritional factors ordinary found in milk products. Algae powder has also been used in the United States to reduce seasonal pollen allergies.[2]

Two animal studies have found that the inclusion of blue-green algae in the diet contributes to a reduction of anaphylactic and immune-type allergic reactions. Serum histamine levels are significantly inhibited in rats administered *Spirulina.* These results suggest that blue-green algae may contain compounds that act to inhibit mast cell–mediated, immediate-type allergic reactions.[3-4] Mast cell mediated allergic reactions include type 1 hypersensitivity responses to allergens (like a common bee sting) and asthma. Many forms of upper dermal (skin) and mucosal allergy are mediated in large part by mast cells; they play a central role in asthma, eczema, itch (from various causes), allergic rhinitis and allergic conjunctivitis.

In another study, when *Spirulina* was ingested by mice along with a known antigen (a crude shrimp extract), enhanced IgA antibody levels in the intestines were evident. According to the scientists, these significantly enhanced IgA antibody levels may protect against allergic reaction. Positive effects were found in both *in vitro* studies with lymphoid cells from the spleen and mesenteric lymph nodes, and *in vivo* studies with live animals. The researchers concluded that, "*Spirulina* may at least neither induce nor enhance allergic reaction such as food allergy dependent on an IgE antibody, and that when ingested both concurrently with antigen and before antigen stimulation; it may significantly enhance the IgA antibody level to protect against allergic reaction."[5]

A pilot study that used a survey developed by the National Center for Health Statistics (1996) reported fewer allergies, skin problems, and asthma among *AFA* consumers. The algae eaters scored significantly better than average on multiple measures of health when scores were compared to normative baseline data.[6]

In my own research a significant reduction was observed in parents' reports of their children's allergy-type symptoms during a three-month trial on *AFA* supplementation. This study included 26 children and used a standardized health symptom checklist as well as a number of other test instruments.[7]

Microalgae may be helpful in Treatment of Patients with Allergic Rhinitis, "Hay Fever"

Allergic rhinitis (AR) is the most common chronic allergic condition in outpatient medicine, associated with substantial health care costs and lifestyle consequences. AR often called "hay fever" is an especially common nasal problem in adolescents and young adults. Allergies to inhalants like pollen, dust, and animal dander begin to cause sinus and nasal symptoms often in early childhood. Infants and young children are especially susceptible to allergic sensitivity to foods and indoor allergens. The prevalence of allergic rhinitis is increasing globally due to various environmental causes. It affects the quality life of a large group of people worldwide. Allergic rhinitis is still inadequately controlled by present medical interventions. The necessity of continuous medical therapy, typically with nasal corticosteroid sprays, creates anxiety about the long term side effects of the drugs. So there is a need for an alternative strategy.

Mao et al. (2005) conducted a randomized, double-blind study on the benefits from blue-green algae, as a dietary supplement, taken by patients suffering from 'hay fever,' or allergic inflammation of their sinuses. Thirty-six allergic rhinitis patients were given either a placebo or a daily dose of either 1000 mg or 2000 mg *Spirulina* for 12 weeks. There was no improvement for the participants taking the placebo. There was also no notable improvement for participants taking 1000 mg *Spirulina* daily. However, the allergy sufferers who took 2000 mg *Spirulina* every day saw significant improvement in their allergy symptoms. The researchers reported that "we discovered that *Spirulina*, administered at 2,000 mg/day, significantly reduced IL-4 levels by 32% from PHA-stimulated cells. These results indicate that *Spirulina* can modulate the T-helper profile in patients with allergic rhinitis by suppressing the differentiation of Th2 cells mediated, in part, by inhibiting the production of IL-4."[8] T-helper cells are a type of white blood cell, which helps to recognize foreign antigens and to activate inflammatory cytokines that mobilizes an immune response. Interleukin 4 (IL-4) is a cytokine that boosts an immune reaction.

Another independent, double blind, placebo controlled follow-up study (2008) evaluated the effectiveness and tolerability of a blue-green microalgae for treatment of patients with allergic rhinitis ("hay fever"). The Turkish researchers reported that "*Spirulina* consumption significantly improved the symptoms and physical findings compared with placebo ($P < 0.001$) including nasal discharge, sneezing, nasal congestion and itching." The investigators advocated that "*Spirulina* is clinically effective on allergic rhinitis when compared with placebo."[9]

In an experimental animal study (2008), blue-green algae was effective in relieving inflammation of the nasal mucosa and decreasing serum histamine and immunoglobulin (Ig) E levels.[10]

Overall, these research findings suggest that blue-green algae (BGA) can modulate the cell profiles in allergic rhinitis patients by suppressing the differentiation of immune cells. In part, BGA helps by inhibiting the production of IL-4 in allergic sufferers.

Algal Oils Reduced Blood Levels of Arachidonic Acid

Researchers at Massachusetts General Hospital, affiliated with Harvard Medical School, found that algal oils significantly reduced blood levels of arachidonic acid in rats.[11] Arachidonic acid produces molecules (leukotrienes) that trigger allergic reactions and contribute to water retention (edema) and puffiness. These molecules may be 1,000 times more problematic than histamine in contributing to asthmatic bronchial constriction.[12]

A daily dose of beta-carotene, from an algae extract, demonstrated a protective effect against exercise-induced asthma. Of thirty-eight patients given 64 mg of algal beta-carotene extract daily for one week, 53% were protected from exercise-induced asthma. All of the patients in the placebo condition showed a significant post-exercise reduction of breathing in a forced expiratory volume test.[13]

Asthma is an erratic disease and various environmental and genetic factors influence asthmatic symptoms and levels of asthma control. Numerous medications are available at present to treat this disease yet, often times drugs may fail to provide a robust and success treatment. Therefore, medical researchers have turned to studying the role of diet and possible nutritional interventions. "In vitro and in vivo experiments have been conducted with extracts or compounds from algae and found that they showed significant activities against asthma. Accordingly, many marine macro- and microalgae have been reported to have potential to ameliorate the effect of asthma."[14]

The time to fatigue after the two hour run was significantly longer among the men who took spirulina. The carbohydrate oxidation rate was decreased by over 10% and the fat burning rate was increased by almost 11% over the placebo group.

A January 2010 research study published in *Medicine and Science in Sports and Exercise* reported that *Spirulina* had a positive impact on exercise performance.[15] In this study nine physically fit men took either a placebo or *Spirulina* for four weeks. The men ran on a treadmill for two hours every day, running at 70-75% of their VO_2 max and 95% of their VO_2 max to exhaustion. Both exercise performance and respiratory function were measured. Blood samples were drawn at regular intervals during the study.

The results were impressive. The time to fatigue after the two hour run was significantly longer among the men who took *Spirulina*. The carbohydrate oxidation rate was decreased by over 10% and the fat burning rate was increased by almost 11% over the placebo group. The study concluded that *Spirulina* really does enhance exercise performance.

Phycocyanin and Bilirubin, Products of Heme oxygenase (HO-1), have Anti-inflammatory Properties

Over a billion years ago, photosynthesis was a remarkable evolutionary break-through that allowed organisms to capture sunlight for energy, along with exchanges of water and air (CO^2), to develop ways to handle "toxic" oxygen. While it may seem incredible that pigments in microalgae, like phycocyanobilin or C-phycocyanin could possibly assist human lung tissue, one key to understanding is heme oxygenase.

Heme oxygenase performs the rate-limiting step in heme degradation (recycling of red blood cells components which carry oxygen) and is activated by cell injury or stress.

It has been substantiated that heme oxygenase is essential to tissue health such as lungs, the interface where oxygen comes into contact with blood. But mammals are not the only creatures that benefit for heme oxygenase. "Heme oxygenase has evolved to carry out the oxidative cleavage of heme, a reaction essential in physiological processes as diverse as iron reutilization and cellular signaling in mammals, synthesis of essential light-harvesting pigments in cyanobacteria and higher plants, and the acquisition of iron by bacterial pathogens. In all of these processes, heme oxygenase has evolved a similar structural and mechanistic scaffold to function within seemingly diverse physiological pathways."[16]

There are reasons to consider the involvement of heme oxygenase (HO-1) in both of the two main pathogenic mechanisms of asthma, bronchospasm and inflammation. Two products of heme oxygenase activity, phycocyanin and bilirubin, have well-documented anti-inflammatory properties. "There is now reasonably good evidence that HO-1 activity is affected by the course of asthma, and it appears probable that HO-1 activity also affects the course of the disease. . . . New experimental data point to a possible therapeutic effect for heme oxygenase in asthma. Heme oxygenase and bilirubin have been shown to decrease airway smooth muscle contractility in an ex vivo model."[17] The major symptoms of asthma are produced by airway smooth muscle contraction.

"Heme oxygenase may also have an inhibitory effect on the process of lung remodeling that is seen in asthma. In a guinea pig model of aeroallergen-induced airway remodeling, activation of heme oxygenase prevented the increase in bronchial smooth muscle area, whereas inhibition of HO had the opposite effect. The investigators attributed this finding to the antiproliferative effect of bilirubin, which they showed was capable of inhibiting human airway smooth muscle proliferation in vitro."[18] The chromophore phycocyanobilin (PCB), found in blue-green algae, is rapidly reduced in mammalian cells to phycocyanorubin, a close homolog of bilirubin.[19] Again, the potential clinical use of C-phycocyanin (bilirubin's counterpart) and additional chromophores found in microalgae, offer promising therapies for asthma and other inflammatory conditions.

HO -1 can Protect the Lungs from an Increased Oxidant Burden Under more Stressful Circumstances

In Japan, blue-green algae are reported "to forestall pancreatic exhaustion and return balance to the flow of enzymatic secretions."[20] Effective digestion requires our body to secrete sufficient hydrochloric acid and pancreatic enzymes into the stomach to process foods. Certain food allergies can be traced to poor digestion combined with "leaky gut syndrome" that allows undigested proteins to enter the blood; the immune system reacts to these large molecules as foreign invaders. "*AFA* blue-green algae contain carotenes and chlorophyll, both of which are able to dramatically stimulate specialized cells around the intestinal walls to secrete lubricating material and thus help to prevent this type of allergic reaction."[21] The omega-3 fatty acids are likely to be helpful as well.

"Many of these enzymes, such as catalase and glutathione peroxidase, are constitutively active and protect the lungs against everyday insults. HO-1 is one of the few inducible molecules that can protect the lungs from an increased oxidant burden under more stressful circumstances."[22]

"As the lung interfaces directly with an oxidizing environment, it is expected that heme oxygenase-1 would be involved in many aspects of lung health and disease . . . The protective effects of heme oxygenase-1 and products of its enzymatic activity, including carbon monoxide, biliverdin and bilirubin, and ferritin, have opened the door to potential therapeutic and disease-monitoring possibilities that one day may be applicable to pulmonary medicine [providing] . . . cytoprotection against oxidative injury and cellular stresses."[23]

Microalgae, Clean Air and Life Support Systems

The effectiveness of microalgae to recycle CO2, release oxygen and purify the air has been considered by biomedical engineers as a way to improve air quality from long-lasting space missions to urban office spaces. Hypothetically with futuristic biotechnology, even a photosynthetic artificial lung might be developed, using microalgae technology, for a human life support system. Microalgae photosynthesis could be used to catalytically convert blood CO_2 to O_2 directly across a semipermeable membrane device, needing only an adequate light source.[24] Because microalgae has such a deep and far reaching global effect on cellular respiration a myriad of therapeutic applications related to lung function are likely to emerge.

CHAPTER

17

Safely Using Edible Microalgae

"Since 1950 small amounts of arsenic as arsanilic acid have been incorporated into poultry feed to stimulate early maturation, increase efficiency of feed utilization, produce more eggs, "improve" skin coloring and feathering, and yield more profits...Currently 90 per cent of all commercial chickens are raised with arsenic in their feed...The arsenic-containing feed must be discontinued long enough before slaughter for the birds to eliminate most—but not all—of it from their meat."[1]

— Hunter B, 1971

Microalgae Hypothesis XII: Microalgae A Safe Food

Many commercial food and cosmetic items already contain algae compounds.

Edible algae products are open to regulatory assessment, using reliable laboratory methods.

Prior confusion about the taxonomy of Aphanizomenon flos-aquae (AFA) microalgae created false safety concerns.

Legal limits for toxic microcystins are far below "no adverse effects" levels.

There are few reported cases of allergic reactions to microalgae.

Buy from microalgae producers with Good Manufacturing Practice (GMP) that meet stricter U.S. Food and Drug Administration standards.

Many Commercial Food and Cosmetic Items Contain Algae Compounds

Most people would be quite surprised to discover how many commercial food and cosmetic items already contain algae compounds. For example, phycocyanin, a blue pigment in blue-green algae, is currently used as a natural coloring agent in "chewing gum, ice sherbets, popsicles, candies, soft drinks, dairy products and wasabi, as well as a pigment for natural cosmetics like lipstick and eyeliners. Microalgae carotenoids are also used as natural food colorants (e.g., orange juice) and as an additive for animal feed and in cosmetics. Microalgae derived DHA is currently used in the production of many foods, such as cheeses, yogurts, breakfast cereals, spreads, salad dressings and beverages. Certain microalgae DHA products are used for pregnant or nursing mothers and even in infant formulas![2] Unfortunately, industrial chemicals and microbial pathogens can contaminate any food source. In recent years, the news media have broadcast many stories of illness and death related to bacteria-contaminated poultry, mercury in fish, aflatoxin in moldy peanuts, and viruses in uncooked shellfish. The Food and Drug Administration reports that food-borne infections caused by *Salmonella* alone are responsible for an estimated 6.5 million cases of human illness and 9,000 deaths annually in the United States.[3]

Edible Algae are Amenable to Regulatory Assessment using Reliable Laboratory Analysis

The chief areas of concern for edible microalgae safety include possible microbiological load, heavy-metal content, pesticides, and extraneous matter and cyanobacterial toxins. The primary safety issue with wild-grown microalgae comes from possible contaminant algal species that, under certain conditions, may grow in the same lake and thus be harvested along with the food algae. Yet, known toxins that might potentially contaminate edible algae are amenable to regulatory assessment, using current techniques for reliable laboratory analysis for toxins, along with safety guidelines and consumption rates for the food.[4-5] A reputable microalgae company will guarantee that accurate and independent tests are done on each batch of harvested algae to ensure purity and safety.

Another concern is how fresh or nutritionally intact the final product is. The best freshness indicators for microalgae are the amount and kind of chlorophyll breakdown products it contains. In one comparison of five companies that harvested wild microalgae, the percentage of intact chlorophyll ranged from 0 to 65% (average of 21%).[6] Breakdown products of chlorophyll include pheophytins and, potentially, pheophorbides. The latter are known to be toxic. Careful harvesting and avoidance of heat will minimize these breakdown products. The same is true of any species of microalgae and also of all chlorophyll-rich "green" foods.

Most Microalgae are Grown in Highly Controlled Environments

Other species of microalgae, such as *Spirulina* and *Chlorella,* are typically grown in man-made ponds where the environment is subject to control. However, "Growing just one kind of algae in a pond is not an easy feat. Hundreds of aquatic organisms can flourish in pond water...Unlike a garden, weeding out unwanted algae becomes a difficult task since this algae is microscopic...An algae farm can keep out weed algae by balancing the pond ecology, however, without resorting to poisons used by conventional agriculture. Each farm has its own trade secrets."[7]

Unlike *AFA,* which flourishes only in highly alkaline water unfriendly to most other algae, *Chlorella* grows in normal water conditions where other undesirable weed algae easily coexist. Also because its cell size is so tiny— only 2 to 8 thousandths of a millimeter in diameter—it cannot be harvested using screens the way blue-green algae can be. Instead, each pond is harvested all at once using centrifuges. Because *Chlorella* has a hard cellulose cell wall, an elaborate drying process is used to crack open the cell wall or mechanically crush it. Though *Chlorella* can only be farmed in a highly controlled environment, strict procedures and safety testing are still necessary.

Likewise, while *Spirulina,* a blue-green alga, is the most easily domesticated and farmed commercial algae, the need for rigorous growing, testing, and processing standards also applies. *Spirulina* is grown worldwide and the standards and purity of harvested microalgae vary considerably.[8] In 1982 US Customs and FDA agents blocked *Spirulina* harvested from Lake Texcoco (the location of the original Aztec floating gardens) due to persistent quality and contamination problems. Additionally, traces of microcystins are also present in *Spirulina* products.[9] It is important to know where any microalgae comes from and the manner in which it is harvested, tested, and processed.

Harvesting Wild Microalgae Takes Extra Safety Precautions

Harvesters of wild algae need to take additional precautions to ensure that the surrounding environment and the microalgae are safe from external sources of contamination. For example, a reputable harvester of *AFA* has its own in-house testing and quality assurance program. Furthermore, the company also uses independent laboratories to conduct purity analyses on every batch of algae it harvests. These tests include heavy metal assays, pesticide assays, extraneous substance detection, and moisture content analysis, as well as toxicity analyses for algal-toxins from dangerous species, bacterial analysis, and pigment analysis to measure chlorophyll intactness. Redundancy tests by two or more independent laboratories are used to compare every batch, to help ensure greater reliability. Finally, the Oregon Department of Agriculture regularly tests algae-harvesting companies for the presence of microcystin toxins (MC-LR) in their products. MC-L is a toxin produced by particular species of blue-green algae.

Fortunately, it appears that *Microcystis* and *Aphanizomenon* taxa infrequently bloom together. Marine environmental scientists (2013) studied the relationship between cyanobacterial composition within (12 locations) and between four eutrophic lakes in a watershed area of Wisconsin on a weekly basis, for five months. Ninety-seven cyanobacterial taxa were identified using genetic methods. "Across all stations, a total of 37 bloom events were observed based on distinct increases in phycocyanin. Out of 97 taxa, a single *Microcystis*, and two different *Aphanizomenon* taxa were the dominant cyanobacteria detected during bloom events. The *Microcystis* and *Aphanizomenon* taxa rarely bloomed together and were significantly anticorrelated with each other at 9 of 12 water sampling stations with Pearson R values of -0.6 to -0.9 ($p < 0.001$)."[10]

Additionally, the scientists were able to correlate environmental variable, like nitrate levels, with the appearance of different types of cyanobacteria (taxa). Significantly there was well-defined "spatial variability" found between different cyanobacteria taxa within and between lake samples. These findings support the practice that harvesters of wild edible species of *Aphanizomenon* can selectively harvest, Aphanizomenon blooms when the risk of contamination from microcystin blooms are least likely. Additionally, regular testing of each batch of *AFA* further assures a pure harvest of microalga species.

Prior Confusion Existed around the Taxonomy of *Aphanizomenon flos-aquae (AFA),* Microalgae

The first report alleging toxicity of *Aphanizomenon flos-aquae (AFA)* was found in a 1960 U.S. Department of Health, Education and Welfare seminar in which the authors Phinney and Peek (1961) referred to a toxic algal bloom that occurred in Upper Klamath Lake during the late 1950's.[11] While a sample of this algal bloom was sent to Dr. Paul Gorhan, then at the National Research Council, Ottawa, Canada, for a toxicological analysis, Gorham determined that the sample was not pure *Aph. flos-aquae*, but nearly half of the algae sample was contaminated with Microcystis – an algae known to produce harmful toxins. Microcystin is a toxin released by peculiar species of blue-green algae, which can be especially damaging to the liver. "Gorham concluded that the toxicity came not from the Aph. flos-aquae, but from the Microcystis."[12]

Confusion about the safety of *Aph. flos-aquae (AFA)*, wild-harvested edible microalgae, from Klamath Lake, Oregon, again raised unfounded concerns when a scientific paper was published by Sawyer et al. in 1968.[12] The authors claimed that samples of *Aph. flos-aquae* were found to produce neurotoxins. Moreover, two teams of researchers (Gentile and Maloney, 1969; Rapala et al., 1993) also published similar toxic *Aph. flos-aquae* allegations based on wrongly classifying their microalgae. Unfortunately, genetic testing was not as developed at that time and the structural similarities of *Aph. Flos-aquae* to another species of microalgae, Anabaena, were still unclear. Better identification between these two algal species resolved the safety concerns. Advances in genetic tools were able to show that all the so-called toxic strains of *Aph. Flos-aquae (e.g.,* NH-5 *AFA*) are genetically dissimilar to the nontoxic, safe AFA, strains and these toxic strains are more similar to the Anabaena genera.[14]

In their (2000) study, Li, R., Carmichael et al., the taxonomy (science of identifying and naming species) of the NH-5 *AFA* was compared to several other *Aphanizomenon* strains, including some that fit the same structure as *flos-aquae* and some that did not. A particular gene sequence was used for comparison between the strains.

Based on the gene comparison results, there were four species identified. They then were classified into four different groups based on their cell grouping. The NH-5 strain was found to be in an atypical nonbundled form, in contrast to the others. It was improperly identified as *Aph. flos-aquae* in 1968 due to its similarities to the *flos-aquae* species. The Li, Carmicheal, et al, study (2000) determined that NH-5 is not *flos-aquae* based on several important findings during the study, including the lack of bundle formation, lack of similarities to the gene comparison, and differences in cell grouping/structure and the location of specific cells.[15]

These results determined that Sawyer's *Aph. NH-5* strain is a distinctly separate species from *Aph. flos-aquae (AFA)*. In fact, the NH-5 strain cannot be matched to any *Aphanizomenon* species. So currently the species of this NH-5 strain is designated as *Aph. sp.* In other words, *AFA* was falsely accused of producing dangerous neurotoxins found in paralytic shellfish poison, anatoxin-a, because of comparisons to a different species of microalgae. Just as accurate taxonomy of mushrooms is critical for safety, precise identification of edible microalgae is crucial.

To date, *Aph. Flos-aquae* wild harvested from Klamath Lake, Oregon have been safely eaten by over one million consumers. Several decades of regular testing by independent laboratories has failed to reveal the presence of any toxins produced by this species of microalgae. In the late 1990s two lawsuits were filed against companies harvesting from Klamath Lake for allegedly containing neurotoxic microalgae. In both cases, a number of expert toxicology laboratories were unable to prove the existence of neurotoxins in *AFA* microalgae. Both legal cases were dismissed after considerable effort to detect toxins proved unsuccessful and the court required both claimants to publish retractions.[16]

Current Legal Limits for Toxic Microcystins are Based on "No Adverse Effects" Limits

Risk assessments have been performed by researchers at the University of Illinois; by Dr. Gary Flamm, former head toxicologist at the Food and Drug Administration (FDA); and by the Oregon Health Division. The proposed safe levels of microcystin-LR toxin (MC-LC) varied between 1 and 15 parts per million (ppm). Because a thousand-fold "uncertainty factor" is used to establish conservative safety limits, the levels where microcystin-LR actually demonstrates any harmful cellular effects are around 10,000 to 15,000 parts per million. Still, "the health risk posed by exposure to cyanotoxins is difficult to quantify, since the actual exposure and resulting effects have still not been conclusively determined, especially for the human situation."[17]

The current legal limit for microcystin is based on the "no adverse effects limit" determined in a 1994 study, and then further reduced by a "safety factor" of 1,000 by the ODA. This conservative limit which is similar to that in force for Canadian public drinking water sources was determined using a procedure in which mice were tube-fed with pure microcystin extract in water. However, the toxic effects of microcystins are especially amplified when taken singularly without food.

Most users of microalgae take their supplemental algae with food. In a study published in *Ecotoxicology and Environmental Safety*, researchers re-examined the toxicity issue, but instead of using microcystins in a water solution, they used a diet of ordinary rodent food combined with *AFA* that was contaminated with twenty times the legal limit of microcystin.[18] That would be at least 20 parts per million. This method of combining microcystin contamination with algae and ordinary food more closely approximates the way humans would be exposed if they were to eat contaminated algae.

The mice tested at the highest consumption rate were fed a quantity of algae that corresponds to a human eating over 2.3 pounds of AFA (4,600 capsules or 1150 grams) per day. Yet no adverse effects were seen at these extreme levels.

The mice tested at the highest consumption rate were fed a quantity of algae that corresponds to a human eating over 2.3 pounds of *AFA* (4,600 capsules or 1150 grams) per day. Yet no adverse effects were seen at these extreme levels. To put this into a larger perspective, if aspirin were used instead of algae tablets, all the rats would have rapidly died the first day. Still the researchers felt that a conservative safeguard would be limiting microcystin food consumption to 10 parts per million, daily.

Three other studies have also determined that the regulatory limit for microcystins in the *AFA* harvesting industry is many thousands of times (ppm) below the acceptable safe limit for humans.[19-21]

However, even within this context it is also important to understand that while most commercial edible microalgae does not show high levels of microcystin (above 1.0 ug MC equiv/gm), rigorous independent testing across various batches, with companies having Good Manufacturing Practice is recommended. Not all microalgae companies do rigorous independent testing. For example, "Although samples of toxin contaminations of more than 10 ug MC-LR equivalents/g are the exception, 8 of 13 blue-green algae products from the German and Swiss markets tested recently have shown more than 1.0 ug MC-LR equiv/g in concurrent analyses."[22] Consumers are advised to best determine whether any grower or producer of edible microalgae products relies upon strict testing and procurement practices. A huge percentage of blue-green algae (e.g., *Spirulina*) are exported from China, where current safety standards and independent testing requirements fall way below that of the United States and Europe.

No Toxic Effects on Human Blood Cells, from Equivalent Ingestion of 15,000 *AFA* Capsules per Day

Dr. Gitte Jensen, an immunologist affiliated with McGill University in Canada, sought to establish a level at which uncontaminated *AFA* algae might exert a toxic effect on human blood cells. She created various dilutions of live blood mixed with *AFA* algae and observed the reactions. Even at extreme concentrations—the equivalent of a human ingesting 15,000 capsules of *AFA* per day—Dr. Jensen found no toxic effects on human blood cells.[23]

Furthermore, she also collected blood from long-term *AFA* algae consumers to look for any signs of liver enzyme impairment. Her concern was to detect any signs of liver damage due to possible contamination with microcystin, a known liver toxin. Dr. Jensen discovered that long-term *AFA* algae consumers showed no signs of liver enzyme impairment. As a further precaution, Dr. Jensen looked for any signs of increased cellular inflammation, which is also associated with microcystin toxins, but found none. Instead she discovered that *AFA* algae exerted an inhibitory effect or even reversed inflammatory conditions.[24]

Beta-N-methylamino-L-alanine (BMAA)

Different species of blue-green algae can produce different bioactive chemicals, similar to edible versus poisonous mushrooms. Some blue-green algae species have been identified to produce a neurotoxin called BMAA (beta-N-methylamino-L-alanine), which has been hypothesized to contribute to certain neurodegenerative diseases, like Lou Gehrig's disease (amyotrophic lateral sclerosis, or ALS).

Ethnobotanist Paul Cox in the 1990s proposed that Beta-N-methylamino-L-alanine (BMAA) – a cyanobacterial neurotoxins found in contaminated seafood (shellfish), drinking water supplies, and recreational waters – may be a contributing environmental factor for understanding specific neurodegenerative diseases, like ALS.[25] Although many medical researchers consider ALS to be caused by a combination of genetic and environmental risk factors, so far the latter have not been firmly identified.

Dan Dietrich, a toxicology professor at the Universität Konstanz in Germany, reported finding unspecified large quantities of BMAA in commercially available blue-green algae dietary supplements, including *Spirulina* and *Aphanizomenon flos–aquae,* a finding that to date has never been replicated.[26]

While BMAA is produced by many genera of cyanobacteria, a prominent American harvester of *AFA* reported no detectable levels of BMAA in their blue-green algae, despite ongoing testing.[27] Dr. Wayne Carmichael (2012), one of the world's foremost algal toxicologists, has found no evidence of BMAA in nutritional algae products (*Spirulina* and *AFA*) that he's independently tested.[28-29] Dr. Carmichael reported that the hypothesis linking BMAA to ALS, as well as the detection methods proposed by Dr. Cox's group, as unsettled. Moreover, many commercial blue-green algae products have been eaten for nearly three decades by millions of people worldwide. No association has ever been reported between the increased risk for neurological problems and daily consumption of nutritional blue-green algae.

Few Cases of Allergic Reactions to Microalgae

Very few reported cases of allergies to microalgae have been reported, typically in mold-sensitive individuals.[30-31] Because molds and some microalgae grow in very similar conditions, individuals sensitized to mold may show a similar symptom pattern of response to microalgae, as is occasionally reported with *Chlorella.* However, these researchers concluded that *Chlorella* is a "weak" allergen even in atopic children with multiple sensitivities. Overall, most children exposed to *Chlorella* do not show any signs of allergic reactions.

Careful temperature-controlled processing of heat-sensitive edible microalgae is critical. Low-grade processed microalgae may pose a risk to photosensitive individuals. Japanese medical researchers noted that five patients who consumed *Chlorella* developed photosensitization with swelling and "erythematopurpuric lesions" on sun-exposed areas of their body. The scientists believe that pheophorbides and their esters in the *Chlorella* tablets are photosensitizing agents.[32] It is interesting, in this regard, to note that pheophorbides are toxic breakdown products of chlorophyll and have been found in some brands of less carefully processed algae. St. John's wort, a popular herb used to relieve mild depression, is also known to contain potential photosensitizing agents.

High Ratios of Nucleic Acids Can Limit the Use of Edible Microalgae as a Food Source

A final consideration relates to nucleic acids, which are produced by all living organisms and are necessary for the maintenance of life processes. A potential excessive accumulation of nucleic acids has been cited as a major limitation in the use of edible microalgae and other microorganisms for food sources.[33] The high content of nucleic acids in microbial cells used as food or feed has been a concern, for when purines are metabolized, high metabolite levels may result in pathological conditions such as gout. Gout, which is characterized by deposits of monosodium urate crystals in tissues, appears more frequently in men than in woman and is more common after the age of 30 years.

Based on typical, health, blood plasma uric acid concentration rates, experts have recommended that the daily intake of nucleic acids from single-cell protein (SCP) should not be more than 2 grams of total nucleic acid from all sources, not to exceed 4 grams per day. Accordingly, a maximum daily intake of up to 30 grams of microalgae provides a sufficient safety margin.[34] For the vast majority of edible microalgae consumers, excessive amounts of nucleic acids are not a realistic concern. Typically, most people eat only a half gram to two grams per day, so there is a plenty of leeway. Thirty grams (within a safe ratio for nucleic acid content) would be the equivalent of eating two hundred and forty (250-mg) capsules, daily.

Based on traditional use of *Spirulina* harvested from Lake Chad region in Africa, 9 to 13 grams per individual were consumed by the local population, up to three times daily. The blue-green algae was often combined with millet cakes and represented between 10 to 60 percent of the local population's daily meals. NASA and other space agencies have studied the use of microalgae as a "space food" for astronauts on extended space expeditions, such as a Mars exploration. Not only can microalgae be used safely for integrated life-support systems, to help re-cycle carbon dioxide and oxygen, but they can provide a useful source of food to supplement the astronaut's diet.

Blue-green Algae Generally Recognized As Safe (GRAS)

"GRAS" is an acronym for the phrase Generally Recognized As Safe. Under sections 201(s) and 409 of the Federal Food, Drug, and Cosmetic Act, any substance that is intentionally added to food is a food additive and is subject to pre-market review and approval by the FDA, unless the substance is generally recognized, among qualified experts, as having been adequately shown to be safe under the conditions of its intended use. The FDA has included blue-green algae, such as *Spirulina*, under its food ingredients whose use is generally recognized as safe. However a food substance can be GRAS even if not listed by FDA. Because it is impractical to list all foods as GRAS, the FDA simply posts reviews on their website from voluntary submissions. Most safe food ingredients have not received GRAS status, like wheatgrass or blue-berries, because of the voluntary and time-consuming nature of the submission process.

Primarily, the maintenance of the quality and safety of edible microalgae products is dependent upon the entire process from the aquatic environment and harvesting to drying and packaging. Strict process control from the initial aquatic medium to the final packaging and product storage are all very important aspects of edible-microalgae production.

Good Manufacturing Practice (GMP) Regulations are Issued By the U.S. Food and Drug Administration

Good Manufacturing Practice (GMP) is a set of regulations issued by the U.S. Food and Drug Administration (FDA) requiring that food and supplement processing facilities follow a set of strict guidelines, minimizing the risk of contamination and ensuring quality. Receiving GMP certification means that a producer has passed a rigorous audit that examines every step of the manufacturing process, from the initial raw material control to release of the finished product, as well as ongoing employee training, validation of testing equipment, product traceability, and maintenance of the entire facility. The manufacturing facility is also audited and inspected for violations of GMP on a regular basis by the Quality Systems Manager (QSM).

Currently, only three edible microalgae companies in the United States have established manufacturing practices in accordance with current good manufacturing practices (cGMPs) created under the United States Federal Food, Drug, and Cosmetic Act as well as applicable state statutes and regulations.

Edible microalgae facilities in the United States are subject to periodic unannounced inspections from federal, state, and local regulatory agencies. Independent product evaluations are also conducted by third-party registrars as part of an ISO 9001:2000 or equivalent program. External laboratories are used to standardize methods and for independent verification of analytical results.

Several Edible Microalgae Products in the United States Have Organic Certifications

A few edible microalgae products in the United States have organic certifications, governed by the U.S. Department of Agriculture's National Organic Program standards. With so many genetically modified foods ingredients on market shelves, the "organic" certification provides additional assurance for health-conscious consumers. During the certification process, a careful review of records and processes is performed by the certifying agency and inspections are made of their production facilities and the ecosystem from which the raw products are extracted.

PART

III

Algae to the Rescue: Planetary Medicine

CHAPTER

18

Global Food Supply and Ecological Restoration

Human health, the primary concern of this research review, is not the only arena in which microalgae shows promise for solving some of humankind's issues. Presented here are a few more areas in which potential applications of microalgae technologies seem to hold major promise.

Indisputably, solar-powered microalgae are Earth's most important transporters of global nitrogen and carbon cycling. "An astonishing 50% of the planet's biological nitrogen fixation is attributable to tropical and subtropical marine cyanobacteria . . . On a global scale, cyanobacteria fix an estimated 25 Giga tons of carbon from CO2 per year into energy dense biomass."[1] Microalgae broadly transformed our planet, being capable of colonizing barren rock, as land emerged through volcanic activity.[2]

Global Hunger Reduction and Ecological Restoration

"The possibility of cultivating crops of vegetation on the vast areas of the open ocean . . . has gripped the imaginations of a growing number of technicians and planners throughout the world. If successfully realized, this technology would enable the planet's oceans to become a huge new source of feeds, foods, fuels, and chemicals — fixed carbon and fixed nitrogen — for the benefit of humanity."[3]

— Wilcox H, 1982

"As we enter the new millennium, nearly 800 million of the World's population will remain chronically malnourished . . . Eradication of nutritional deficiencies among woman and children on a global scale are needed to ensure improved quality of life for the next generation of citizens."[4]

— Iyengar, Nair, 2000

The major advantage of algae as a protein source is its economical land and water use. A one-quart jar may hold billions of microalgae cells that can double on a sunny day. Best yet, microalgae can be sustainably cultivated on marginal (agriculturally poor) lands or in the ocean.[5] This ability is significant, as the UN Environmental Program (UNEP) reports that "11 billion acres — 35% of the earth's land surface — are threatened by desertification and, with them, fully one-fifth of humanity."[6] Furthermore, it is estimated that the world population will increase by nearly 30% in the next two decades, while expanding fertile crop lands will be increasingly difficult. Between 1961 and 2010 farmers world-wide were able to keep crop yields ahead of rising worldwide food demand. Although the global population grew by nearly 4 billion people between 1961 and 2010, per capital food production stayed ahead of the 123% population growth. Advances in crop varieties, increase of agricultural lands (434 million hectares), along with cost-effective energy and water sources; all kept global food prices at historic lows.[7]

Yet, how long will food production remain ahead of population growth? What happens when we run out of new fields for crops? Will wealthy nations continue to use food sources as biofuels, when poor countries struggle to afford rising food prices? While currently our world produces enough food to feed everyone; too many people do not have sufficient land to grow or income to purchase enough food. Poverty is the principal cause of hunger. Many experts believe that food commodity prices will be both higher and more volatile in the decades to come.

Figure 25. Malthus Theory of Hunger

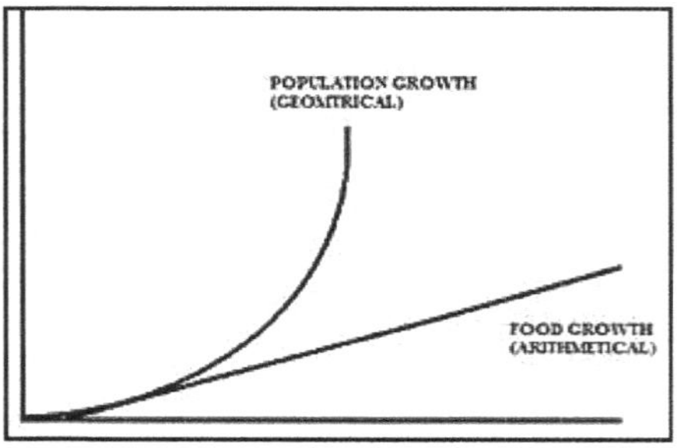

The straight line shows projected increase in food supply, whereas the curved line anticipates population gains

Malthus Theory of Hunger

The problem was first highlighted by Thomas Malthus, a British economist in 1798. Malthus said that human population grows at exponential rates, whereas food supplies only increase at a linear, arithmetical, growth rate (See diagram). This theory highlights a major problem: when population expansion exceeds food resources, then mass starvation results.

Fortunately, Malthus' theory has not come true. So far, on a global scale the world's population has increased at an exponential rate as predicted by Malthus. Scientific advances however have allowed food production to stay ahead of population growth, something that Malthus couldn't have predicted. Agricultural advances have included the wide use of artificial fertilizers and the rapid developments in mechanization (the use of machines in the production of food).

Figure 26. Exponential Rise in Human Population

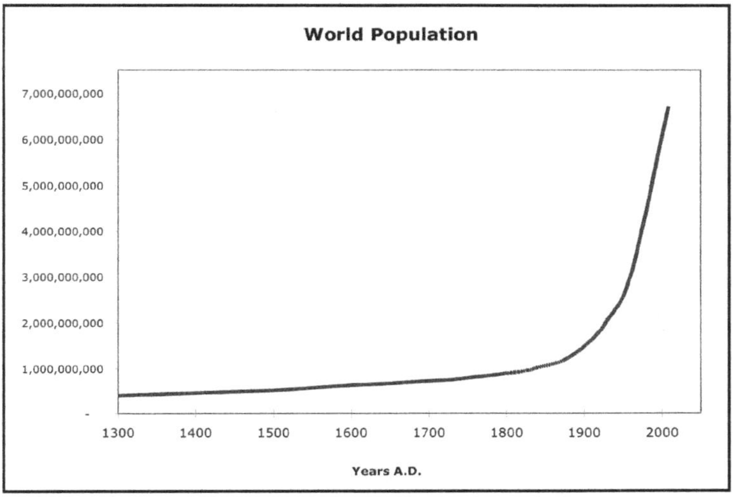

Yet, it is widely believed that agricultural advances may be slowing, while population continues to rise exponentially (see Figure 26). While international trade, "globalization" of food and agriculture, has helped reduce poverty in Asia; multinational food companies disempower local farmers and lead to an increased monopolization and ownership of the world's food supply. Food has increasingly become a global commodity transported across far distances for the highest price. While vast numbers of undernourished people in Asia declined; tens of millions of hungry in Africa and even malnourishment in more developed nations climbed.

One of the greatest challenges facing humanity in the 21st century will be the competing demands of increasing food production, urbanization, and restoring a healthy environment. Between 2000 and 2012, the World Bank global food price index increased 104.5 percent, at an average annual rate of 6.5 percent. The upward trend in food prices and volatility can be traced to additional factors including "climate change, policies promoting the use of biofuels, rising energy and fertilizer prices, poor harvests, national export restrictions, rising global food demand, and low food stocks.

Perhaps most significant has been an increase in biofuels production in the last decade. Between 2000 and 2011, global biofuels production increased more than 500 percent, due in part to higher oil prices and the adoption of biofuel mandates in the United States and European Union (EU). According to a 2012 study by the University of Bonn's Center for Development Research in Germany, if biofuel production continues to expand according to current plans, the price of feedstock crops (particularly maize, oilseed crops, and sugar cane) will increase more than 11 percent by 2020."[8]

Algo-cultivation more Protein per Acre

The loss of prime agricultural lands may occur as population pressures increase and the human-related climate de-stabilization influences remain unchecked. Modern agriculture is highly dependent on inexpensive petrochemical fertilizers, topsoil, and groundwater. In comparison to beef, microalga protein is hundreds of times less wasteful in terms of topsoil, water, and land use. Another alternative is the increasing use of genetically modified food species that pose unknown risks to our environment and health.

Land management studies illustrate that it takes hundreds of times more land and water to produce a pound of beef than a pound of vegetables.[9] When microalgae protein levels are compared to vegetable crops still greater gains are realized. "Algo-culture systems can produce up to 15,000 kilograms of proteins per acre per year. This is almost 20 times the per-acre yield of soybeans, which is the highest yield obtainable through conventional agriculture. Through algo-culture, in theory, a protein supply adequate for the entire planet could be produced in an area the size of the state of Maine...From the standpoint of food energy, our algal cultures have yielded dietary energy on an area basis at rates 8 times as great as that of sugar beets, 22 times as great as that of corn, and 45 times as great as that of potatoes...[and could supply] up to two thirds of the protein supplied by fish for adult humans, without impairment of nitrogen retention."[10]

In equatorial regions where malnutrition, over-population, and climate impacts are most intense, microalgae cultivation for food and fertilizer is very promising. "Once scientists better understand how to cultivate microalgae in seawater, new food producing areas can open up on the more than 10,000 miles of accessible desert coastline in hot climates. Coastal farms could produce food and support new ecological industrial zones, recycling industrial carbon dioxide and mineral wastes into food. Marine algae farms could be built along the coastline of Baja California, Peru and Chile, West Africa, Mediterranean North Africa, Indian Ocean coasts of East Africa, Egypt and the Arabian Peninsula, India and Australia. Many of these coastlines are close to famine threatened areas with millions of people."[11]

Many people associate the fetid odor and unpleasant sights of swamps and bogs, or stagnant ponds with algae and therefore reject it as a food source. Often standing waters, depleted of oxygen by unsustainable algal blooms, have decaying biomass. However, in a moving, coursing, body of water where oxygen and nutrient levels remains optimal, algae is fresh and the air around the algal pool of water will actually be very fresh from all the oxygen produced. There will not be the fishy or fetid smells associated with rank and decaying algae swamps.

Given our current course, the time will come when microalgae cultivation could play an essential role in meeting the food needs of a protein-hungry world. Much of the developing world currently depends upon fish as a protein source. However, our global fish reserves are being rapidly depleted, while animal protein is too inefficient and environmentally wasteful but significant economic and distribution barriers exist. The hope for blue-green algae cultivation in the early 1970s was meant to provide the cheapest sources of protein for the world's poor and hungry. Presently instead of feeding the world's neediest population, blue green algae are more often consumed by the educated and wealthy.

Microalgae to Combat Malnutrition

Though people can exist on algal protein exclusively for long periods of time, microalgae are typically eaten in concert with other foods. Based on traditional use of *Spirulina* harvested from an African lake in Chad, 9 to 13 grams per individual were consumed by the local population, up to three times daily. The blue-green algae was often combined with millet cakes and represented between 10 to 60 percent of the native's daily meals. The optimum dietary results are achieved when algae is used to supplement a varied diet. Used in this manner, even a single gram or two of microalgae daily go a long way. "If every child could get one spoonful of algae each day, we could reduce malnutrition by 50%. Algae can remediate the principal nutrient deficiencies."[12]

African relief workers used a food combination called "Misola," made of millet, soy, and peanut to combat malnutrition in children. In order to assess the benefits of *Spirulina*, Medical workers divided 550 malnourished children, under the age of five, into four equally sized random groups. One group of children received only traditional meals and acted as a control. The other groups were given the following: *Spirulina* plus a traditional meal; *Spirulina* plus Misola; or Misola alone. Total calories were kept nearly equivalent in all groups (around 800 calories per day). Physical body measurements, along with blood tests were used to evaluate the effectiveness of the diets. The study results indicated that "Misola, *Spirulina* plus traditional meals or *Spirulina* plus Misola are all a good food supplement for undernourished children, but the rehabilitation by *Spirulina* plus Misola seems to synergically favor the nutrition rehabilitation better than the simple addition of protein and energy intake."[13]

In contrast, another African nutrition study was carried out among 182 infant children (from three months to three years old; mean age was 14 months) diagnosed with malnutrition. Within five relief centers, three groups were defined by randomization: group 1 - usual nutritional rehabilitation program; group 2 - usual nutritional rehabilitation program, plus 5 grams of *Spirulina* daily; group 3 - usual nutritional rehabilitation program, plus 5 grams of *Spirulina* daily and fish. Tragically during the three month study, six children died (3.3%) and 11 children were hospitalized. In the investigator's final analysis the only criteria used were height per age and weight per height at 60 and 90 days. The researchers concluded that "*Spirulina* dose does not bring any benefit over 90 days, compared to traditional nutrition." Surprisingly, neither did the addition of fish make any measurable differences. Perhaps, due to the shortness of the study (2 to 3 months) and limiting measures to only body weight and size, overlooked the other possible health benefits of blue-green algae (e.g., cognitive assistance). Also, a larger program of study may have reduced death and hospitalization rates in severely malnourished children using blue-green algae drinks as part of nutritional rehabilitation.[14]

One large study in Central America examined cognitive measures, school attendance and health status, along with dietary AFA microalgae supplementation with malnourished children. "In Nicaragua, malnutrition is a serious public health problem, associated with 72 percent of hospital mortality among children and with high rates of ill health and mortality due to infectious diseases in children. Of the 2,500 children who visited the Health Care Center in Nandaime, Nicaragua in 1995, 1,375 exhibited some degree of malnutrition. Among the many consequences of malnutrition in children is its direct impact on the learning process and school achievement."[15] In 1992 a program was started which provided 1 to 2 grams of AFA algae supplementation to 2,000 primary school children and 500 adults from Nandaime, Nicaragua. Due to the dramatic health benefits reported, the director of the Carmela Noguera School in Nandiame requested AFA algae be provided for their students, as well. After AFA supplementation was provided for all students during the 1993 academic year, the school's national standardized test scores soared and the school was awarded first place among all primary schools in Nicaragua, an outstanding achievement!

Two doctoral students, Irma Sevilla and Nereyda Aquirre impressed by the school's improvements, conducted a scientific, controlled evaluation of AFA microalgae's effects on children. They chose a different school, in the same town, which had not yet received the microalgae. From June through December, 1994, under the auspices of the Centro Comunitario in Nandaime and the Central American University in Nicaragua, these doctoral students assessed the health, behavior and academic status of first-, second-, and third-grade children attending the Monseñor Velez School in Nandaime, Nicaragua.[16] One hundred eleven children were selected in the experimental group who presented a malnourished, deteriorated physical appearance, along with "disheartened behavior." These families were from low socioeconomic, impoverished, backgrounds. The children in the study group were given a total of one gram of AFA algae, five days a week, both at school and at home, for a period of six months. Children presenting a more severe state of malnutrition received one gram of the algae seven days a week during the same six-month period.

A control group, also consisting of 111 first-, second-, and third-grade children attending the Monseñor Velez School in Nandaime were not provided AFA microalgae, and were evaluated pre-, mid-, and post-study using the same criteria and study variables.

It was not possible to create an absolutely equivalent control group, since the 111 children selected for the study's sample group (those receiving the algae) were those who generally suffered the greatest degree of malnutrition and its associated problems. In other words, while the age group and gender were the same, the health status was initially better in the control group.

The children consuming the AFA in the study group experienced significant improvement in their nutritional status. At the onset of the study, only 14% had a normal nutritional status. After six months of eating the algae (with no other changes in eating patterns), the percentage of children with a normal nutritional status increased dramatically to 79%.

There also was a notable change in the children found to be malnourished in the study group. At the beginning of the study, 86% of the children were found to be in a state of malnutrition; this fell significantly after consuming the AFA microalgae for six months to 21%, or four times lower than the original figure. It was noted that the children in the remaining 21% category were among those with the greatest nutritional deficiencies at the beginning of the study.

Significant improvements were observed in the hair, skin, eyes and oral health of the AFA experimental group. For example, an initial paleness and dermatitis were identified in 27% of the experimental children, which disappeared completely in the course of the six months, a 100% improvement. Similarly an 87% improvement of eye problems (conjunctiva) and a 90% decrease of bleeding gums were evident.

Not surprisingly, malnourished children experience elevated rates of illness and associated school absences. At the initiation of the study, only 72% of the students had a good school attendance record (0-3 absences per month), this increased to 95% in the following six month AFA supplementation period. The percentage of children with an unsatisfactory school attendance record (4 or more absences per month) dropped from 28% to 7%. The percentage of children in the malnourished (AFA-fed) group who were active and participating in class when the project began (11%), increased by almost five times, to 54%, by the end of the study.

Nutritional status is known to have an impact on academic performance, and this relationship was strongly evidenced in the study's findings as well. At the outset of the study, 52% of the sample group was evaluated as showing unsatisfactory to fair performance. By the end of the study, this low performing group had decreased to only 20%. Conversely, at the beginning of the study, 48% of the AFA group tested as good or excellent, and by the end of the six months, this group had increased to 80%. Included in this group were those who scored in the top 10% of national standardized test scores: this top-performing segment tripled in size over the course of the study, growing from 7% to 21% of the total. In the control group (those not eating AFA microalgae), the percentage of children who achieved an excellent or good school performance rating dropped from 70% to 58% during the same six-month period. The dramatic physical, behavioral and academic improvements assessed in the malnourished children were achieved with only the addition of a single gram of AFA microalgae, daily!

CHAPTER

19

Microalgae as Fertilizer and Animal Feed

"We have noticed already that the dogs and cats are not shedding nearly as much since we started them on microalgae."

— Peterein C, 1993

Algae have been used as a fertilizer for thousands of years in Japan. In fact, the initial domestication of rice and the use of algae as a fertilizer occurred close together. "Nitrogen-fixing blue-green algae have been cultivated in Southeast Asian rice paddies for a thousand years to increase rice production. A 1981 UN FAO [Food and Agriculture Organization] report, "Blue-Green Algae in Rice Production," documented the possibilities of blue-green algae replacing chemical fertilizers and rebuilding the structure of depleted soils."[1]

Algae are already nearly universal members of the soil micro flora community, and finding better ways to increase or to supplement their inclusion in soil content can provide greater amounts of fixed nitrogen and important minerals to crops. Algae fertilizers may offer advantages over standard chemical fertilizers. They can be locally grown, environmentally friendly, renewable sources of plant nutrients that are not dependent on fossil fuels for production. Algae fertilizers can either replace or supplement traditional chemical fertilizers. They can be classified as organic fertilizers and used to produce certified, organic food crops.

Soil is a living matrix of organic carbon materials, living microbes and other organisms (e.g., worms & microalgae) and mineral elements. The process of converting raw organic matter into soil humus feeds the soil population of microorganisms and worms, thus supporting a healthy soil. Like the human gut, microbes and living organisms help to break down the nutrients and make them available to plants.

In soil science, the term soil humus refers to any organic matter that has reached a point of relative stability. Thus soil humus might last for hundreds or even thousands of years, if not disrupted. Humus adds to good soil structure and significantly contributes to holding moisture and nutrients within the soil. Humus contributes to the fertility of the soil.

Humus can hold the equivalent of 80-90% of its weight in moisture, and therefore increases the soil's capacity to withstand drought conditions.[2] "Increasing humus content of soil by 1 percent can improve the soil's ability to hold water by a significant average of 20,000 gallons per acre."[3]

Farmers have known for centuries that adding organic residues to soils improve structure or "tilth." Tilth is the composition of the soil in relation to its suitability for crop growth. Martin (1971) reported the most active binding substances synthesized by effective microbial species were polysaccharides. Microalgae and microbial polysaccharides in concentrations as low as .02%-.2% exerted a marked binding action on soil particles.[4] Microalgae and bacterial polysaccharides are more water stable than plant polysaccharides, resisting decomposition long enough to help hold soil particles together in soil aggregates.

Soil aggregates are 'clumps' of soil particles that are held together by organic compounds (polysaccharides from soil algae, bacteria and fungi), other organic matter, like roots, or held together by moist clay. Bailey, Mazurak, and Rorowski (1973) found water stable soil aggregates were produced within 6 weeks by three algal species, which led them to conclude "...the fact that algae are frequently important pioneer vegetation, it is believed that algae are important in stabilizing and improving the physical properties of soil as a medium for plant growth."[5] Thus, algae is not only found in marine and freshwater environments, but is an important aspect of soil, underground as well.

Even in dry desert environments, a Chinese research team (2006) through microscopic inspections of soil crusts discovered "an intricate network of filamentous cyanobacteria and exopolysaccharides, which binds and entraps sand grains and conglutinate fine particles with each other." Using a portable wind tunnel, they found that even "maximum wind tunnel velocity tests did not lead to any wind erosion on the surface of undisturbed microbiotic soil crusts, i.e. 100% covered by microbiotic [cyanobacteria rich] soil crust."[6] Understanding how cyanobacteria support desert ecosystems will be important to understand the complex process of "desertification." Desertification is a form of land degradation, turning more hospitable dry lands into deserts. Dry lands occupy an estimated forty percent of the Earth's land surface, home to nearly a billion people, who live under threat from desertification.

Microalgae "green manure" has been field tested and applied to thousands of farms over decades to improve the soil, produce beneficial organic metabolites, while fixing atmospheric nitrogen. Reported benefits include improving aeration, increasing soil aggregation, reducing fertilizer needs, and improving crop yields. "Soil algae constitute the initial successional stage on substrates which are poor in plant nutrients, for example, recent volcanic deposits, eroded hillsides, and disturbed soils. Soil algae can have dramatic effects on soils that have mechanical problems caused by intensive agricultural use as well."[7]

Microalgae can help to stimulate the production of plant growth hormones, which accelerate cell division and size, providing higher crop yields. Microalgae are reported to "stimulate plants to secrete compounds that repress harmful bacteria, fungi and other pests. In some cases, algae operate as a catalyst that helps plants manufacture natural insect repellent on their leaves."[8]

Figure 27. Benefits of Using Microalgae as a Biofertilizer for Agriculture

Improves Soil:	**Reduces Production Costs:**
Benefits Soil Microbes Increases Soil Porosity Decreases Soil Erosion Enhances Nutrients Aids Moisture Retention Adds Organic Materials	Less Soil Tillage Required Less Diesel Fuel Needed Less Irrigation Water Needed Less Fertilizer Required Less Pesticides/Herbicides/Fungicides Improved Waste Recycling
Improves Crops:	**Helps Market Value:**
Faster Germination Rate Faster to Maturity Greater Yield and Quality Better Size and Weight Better Crop health	Improved Color & Appearance Better Smell & Taste Increased Vitamin Amounts Greater Trace Minerals A Longer Shelf Life

Table Adapted from: Henrikson, R. and Edwards, M. (2012) Imagine Our Algae Future

Additionally, some scientists have proposed that marine and freshwater microalgae might be a source of novel herbicidal compounds. Microalgae species are well-recognized to produce a remarkable diversity of biologically active metabolites, as part of their survival strategies. The term, allelopathy means that an organism releases one or more biochemicals that prevents or influences the growth, survival, and reproduction rates of other organisms nearby. Marine biologists are just beginning to appreciate how bioactive metabolites from certain algae species, exhibit allelopathy as an advantage over other photoautotrophic competitor species.

"Moreover, given the overlapping targets of action between allelopathic metabolites and herbicides (e.g. inhibition of photosynthesis), there is an emerging literature on the potential – including possible commercial potential – of algal metabolites as herbicides . . . Other compelling feature of these algal metabolites as herbicides include a presumptive biodegradability of many of these natural products, as well as water solubility of many (e.g. peptides) which might minimize concerns of bioaccumulation, and associated environmental concerns."[9]

Cyanobacterial Electrogenic Activity

Microalgae utilize the energy of sunlight to drive photosynthesis, in which the energy of light is used to split water molecules into oxygen, protons, and electrons. While most of the high-energy electrons derived from water are utilized by the cyanobacterial cells for their own needs, a fraction of these electrons are donated to the external environment via electrogenic activity. "Cyanobacterial electrogenic activity is an important microbiological conduit of solar energy into the biosphere."[10] While any impact from cyanobacterial electrogenic activity on planetary environment has yet to be determined, the sheer magnitude of cyanobacterial electron discharges into the environment, over billions of years, might have provided novel adaptive evolutionary advantages for other life forms.

For example, electrogenic enriched atmospheric conditions have been reported to benefit plant development. In the 1970s, A. Zaderej and C. Corson formed Intertec, Inc., to develop and market their "Electrogenic Seed Treatment". Electrogenic seed treatment was claimed to result "in more rapid germination and increased yields . . . increase the metabolism of ATP. . . [and to reduce] the mortality rate of seeds by providing a source of electrons to buffer the reaction with free-radical nutrient ions."[11] Such alternative "electro-culture" farming methods that use forms of electricity, magnetism and sound to stimulate plant growth will strike many as pseudoscientific. Yet, because photons and electron exchanges are integral to cellular life; cells are "dynamically wired" to often respond to subtle energy influences.

Microalgae an Effective Animal Feed Supplement

Numerous studies have documented the effectiveness of algae as a high-quality feed for animals. Microalgae have proven to be an effective animal feed supplement, providing higher quality protein and better trace mineral uptake than many conventional feeds.

Benefits have included more rapid weight gain, enhanced immunity, and larger litter sizes.[12-19] The variety of animal species successfully given algae feed are numerous and include, fish, rodents, birds, cats, dogs, swine, horses, and cattle.

"In Europe for a number of years people have been feeding domesticated animals on seaweeds when regular fodder is insufficient. They probably got the hint for this policy from watching wild animals feed on sea vegetables. Incidentally, sea vegetables stimulate animal appetites...The history of seaweeds, especially kelp, as fertilizer is very old. In Europe, kelp meal was effectively put to this use in the twelfth century."[20]

A series of experiments conducted on laying hens, found that the alga *Antrhospira platensis* produced "a positive effect on the number of laid eggs, their measurement and their quality (Anonym, 1998). Likewise, in 1960, when Arakawa et al., fed chicks diet supplemented in *Chlorella* powder, the total number of laid eggs increased and improved coloration of yolk was observed. Additional studies have shown the microalgae could have a beneficial effect on reproduction. Ishibashi, 1971, stated *Chlorella* influence on estrus in rats that could have been caused by estrogenic substances present in the algae cells."[21] Another green alga, *Spirulina,* was shown to increase the number of mice pups in litters in comparison to mice fed casein (Kapoor and Mehta, 1993).[22]

Similar effect was received when pigs were fed a diet containing 5% of *Spirulina maxima.* The effect was only observed in the first reproduction cycle, whereas in the second cycle the sows fed this diet delivered fewer piglets in their litters (Fevrier and Seve, 1975).[23] Pabst et al., 1978, fed mice diet containing 20% of another alga, *Scenedesmus acutus,* over 6 generations, and found a decrease in the number of pups with accompanying increase in the birth weight.[24] However, a doctoral dissertation by Michael Harrison (1986) found that algal products were "not a satisfactory protein for swine diet."[25] Different species of algae will likely have differing nutritional effects for various animal species.

"Dr. Combs fed chicks a diet containing *Chlorella* as a substitute for soybean meal. He found that when algae were substituted for 10%

of the soybeans in the diet, a more efficient utilization of protein resulted and led to a significant improvement in growth."[26] Reportedly the Grain Processing Corporation of Muscatine, Iowa produced about two million pounds of algae, grown on corn liquor, which was then added to chicken feed to enhance formation of yellow pigment in their egg yolks. Currently consumers can obtain DHA enriched eggs in many supermarkets knowing that, microalgae enriched diets were used to increase DHA levels in eggs.

One of the central problems in feeding livestock animals high quantities of powdered algae was most pure bulk algae was not very palatable to many animals. This challenge was tackled by the University of California at Davis research group led by, (Dr. Harold Hintz and others), "when they made pellets from processed algae and steam-rolled barley. The cost of pelletizing algae is about ten dollars per ton of final feed or about one dollar per ton of algae."[27]

Algae are also useful in waste treatment, both directly and in conjunction with its role as animal feed. In a simple waste treatment system, such as the one used in the town I live in (Pacifica, California), the municipal sewage treatment plant uses marsh plants and algae to help treat the discharged water. In an integrated biological system, algae are harnessed to break down fecal waste and is later collected and used for animal feed. Importantly, the algae can be tested for any signs of contamination, prior to use as animal feed. In an integrated algal system, algae plays two roles—one as a direct means of waste treatment and the other as a commercially valuable by-product. The first effective prototype of this kind of system was designed by Dr. Oswald at the University of California at Berkeley (of "Algatron" fame) in cooperation with the Davis campus research team mentioned earlier. They cultivated algae in a one-million-liter-production pilot pond and then fed large animals with algae grown on waste products.[28]

World-wide Microalgae Demand for Aquaculture, Livestock and Pets

"Today, about 33% of the algae grown globally go to aquaculture. Recent studies show fish grow faster, healthier, and more stress tolerant with algae than food grain diets. This makes sense since algae are the natural diet for fish in nature. Taste tests show algae feeds improve color, taste and sensory appeal of seafood."[29]

— *Henrikson R, Edwards M, 2012*

World-wide demand for microalgae additives is growing for aquaculture feeds, which increase disease resistance and increase growth rates for factory farmed (aquaculture) fish and prawns. Tropical fish, ornamental birds, animals and pets of all kinds also consume a portion of global microalgae production.

Aquarium Fish Food. For many species of exotic tropical fish, algae are an essential part of the diet. Microalgae promises five benefits for healthy aquarium fish: 1) great profile of natural vitamins and minerals; 2) rich in proteins for healthy skin; 3) phycocyanin for reduced obesity and better health; 4) essential fatty acids for proper organ development and; 5) richness in natural coloring agents such as carotenoids. In brief, feeding will result in beautiful, healthy and longer-lived fish.

Highly prized fish have distinctive bright red, yellow, orange, white and black markings, and are often seen in ornamental pools and fountains. Many koi feeds include 5 to 20% *Spirulina* for its rich carotene pigments that enhance red and yellow patterns, while leaving a brilliant pure white. Their improved clarity and color definition, from eating microalgae, increases their value.

Aquaculture Farming. The fish farming industry is rapidly expanding. In the United States alone, aquaculture farming produced nearly $1.1 trillion dollars in sale, in 2005. But aquaculture is even bigger worldwide and the USA is a net importer of seafood. Fish start life as tiny hatchery fry and their survival rates may be quite low. *Spirulina* added to the feed ration at 1 to 10% levels has been shown to increase survival rates. Furthermore microalgae help to increase fish growth rates and body size, reducing fish production time.

Furthermore, adding microalgae to fish feeds helps solve the two biggest problems. First, farmed fish are susceptible to infection and disease. Second, the flavor, texture and skin color may be inferior to wild fish. Fish farmers report that adding microalgae to their fish feeds helps to: 1) reduce medication requirements; 2) improve quality and coloration; 3) increase survival rates; 4) improve growth rates and; 5) reduce waste runoffs in the effluent.

Health, Beauty and Color for Ornamental Birds. Zoos feed their flocks of flamingos and ibises a diet rich in blue-green algae to support the birds' health and color. Beta-carotene microalgae have shown increase, the feather color and shine, health of beaks and skin, and promotion of beneficial bacteria in the digestive tract.

Enhanced Fertility for Bird Breeders. Breeders of canaries, finches, parrots, lovebirds and other bird varieties, use of microalgae to increases coloration, accelerates growth and sexual maturity and improves fertility rates. Ostrich and turkey breeders use it to increase fertility and reproduction rates microalgae also enhances the deep yellow color of their egg yolks. Studies with chickens suggest that adding a small percentage of microalgae in their diet stimulates macrophage production, improving immune performance and disease resistance without side effects.

Healthy Food for Cats and Dogs. There is evidence that microalgae have health benefits for domestic cats and dogs, at all of their life stages. Appetites of finicky cats sometime improve with the addition of algae sprinkled on their food. Pet owners report microalgae fed pets have a fresher breath odor, healthier skin and more lustrous coats.

Tonic for Horses, Cows and Breeding Bulls. Highly valued racehorses are given microalgae in their feed ration for improved performance and recovery, but in that competitive environment trainers tend not to share results. Dairy farmers use it to maintain herd health as it improves intestinal flora that is so vital in ruminants. It has been reported that microalgae also increases the sperm count of breeding bulls and fertility in females."[30]

Nourishment for Zoo Animals and Wildlife Conservation Centers. *AFA* microalgae have been used to supplement diets of captive wildlife at the Rocky Mountain Wildlife Conservation Center in Keenesburg, Colorado. For more than five years the center incorporated *AFA* microalgae into their food plan for the large cats and bears that had been rescued by the sanctuary. The center director, Pat Craig, described how the addition of *AFA* microalgae has been a boon to the animals and that, "We will continue to make it an integral part of the animal's diets, and we highly recommend it to our colleagues in the world of animal care." Benefits reported include; assisting in the reduction of metabolic bone disease in older animals and reversing this condition in cubs. Helping to improve skin and coat condition. Improving animal's appetites and food absorption, and improving the overall metabolism of the animals. Helping to relieve pain associated with arthritis and other joint conditions in the captive animals.[31]

CHAPTER

20

Microalgae in Water Purification and Conservation

"When the well is dry, we know the worth of water."
— *Franklin B,* Poor Richard's Almanac, *1746*

Water plays a Primary Role in Health and Society

Water is the most plentiful and a most essential element within the human body. Yet, one of the greatest challenges facing humankind is scarcity to clean water for drinking and sanitation. An estimated 780 million people in developing countries have no easy way to get hold of uncontaminated water. It has been estimated that world-wide, women spend 200 million hours a day just collecting water for their families. Lack of adequate water sanitation is the world's leading cause of infections. Nearly a billion people world-wide don't have ready access to tap water, much less a sink, shower or a toilet.[1] The World Health Organization estimates that safe water could prevent 1.4 million child deaths from diarrhea each year.[2] Furthermore, the poorest people living in deplorable slums, often pay five to ten times more per liter of water than wealthy people living in the same city, who have cheaper access to public water supplies.[3]

Meanwhile, most Americans do not realize just how much water is used to support our modern lifestyles. Water plays a principal role in today's economy. Water is essential to grow crops, wash or clean-up wastes, produce energy and fuel, required in manufacturing and packaging, used for transportation, and consumed in a myriad of unseen ways. According to Professor James Shine (2013) Harvard School of Public Health, the average American uses about 1,600 gallons of water per day. How? "It takes 52 gallons of water to make one glass of milk. It takes more than 600 gallons to produce a quarter-pound hamburger. It takes 1,800 gallons to make a pair of jeans." [4]

Our water usage adds up quickly. Even without drinking water, which is impossible to live, we cast a huge water footprint. It takes 31 gallons of water to make a glass of wine; 37 gallons of water to make a cup of coffee; 49 gallons of water for one bag of chips; 72 gallons of water to produce two chicken eggs; and a staggering 2,847 gallons of water to make a pound of chocolate.[5] Far more water is used to create a plastic bottle, than the amount of water it can hold. Bigger ticket items like a car might require upwards of 39,090 gallons of water for manufacturing; 518 gallons of water per tire.

Not only do we require fresh water to survive, but we use huge amounts of water to support our highly digital electronic world. Annually numerous billions of gallons of water are required to support electrical power generation plants. An estimated two gallons of water is evaporated by a typical power plant to just produce one kilowatt hour of energy for end-use electricity.[6] In addition to massive amounts of water needed to cool power plants, water is vital to generate the steam that spins turbines. Water is essential throughout the fuel cycle – extracting, processing and transporting fuel – as well as during the construction of power plants and all other energy devices.

The titanic growth in U.S. natural gas production using hydraulic fracturing is presenting a new stress on water supplies. Hydraulic fracturing can quadruple the amount of water consumed when extracting natural gas from the ground, compared to more conventional methods. "Fracking" requires millions of gallons of water per well, annually, which then gets removed or sometimes contaminates underground fresh water resources. 'Frac' flowback water is around three to five times saltier than seawater and far more difficult to purify.

Animal Factory Farms Use Vast Quantities of Water

An estimated seventy percent of human water use goes for agriculture and animal horticulture. Modern animal factory farming, wastes enormous quantities of water and is the leading source for water pollution in the United States. Our heavy consumption of meat, made popular by food fads like the "Paleo Diet," threatens our water supply. "Every 24 hours, the animals destined for America's dinner tables produce 20 billion pounds of waste. That is 250,000 pounds of excrement a second. The livestock of the United States produce twenty times more excrement than the entire human population of the country. Over half this staggering [manure] production comes from confinement operation from which it cannot be recycled . . . the result is their waste tends to end up in our water."[7] That is a lot of polluted and wasted water coupled to all that animal feces.

North Americans are headed for an ecological and economic disaster based on the inefficient use of water used for livestock. The Ogallala Aquifer system, the largest freshwater aquifer in North America is becoming rapidly depleted because of the 13 trillion gallons of water used, annually, primarily for meat production. The Ogallala aquifer which took millions of years to gather its enormous stores of water, covers an area of nearly 174,000 miles, and provides drinking water for 82 percent of people living in the Great Plains region. This vast underground water table extends below areas of South Dakota, Nebraska, Kansas, Oklahoma, Wyoming, Colorado, New Mexico, and Texas.[8] According to a recent (2013) four-year study out of Kansas State University, "the Ogallala Aquifer will see a 39 percent depletion of its available groundwater over the next five decades, which is in addition to the 30 percent reduction that the water supply has already recently faced."[9]

Simply put, 70 % of the ground water used to support some of America's most productive agricultural regions will simply no longer exist. Experts, say it could take 500 to 1,300 years to replenish the Ogallala aquifer, naturally.[10] The lands overlying the Ogallala aquifer are some of the most productive in the United States and are used for livestock, corn, wheat, and soybean production. Food costs will continue to rise dramatically in the future, unless strict ground water conservation measures are implemented, now.

"Reports by the General Accounting Office, the Rand Corporation, and the Water Resources Council have made it clear that irrigation water subsidies to livestock producers are economically counterproductive. Every dollar that state governments dole out to livestock producers, in the form of irrigation subsidies, actually costs tax payers over seven dollars in lost wages . . . most of the water goes to produce livestock, either directly or indirectly. Thus, current water use practices now threaten to undermine the economies of every state in the region."[11]

Cost of Meat Dependent upon Cheap Water

Meat prices are especially dependent upon cheap water sources. Over half the total amount of water consumed in the United States is used to irrigate crops that feed livestock.[12] If water and grain were not subsidized by tax payers, the cost of beef would likely rise astronomically. Daily consumption of meat might well become unaffordable to all but the wealthiest Americans. One professor at the University of California, Irvine estimated that beefsteak would cost nearly $100 dollars a pound, if government subsidies for water and grain, along with environmental costs were included.[13]

Based on calculations by David Pimentel, professor of ecology and agricultural science at Cornell University (2001) it takes about 12,009 gallons of water to produce a single pound of beef.[14] That is about fifty times more water to produce a single calorie from beef than a calorie from potatoes. In contrast, it takes only 25 gallons of water to produce one pound of wheat. Not only is meat water demanding, but extremely fuel intensive. "Since so much fossil fuel is needed to produce it, beef could be considered a petroleum product. With factory housing, irrigation, trucking, and refrigeration, as well as petrochemical fertilizer production requiring vast amounts of energy, approximately one gallon of gasoline goes into every pound of grain-fed beef."[15]

To conclude, if we consider world-wide increasing population growth, coupled with the spread of current energy industries, and the taste for animal agriculture we are headed for a fresh water shortage. People worry about rising fuel costs, yet most people are ignorant of the biggest challenge ahead; lack of clean, cheap, fresh water! At projected rates of water consumption, the majority of people on Earth will face severe water shortages within a generation or two if pollution and waste continues unabated.[16] At which time, the population living in water stressed areas may well reach nearly 4 billion people. Certainly desalination of ocean waters could provide one answer to help meet the world's future fresh water needs. Yet, desalination methods are energy intensive, thus expensive, and how to afford a vast network of water distribution for people living far from coastal areas?

Water Conservation: Earth's Precious Gift

Our immediate choices appear clear. First, fresh water must be considered one of Earth's most precious commodities and water management practices need to be a top priority across all sections of society. Since consuming a protein diet from eating animals uses over twenty times more water than a vegetarian diet, people need to be made aware of both the environmental impacts of meat consumption, as well as potential health risks. Legislatively, animal factory farms need to be required to use integrated wastewater treatment systems, such as microalgae with a goals of: (a) recycling waste water to better conserve water; (b) minimize water pollution run-off that is tainted by animal excrement; (c) use algae by-products for animal feed supplement, where safely possible, to further reduce land and irrigation requirements for growing crops for animals; and d) the actual costs (without water and grain government subsidizes) need to be added to animal farming economics. If assisting in recycling waste water were the only contribution microalgae made to global restoration that would be valuable enough.[17-18]

Microalgae aquaculture wastewater treatment system efficiencies vary, but have reached as high as 100% in their removal of phosphorus, nitrogen, and ammonia by the end of the treatment period. Additionally, applied aquatic systems have demonstrated a reduction of heavy metals in the range of between 52.3% and 100% in the batch system and 64.2% and 100% in continuous system designs.[19]

Most remarkably many valuable byproducts can be produced utilizing recycled waste water grown algae including energy applications such as:

- Biodiesel
- Ethanol
- Hydrogen
- Methane
- Hydrocarbons
- Gas Production
- Power generation

Nonfuel applications of wastewater grown algae include:

- Chemicals / and Fertilizers
- Animal Feed Supplements

Additionally, the recycled water could be used for irrigation purposes, watering landscapes.

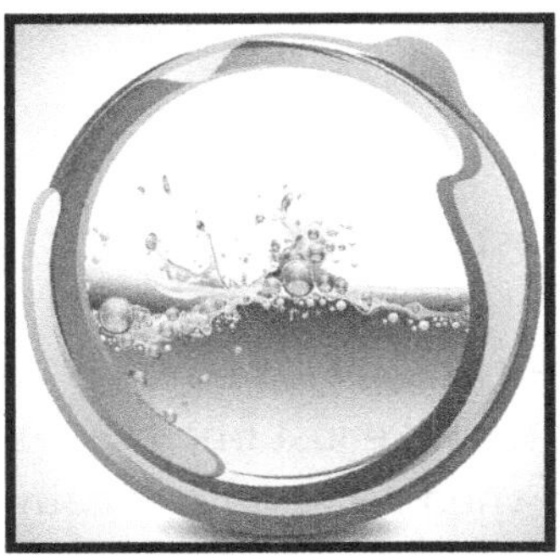

CHAPTER

21

Microalgae Chemical Know-how and Toxic Clean-Up

"If you look at the ecological circuitry of this planet, the ways in which materials like carbon or sulfur or phosphorous or nitrogen get cycled in ways that makes them available for our biology, the organisms that do the heavy lifting are bacteria. For every cycle of a biologically important element, bacteria are necessary; organisms like ourselves are optional."

— Knoll A, 2004

"Some things are clear. We must begin by understanding that there are real costs that we pay for pollution, even if they are not printed on each object's price tag."

— Needleman H, Landrigan P, 1994

Industrial and human waste steadily pollutes our world. The amount of toxic by-products discharged into our air, water, and landfill is staggering. Unfortunately, it appears that the task of clean-up is being left for future generations. Take just one example, "Today there are about 15,000 publicly owned wastewater treatment works in the United States, discharging approximately 26 billion gallons per day of treated wastewater into lakes, streams and waterways . . . [about] fourteen percent—approximately 28 million pounds per year—winds up in sewage sludge . . . Over 60,000 toxic substances and chemical compounds can be found in sewage sludge, and scientists are developing 700 to 1,000 new chemicals per year."[1]

Stagnant microalgae blooms are often blamed for fish kills, beach closures and other maladies that harm the Great Lakes and other key waterways. That's why algae blooms are often considered a threat to the Great Lakes and its tourism industry. This is not only an issue in the Great Lakes, but also in the Chesapeake Bay, the Everglades and other ecologically important bodies of water. Algae blooms are the result of warm temperatures, ample light and an abundant supply of such nutrients as nitrogen and phosphorus. The excess nutrients typically come from fertilizer, manure and sewage that rainwater washes into the watershed.

Bioremediation using Microalgae

Now researchers from the University of Buffalo, the University of Maryland, the Western Michigan University and other institutions are gathering microalgae from the Great Lakes and other bodies of water, to investigate the potential uses of this often maligned vast resource. One of the main goals is the use of microalgae for pollution recovery. "By using the algae beds to remove excess nutrients from the lake, we can improve water quality," says Blersch, research assistant professor in University of Buffalo's School of Engineering and Applied Sciences. "The other aspect is studying its properties; is it viable to turn algae into biofuels, fertilizer or other commercial products? . . . Turning Great Lakes algae into biofuels, fertilizer or other products may become a reality someday." For now however; Blersch is focused concentrating on the water quality around Buffalo New York and beyond."[2]

Bioremediation generally uses microbes (bacteria, fungi, yeast, and algae) to remove heavy metals and break down chemical toxins. New bioremediation techniques are emerging based on dramatic advances in biotechnology. Algae will be increasingly relied upon as environmental garbage collectors in the new millennium.[3-6] In fact; the oldest microalgae technological usage in North America is wastewater remediation. When microalgae are added to wastewater, they rapidly break down pollutants, absorb organic nutrients, and assist in water purification after the microalgae are removed.

Microalgae can Decontaminate Heavy Metals

A wide-range of human activities contribute to the discharge of heavy metals into the environment these include: use of intensive agriculture, pharmaceuticals, mining/metallurgy, electronic manufacturing, energy production, and sewage sludge. "Heavy metals are stable and persistent environmental contaminants since they cannot be degraded or destroyed. Therefore, their toxicity poses major environmental and health problems and requires a constant search for efficient, cost-effective technologies for detoxification of metal-contaminated sites,"[7] states W. Kaplan in his article. A growing number of publications advocating the use biological methods, like microalgae, for toxic metal removal and detoxification of environmental wastes have appeared since the 1990s. All of them support Kaplan's contention.

Heavy metals, with a hefty atomic density, exert toxic effects on living organisms, and are thus are termed "toxic heavy metals." Microalgae and other microorganisms play a significant role in the transformation of heavy metal ions in the environment. Microalgae especially have a wide spectrum of absorption (extracellular) and adsorption (intracellular) mechanisms for coping with heavy metal toxicity. There are several potential advantages in using prokaryotic and eukaryotic microalgae living cells or their dead cell biomass over traditional methods for removing toxic heavy metals. Some advantages with microalgae bioremediation can include reduced costs, improved specificity for targeting metals, and better absorption rates when wastewater metal concentrations are low. "In conclusion, many species of prokaryotic and eukaryotic microalgae, as well as their inactive cell biomass, can be used for bioremediation of metal-polluted sites."[8]

Bioremediation is the favored approach for processing biological wastes and may play an increasing role in concentrating metals and radioactive materials to reduce toxicity and to recover metals for reuse.[9-15] Microalgae and bacteria used to degrade organic chemicals can then be used in waste-site cleanup operations, or removing pollutants from agricultural wastewater, as well as a variety of other pollution management uses.[16-19]

Cleaning up Arsenic with Microalgae

Arsenates are salts or esters of arsenic acid. Arsenic is a naturally occurring metalloid element, commonly found in ground water. It is the 20th most abundant element in the earth's crust. Arsenic is widely used as a wood preservative and as a poison for killing weeds, rats, and insects. It is toxic to humans in fairly low doses and its poisonous effects are cumulative, over time. In drinking water, it is tasteless and odorless. "In the last few years, researchers have concluded that the health risks from arsenic are greater than originally thought. Studies indicate that the MCL for arsenic should be 3-5 ppb. The cost of achieving this level for all public water supplies would be exorbitant, so a 10 ppb MCL was adopted."[20]

A green microalga (*Chlorella vulgaris*) was shown to be "capable of consistently removing around 70% of the $As5^+$ [arsenate] present in growth media."[21] The scientists discovered that rather than just soaking up arsenic, the microalgae produced increased amounts of glutathione, which is a powerful detoxifying agent, which accounted for much of the detoxification effect. The scientists proposed that microalgae offered "an increasingly promising low cost arsenic phytoremediation method for developing countries." Microalgae bioremediation methods have already proven themselves to be cost-effective and beneficial additions to chemical and physical methods of managing waste and environmental pollutants.[22]

Microalgae Biotechnology to Enhance Algal Uses

Recently developed rapid-screening assays may help in the future to identify organisms, like strains of microalgae, which are capable of degrading specific wastes. Advanced biotechnology techniques can potentially be used to increase these particular algal metabolic pathways. Of course, a danger implicit in using biologically modified organisms for environmental clean-up is that their release into the environment may inadvertently upset the balance with other life forms. Yet, because microalgae readily grow inside isolated, controlled bodies of water, environmental factors could be created that are quite different and more easily controlled than those in the world at large. For example, a culture might be produced that only grows in water with a high level of cadmium. Conceivably, use of microalgae can avoid problems associated with higher plant sources. Because higher plants spread pollen over many miles, as part of their reproductive cycle, they can be extremely hard to control once released into the environment. But microalgae are less likely to present this particular problem. Finally, because microalgae rapidly regenerate, they offer a novel biosorbent material that can be repeatedly used.

Figure 28. Major Water Pollution Sources

Pharmaceutical factories, municipal sewers, hospitals, wastewater treatment plants, agricultural and horticultural and rural septic systems are major sources of water pollution[23]

According to a recent governmental report (2011) by the General Accountability Office, most drinking water in the U.S. is contaminated with pharmaceuticals; especially common pollutants are estrogen-based contraceptives, prescription drugs and antibiotics. The Environmental Protection Agency funded a 2010 analysis of 48 research publications and found that 54 active pharmaceutical ingredients and 10 metabolites have been detected in treated drinking water. Another (2008) study tested 19 drinking water treatment plants across the United States. The study found pharmaceuticals and metabolites at all of the locations tested. These water treatment plants provide drinking water to well over 28 million Americans.[24]

Microalgae can offer a new generation of green chemicals, bio plastics, pharmaceuticals, which are friendlier to the environment, and require no petrochemicals in their production. Moreover, bioactive algae compounds are more biodegradable than traditional petroleum based pharmaceuticals. Microalgae and microbes have a most remarkable potential to be used as "micro-factories," which can produce a wide range of useful chemicals. Fuzhong Zhang, Ph.D., assistant professor of Energy, Environmental & Chemical Engineering at Washington University in St. Louis, MO, says "My goal is to engineer microbes [including microalgae] and turn them into micro-factories that produce useful chemicals . . . Traditional chemical production requires high pressure and temperatures and lots of chemical solvents, but the microbial approach is very eco-friendly: Once the engineered cyanobacteria start to grow, all they need are water, basic salts and the CO2."[25]

Microalgae will eventually become part of our consumer packaging. Americans discard about 200,000 pounds of packaging daily – only about half of which is recycled. David Edwards, a Harvard bioengineer, proposes that someday we could eat more of our packaging. He has designed a two-layered, edible membrane that he calls the WikiCell. "It can be fashioned into attractive packaging that is capable of keeping food such as yogurt, cheese and juice fresh for up to six months. The layers function as a wrapper and box. A soft inner skin is made from tiny particles of food, mixed with either algae extract or chitosan (a biochemical polymer), and held together with calcium or magnesium ions."[26] Its all biodegradable and edible. WikiCell ice cream and frozen yogurt packages are expected to make their debut in some U.S. grocery stores soon.

CHAPTER

22

Future of Microalgae as a Sustainable Energy Source

"Low-level plants like the algae are a greater source of natural fuel energy than more elaborate, sophisticated plants. Coal, which is produced largely from ferns, is a better fuel than firewood. And petroleum, which is formed from living organisms, including unicellular, low-level algae...is superior to coal as fuel...the potential power of the lower-level plants is greater than that of higher-level plants; ... Probably the utilization of sea vegetables [algae] can have the same revolutionizing effect...that the use of petroleum has had on human industry and civilization."[1]

— *Arasaki S, Arasaki T, 1983*

Greenhouse Gases, Fossil Fuels and Microalgae

Since the 18th century, the start of the industrial revolution, fossil fuels in the form of coal, oil, and natural gas have powered transportation, manufacturing and supported advances in technology, which have shaped modern society. Yet expanded use of fossil fuels threatens our energy supply and puts enormous strains on the environment. The world's demand for energy is projected to double by 2050. Population growths, along with the rapid industrialization of developing countries will likely lead to restrictive shortages of oil and gas within our lifetimes. Ominously, the unrestrained use of fossil fuel puts our health at risk because of the pollution it creates. Carbon dioxide and other greenhouse gas emissions linked to fossil fuel use contribute to climate change.

Figure 29. Rising Atmospheric CO_2 Concentrations

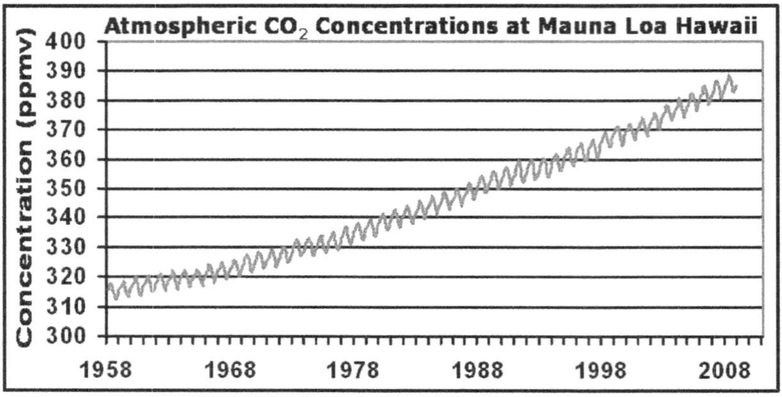

The Mauna Loa Hawaii data constitute the longest available in the world. Carbon Dioxide Research Group, Scripps Institute of Oceanography

According to the University of Delaware's College of Marine and Earth Studies, "There is good evidence that the level of atmospheric carbon dioxide (CO_2) today is more than 25% higher than at any time in the past 420,000 years. The recent increase in atmospheric CO_2 has occurred since the beginning of the "industrial" era (defined as since 1750) and most of that increase has occurred over the past 50 years. The increase in atmospheric CO_2 is primarily from burning of fossil fuels (land-use changes and cement manufacturing also contribute) with half of this increase having occurred since the mid-1970s."[2]

The most important difference between fuels derived from fossilized-forms of sunlight (oil reserves are the ancient remains of organic matter) and those created from current sunlight (e.g., algae) is that one is limited and the other is renewable. Microalgae are by far one of the most profuse living biochemical converters, photo-synthetically changing light energy and carbon dioxide (CO_2) into biomass such as carbohydrates, proteins and lipids. During microalgae blooms, cell densities increase exponentially. Yet, if light is plentiful, but nutrients are not, microalgae may accumulate large amounts of lipid as a survival strategy, followed by large numbers of cell deaths. Then the micro-algal biomass is further degraded by teeming microorganisms, which also consume large amounts of oxygen. In extreme cases, massive algal blooms and associated decomposition, leads to the absence of aquatic oxygen levels. Large bodies of water become deficient of oxygen, with a massive deposit of algal cells and microorganisms. Remarkably, "this process is also believed to have been the key factor for large-scale oceanic anoxic events that led to fossil mineral oil deposition."[3]

Moreover, when fossil fuels are burned they release CO2 that has been stored for millions of years. "Any fuels made from fossil fuels can be made from algae because crude oil, coal and shale are simply fossilized algae. Algae biofuel growers are attempting to produce gasoline, ethanol, green diesel and jet fuel in a matter of weeks rather than the 400 million years required by nature."[4]

The Aquatic Species Program (ASP): To Develop Biodiesel from Algae

In 1978, President Jimmy Carter established the Aquatic Species Program (ASP), a research initiative charged with developing biodiesel from algae as a clean, homegrown alternative to gasoline. "Yet some two decades and $25 million later, the team had failed to produce any significant amount of oil from algae, and in 1996 the Clinton administration axed the program. Still, the researchers hoped their work would not go to waste. "The directors were adamant that we make available a detailed summary of what we'd done, because they knew that in the future someone would be interested," says John Sheehan, a former ASP project scientist. Sheehan and his colleagues compiled a 328-page report on their work and uploaded it to a Department of Energy Web site."[5]

Since 2007, the USDA allocated a budget of 1.6 billion dollars for research into biofuels. The Federal law of 2007 on security and energy independence set ambitious targets from a production volume of biofuels at 9 billion gallons for 2008, to a goal of 36 billion gallons of biofuels by 2022. Of these 36 billion gallons, 21 billion will come from "advance" biofuels, like microalgae, while 16 billion are targeted to come from cellulosic biomass. Yet subsidies for biofuels like corn in effect act like a tax upon basic foods, which competes against rising commodity prices. For the average American more money is being spent at the grocery stores; while the poorest people in developing countries go hungry, unable to afford adequate amounts of grains. Alarmed, the World Trade Organization (WTO) has encouraged new classifications of "sustainable bioenergy," like microalgae, and is seeking lower tariffs on biofuels like corn, which raise the cost of basic grains for everyone.[6]

Algae-derived biofuels help to recycle current CO_2, adding far less greenhouse gases to the atmosphere. However, there are added environmental and social costs to most traditional biofuels. For example, most biofuels require land, water, fertilizers, and even pesticides, as well as processing and transportation costs. "The United States has developed two main branches of the 1st generation biofuels to reduce their dependence of conventional oil. Bioethanol is produced from sugar and grains (e.g., maize, wheat, and beet), biodiesel from oil seeds (e.g., rapeseed oil and soybean oil). Both channels are now faced with constraints of availability of raw materials and competition of food production; as well as the problems of deforestation, abuse of biocides and fertilizers."[7-8]

Producing biofuels from microalgae is estimated to be hundreds of times more efficient than using traditional irrigated crops to produce biofuels (Solix Biofuels). Furthermore, a wider range of suitable climates are available to grow microalgae biofuel, especially closer to the equator, where many of the poorest people live. The great wealth gap between the richer Northern countries and poorer Southern countries might be reduced by shifting towards a microalgae-based energy economy.

One of Earth's Most Productive Life-forms: Microalgae

Given the right conditions, microalgae can practically double its volume overnight. Microalgae are the earth's most productive plants, 10 to 15 times more productive in biomass than the fastest growing land plants. Oil from microalgae can be converted to biofuels such as biodiesel using a similar technology that is currently being used to convert vegetable oil into biodiesel. In addition, it is possible to use other technologies to convert algae byproducts directly into hydrogen and butane fuels, jet and marine fuels, gasoline, ethanol, methane, biodiesel or even run electric power generators. "In the past five years, more than a billion dollars has flowed into algae biofuel development."[9]

Microalgae can yield far more gallons of biodiesel per acre than any other botanical source, if technological and commercial applications made the process cost effective. For example, while soy might produce some 50 gallons of oil per acre per year; canola can produce 150 gallons; and palm can produce 650 gallons; certain microalgae could yield 5,000 or more gallons per acre per year. Higher percentages of oil are available in biomass of algae than most any other botanical sources, as well.

According to Robert Henrikson (2013) microalgae can be grown on marginal farm lands using recycled water sources or wastewater from sewage systems to grow fuel byproducts.[10] The waste water could be reused for irrigation purposes, cleaner as a result of growing the algae, which could absorb the extra nitrates and some of the heavy metals or break down pesticides.

While a special oil-rich microalga, *Botryococcus braunii,* occasionally contains up to 85% hydrocarbons of its dry weight under natural conditions, considerable basic research will be needed to reliably achieve these levels in mass-production.[11] As yet there has not been a suitable method developed for microalgae hydrocarbon production on a commercial-scale.[12] However, with increased biotechnology and understanding of the genetic regulation and environmental parameters that control hydrocarbon-producing capacities, the day may come when the commercial production of microalgae hydrocarbons becomes a reality.

"The exploitation of photosynthetic microorganisms affords the only feasible approach to bio production of hydrocarbon oils. Photosynthetic microalgae: (a) use sunlight as energy for biochemical synthesis from inorganic compounds; (b) can be intensively cultivated in bodies of water generally considered unusable for domestic purposes; and (c) accumulate trace metals and synthesize proteins, carbohydrates and vitamins thereby making excellent livestock feed or fertilizers. Particularly important is the fact that the cultivation of microalgae for the purpose of producing fermentation products is unlike other biomass programs in that all end-products increase our biomass rather than decrease it. Such rewards make the search for the hydrocarbon producing algae a most exciting one."[13]

Bio hydrogen, Bioethanol, Biodiesel and Biogas

Biofuel-bioenergy production has gained serious interest due to growing concerns regarding limited fossil-fuel reserves and their impact on atmospheric CO2 levels. Due to the many harmful environment impacts and limited reserves, petroleum-derived fuels need to be replaced by renewable biofuels. "Several aspects of cyanobacteria and microalgae such as oxygenic photosynthesis, high per-acre productivity, nonfood-based feedstock, growth on nonproductive and nonarable land, utilization of wide variety of water sources (fresh, brackish, seawater and wastewater) and production of valuable co-products along with biofuels have combined to capture the interest of researchers and entrepreneurs. Currently, worldwide biofuels mainly in focus include bio hydrogen, bioethanol, biodiesel and biogas."[14] The main challenge of biofuel development is making microalgae biofuel-bioenergy production economically practical on a large enough scale. The basic science of converting bioenergy sources out of microalgae, such as oil or hydrogen extraction, has already been proven.

"Recently several methods have been developed to use *Chlorella* as a source of energy. One of these (developed in Canada) grows

Chlorella in conjunction with a yeast or bacterium which feeds on the Chlorella. The yeast then develops fat cells which contain hydrocarbon oil almost identical to crude oil. Another method produces algae which is almost 50% oil (by dry weight). This oil is then extracted to form fine grade crude from which gasoline as well as all the other products derived from petrochemicals can be produced. (Petrochemicals are the source of all plastics, many dyes, paints, pharmaceuticals, etc.)" [15]

A Microalgae Hydrogen Economy

Many energy experts predict that hydrogen gas might be the world's best alternative to fossil fuels. A hydrogen-based economy has been promoted to power our future. Hydrogen through a reaction with oxygen can release explosive energy in heat engines or silently within fuel cells; a clean fuel source, producing water as its only byproduct. Hydrogen is so clean burning that the liquid which comes out of a car's exhaust pipe could conceivably be consumed, as pure water. Hydrogen could be locally generated and distributed worldwide without regard for national borders. Creating a hydrogen economy — a clean energy system based on hydrogen gas and hydrogen-generated electricity— is an attractive replacement for fossil fuels. Yet there is one huge problem; hydrogen does not occur in nature as the fuel H_2.

"At present, most of the world's hydrogen is produced from natural gas by a process called steam reforming. However, producing hydrogen from fossil fuels would rob the hydrogen economy of much of its raison d'être: Steam reforming does not reduce the use of fossil fuels but rather shifts them from end use to an earlier production step; and it still releases carbon to the environment in the form of CO_2. Thus, to achieve the benefits of the hydrogen economy, we must ultimately produce hydrogen from non–fossil resources, such as water, using a renewable energy."[16]

Remarkably, microalgae can be grown to produce hydrogen. In 1939 a German researcher named Hans Gaffron, while working at the University of Chicago, discovered that a green microalgae

(*Chlamydomonas reinhardtii*) would sometimes switch from the production of oxygen to the production of hydrogen. However, Gaffron never discovered how the microalgae made this switch possible. In the 1990s UC Berkeley professor, Dr. Melis and his team discovered a "metabolic switch that triggers algae to turn sunlight into large quantities of hydrogen gas, a valuable fuel." "I guess it's the equivalent of striking oil,'" quipped the plant and microbial biology professor.[17]

Professor Melis and his postdoctoral associate Liping Zhang of UC Berkeley made the discovery, funded by the U.S. Department of Energy (DOE) Hydrogen Program - with Dr. Michael Seibert, Dr. Maria Ghirardi and postdoctoral associate Marc Forestier of the National Renewable Energy Laboratory (NREL) in Golden, Colorado. The two institutional teams were awarded a patent on this new technique. To switch microalgae production from oxygen to hydrogen is simple. It depends on removing sulfur from the growth medium. To survive the microalgae are forced to activate an alternative metabolic pathway which generates hydrogen, instead of oxygen. Hydrogen production appears to be universal in many types of algae, as a sort of "ancient strategy that the organism developed to live in sulfur-poor anaerobic conditions." According to Dr. Melis,"In the future, both small-scale industrial and commercial operations and larger utility photobioreactor complexes can be envisioned using this process." For example, " it is conceivable that a single, small commercial pond could produce enough hydrogen gas to meet the weekly fuel needs of a dozen or so automobiles," Melis said.

Microalgal CO2 Absorbing Lighting

French biochemist Pierre Calleja invented a CO2 absorbing microalgal lamp. The microalgae street lamp is a tube-shaped tank filled with green algae. The first installed lamp is estimated to capture one ton of CO2 a year in an underground parking lot. As 25% of carbon emissions come from car exhaust, placing these green street lamps on roads and parking lots could have an enormous impact on CO2 emissions. Reportedly the lamp runs on a battery that charges during the day by "microalgal photosynthesis." These lamps could be used everywhere in homes, streets, and in factories to absorb CO2. Recently, Pierre Callja (2014) received a 12 million euro grant to carry out his project.[18-19] Each algae street lamp is capable of collecting 150-200 times more CO2 in a year than tree. Imagine world-wide if all street lamps were using microalgae!

Diverse species of cyanobacteria all donate electrons to their surroundings. John Pisciotta and his team at the University of Maryland, Center for Biomedical Engineering and Technology (2010), have proven that cyanobacteria possess light-dependent electrogenic activity (i.e. emit electrons into their environment in response to illumination). These researchers state, that "The electrogenic pathway of cyanobacteria might be exploited to develop light-sensitive devices or future technologies that convert solar energy into limited amounts of electricity in a self-sustainable, CO2-free manner."[20]

Collaborating on these findings, Stanford engineers actually generated electrical currents by tapping into the electron activity of individual algae cells. Photosynthesis excites electrons, which can then be turned into an electrical current using a specially designed gold electrode. This study could be the first step toward carbon-free electricity directly from living chloroplasts.[21]

The United States Navy Green Fleet

The "Great Green Fleet" Demo, took place on June, 27, 2012 featuring a carrier strike group of ships and 71 aircraft onboard the deck of the Nimitz, including fighter jets, helicopters and transports, all running on a 50-50 mix of petroleum and biofuel derived from algae and used cooking oil. The supersonic F/A-18 jets launched from the deck of an aircraft carrier were powered by jet fuels fermented from algae. The Australian Navy and the U.S. Navy have signed a statement of cooperation pledging the two nations' navies to collaborate on biofuels research and deployment. The United States Navy aims to deploy a permanent Green, Biofuel, strike force in 2016.

When Mabus took over as Secretary of the Navy, he declared that the service would get half of its energy from sources other than oil by

2020. The U.S. military reportedly ordered 450,000 gallons of biofuel, half of which is algae-based. Secretary Mabus proclaimed, "In history, the Navy has always led in changing fuel types. We went from sail to coal in the 1850s. We went from coal to oil in the early part of the 20th century, and we pioneered nuclear in the 1950s. And we're going to lead once again by establishing a market, by helping establish a market for biofuels now."[22]

Recently, President Barack Obama referred to the energy potential of algae in an energy speech, which drew quick, initial scorn from Republican rival, Newt Gingrich, "who called it "weird." After further consulting with his scientific and political advisors, however, he admitted that algal biofuel is "a terrific concept." President Obama at a University of Miami address, announced 14 million dollars in grants to develop biofuel from algae. "We're making new investments in the development of gasoline and diesel and jet fuel that's actually made from a plant-like substance — algae". Obama continued,"You've got a bunch of algae out here, right? If we can figure out how to make energy out of that, we'll be doing all right."[23] It is becoming increasingly clear even to our politicians that there just isn't enough agricultural land in the U.S. to gain self-sufficiency with bio fuels using conventional crops, like corn or soybeans. However, all of our energy biofuel needs could be attained using micro-organisms and algae-derived methods, upon a total land-mass about the size of Connecticut.

Meanwhile, NASA's space agency scientists are experimenting with making fuel from microalgae grown with sewage 'waste water' in massive, floating plastic bags. "NASA's $10 million, two-year project has at least shown that it's possible for floating plastic bags full of algae and wastewater to yield as much as 2,000 gallons of biofuel per acre annually under favorable conditions. This solves a few problems associated with land-based bioreactors, lowers costs, and takes advantage of open waters to grow the fuel . . . The next step, says NASA on its website, is for other organizations to deploy larger-scale systems offshore in protected bays to determine if [Offshore Membrane Enclosures for Growing Algae] OMEGA can be used commercially for biofuels production, environmental remediation, wastewater treatment or carbon sequestration."[24]

U.S. Coal Energy Plants Release nearly 800 Million Tons of Carbon Dioxide, Every Year

A typical (500 megawatt) coal plant burns 1.4 million tons of coal each year. As of 2012, there were 572 operational coal plants in the U.S., collectively releasing nearly 800 million tons of carbon dioxide into our atmosphere, annually. Carbon Dioxide (CO_2) is the principle heat-trapping gas generating climatic changes. If the United States continues to burn coal the same way, without carbon recapture systems, dangerous increases in global warming is almost a certainty. Several approaches can both increase the productivity and reduce the costs of growing microalgae. For example, carbon dioxide pollution can be sequestered from industrial power generation plants offering cheap carbon sources. Also wastewater from sewage treatment plants or factory farms can offer cheap streams of nutrients, all of which can be used to create microalgae by-products for other uses like fertilizers or animal feed. Green Fuel Technologies, based in Massachusetts, is field testing algae capture technology at coal-fired plants. Industry experts say that it takes about two million tons of algae to capture about one million tons of carbon dioxide. The greening of coal-burning utilities requires no re-engineering of the actual power plant. Since about 40 percent of electricity in the U.S. comes from coal, finding ways to capture carbon dioxide, like feeding microalgae, from coal burning plants is essential.

A waste treatment pilot project in Holland, Michigan in collaboration with Michigan State University is currently underway. Holland offers a location ideal for growing algae in the "open" as it is purposely located near a water treatment plant in order to reuse the carbon dioxide released for the growth of algae. Petrosun Company is working on projects to build plants in Arizona and in Texas. Their facility, located in Rio Hondo Texas, would produce an estimated 4.4 million gallons of algal oil and 110 million lbs. of biomass per year off a series of saltwater ponds spanning 1,100 acres. Twenty of those acres will be reserved for the experimental production of a renewable JP8 jet-fuel. The Federal Aviation Administration, NASA, the U.S. Air Force, and several commercial airlines are all on board with the idea of using algae-based jet fuels. Powerful corporate and national interests are getting behind algae biofuel.

The "Pentagon says it is just months away from producing a jet fuel from algae for the same cost as its fossil-fuel equivalent. The claim, which comes from the Defense Advanced Research Projects Agency (DARPA), has taken industry insiders by surprise. DARPA's research projects have already extracted oil from algal ponds at a cost of $2 per gallon. It is now on track to begin large-scale refining of that oil into jet fuel, at a cost of less than $3 a gallon, according to Barbara McQuiston, special assistant for energy at DARPA."[25] For DARPA, the support for algae fuel is part of a broader mission for the U.S. military to obtain half of its fuel from renewable energy sources before the end of the decade.

Airlines enter the Microalgae Biofuel Age

Even United Airlines officially entered the biofuel age (Nov. 2011) by launching its first U.S. commercial flight from Houston to Chicago powered by a 40% algae-derived jet fuel blend. A fermentation process, similar to those developed by brew masters, is used to create algae-jet fuel blends. According to Harrions Dillion, president of Solazyme, "We put [algae] into a big steel fermentation tank, similar to what you would use to make beer. We put in plant sugars and then we infuse the algae, and the alga converts the sugar into crude oil in about two days. And once we have that oil, we can then refine it to diesel or jet fuel, any type of fuel we want to make, at a regular oil refinery."

Dillion says sugar plus a small vial of algae (the size of one's thumb) and a little time can help fly aircraft. "We put it [algae] into the container, and it divides exponentially until you have enough algae, and then it converts all of the sugar into oil. So the vial that you start is just a seed, and then each vial can be grown to an indefinitely large size, depending on how big your tank is." Then the algae oil is put through a refinery, pipelined, transported, and then used as fuel to power existing jet engines.[26]

International industry association Air Transport Action Group (ATAG), based in Switzerland, says alternative fuels have been identified as candidates for helping to achieve the industry's target of reducing net aviation carbon emissions by 50% in 2050 from the 2005 baseline, and limiting its exposure to volatile oil and jet kerosene prices worldwide. United Airlines, Continental, Virgin Atlantic, Aero Mexico, Air New Zealand, and Air China, are just a few airlines all working to fly greener in the air. An estimated 85,000 commercial aircraft take off and land every day, contributing 3 to 4 percent of all greenhouse gas emissions. In 2006 Richard Branson, founder of Virgin Atlantic promised to invest all of his profits from his airline and train transportation empire, up to 3 billion dollars, into a new company called Virgin Fuels, to develop biofuels. Branson claims after wrongly investing money into corn ethanol, his focus is now on micro-organism produced isobutanol and algae-derived biofuels.[27]

Multiple start-up microalgae energy companies were founded in the United States over the past decade(s), including OriginOil, Aurora BioFuels, and Algae BioFuels, to name a few. While even a few big oil companies like Exxon Mobil Corp., sense the turning tide of biofuels and have invested millions of dollars into algae biofuels research; though further expansion of production scales are required before algae fuels are commercially viable.

Biodiesel from Microalgae to Power Cars & Trucks

Creating biodiesel from microalgae especially offers significant benefits. Consumption of diesel fuel in the U.S. for on-highway use is estimated at around 40 billion gallons per year. While diesel fuel is chiefly used by trucks in the United States, interest in cleaner air and higher mileage efficiency diesel cars is growing. Currently, the diesel fuel market represents nearly a quarter of sales in the total U.S. transportation fuel market. Biodiesel can be run in any modern diesel engine which is designed to use ultra-low sulfur diesel, the new diesel fuel standard for the United States that went into effect in 2006.

"Producing biodiesel from algae has been touted as the most efficient way to make biodiesel fuel. The advantage being that the land requirement for growing the biodiesel is very small. Independent studies have demonstrated that algae is capable of producing 30 times more oil per acre than the current crops now utilized for the production of biofuels. Algae biofuel contains no sulfur, is nontoxic and highly biodegradable. Some species of algae are ideally suited to biodiesel production due to their high oil content, in excess of 50%, and extremely rapid growth rates."[28]

Feasibility Studies on Microalgae to Fuel America

An analysis (2013) shows that land and water resources in the United States could likely support the growth of enough microalgae to produce up to 25 billion gallons of algae based fuel a year, one-twelfth of the country's yearly needs.[29]

The research team looked in-depth at how much water and land resources might be needed to grow vast quantities of microalgae in the United States for biofuels.

First author Erik Venteris of the Department of Energy's Pacific Northwest National Laboratory said that, "While there are many details still to be worked out, we don't see water issues as a deal breaker for the development of an algae biofuels industry in many areas of the country." The Gulf Coast was identified as an especially favorable site because of a combination of warm temperatures, low evaporation, and access to an abundance of water with plenty of fuel-processing facilities. To produce 25 billion gallons of algae oil – equal to the 2022 advanced biofuels goal set out by the Energy Independence and Security Act - the team estimated based on current technology, about one-quarter of the water now used each year in the United States for agriculture would be needed. The amount of land required would be roughly the size of South Carolina.

The team based its analysis on the assumption that no more than 5 percent of a given watershed's mean annual water flow could be used in microalgae production. This projected figure is the same percentage allowance used by the U.S. Environmental Protection

Agency, for power plants to use for cooling. Mark Wigmosta, hydrologist, and leader of the team reported that, "I'm confident that algal biofuels can be part of the solution to our energy needs, but algal biofuels certainly aren't the whole solution."[30]

Currently, the cost of making algae biofuels exceeds the extraction cost of traditional gasoline-based products. However, algae biofuels are gaining attention from major energy companies like Exxon-Mobil who launched a $600 million research effort four years ago. Securing rights to all the necessary resources for algal biofuel production would generate a sustainable, long-term, profitable return. Yet, advances in microalgae fuel production could lead to de-centralized, local bioreactor sites in homes or urban centers.

Another 2013 published study on algae-based fuels in the peer review journal, Bioresource Technology, found that "Algae-derived biofuel can reduce life-cycle CO_2 emissions by 50 to 70 percent compared to petroleum fuels, and is approaching a similar energy return on investment (EROI) as conventional petroleum." Mary Rosenthal, ABO's executive director states, "This study affirms that algae-based fuels provide results without compromise." "With significant emissions reductions, a positive energy balance, nutrient recycling and CO_2 reuse, algae-based fuels will be a long-term, sustainable source of fuels for our nation."[31]

The authors of this study examined field data from two facilities operated by Sapphire Energy in Las Cruces and Columbus, New Mexico, which grow and process algae into green crude oil. Sapphire Energy's green crude can be refined into drop-in fuels such as gasoline, diesel and jet fuel. Currently algae-based fuels have an environmental profile that is comparable to existing conventional biofuels. Yet, likely improvements in the microalgae biofuel industry means surpassing established biofuels such as cellulosic ethanol in terms of both energy returns and reduction in greenhouse gas emissions.

CHAPTER

23

Microalgae, Carbon Fixation and Climate Change

"Algae fix around half of the world's carbon so it is vital that we can understand what affects their growth and wellbeing. Research into fundamental relationships and microscopic bacteria [including microalgae] may not seem important at first but it is only by improving our understanding at this level that we can discover how to maintain the health of ecosystems at a global level."[2]

— Goodfellow J, 2006

"The evolution of O^2 producing cyanobacteria that use water as terminal reductant transformed Earth's atmosphere to one suitable for the evolution of aerobic metabolism and complex life. The innovation of water oxidation freed photosynthesis to invade new environments and visibly changed the face of the Earth."[3]

— Dismukes G, Klimov V, et al., 2001

Microalgae is Essential to the Global Carbon Cycle

"Life in the Earth's biosphere depends on the balance of gases in the atmosphere . . . with two-thirds of the world's biomass, algae has an enormous effect on the biosphere."[1] *— Henrickson R, 1989*

Because algae were the first organisms to extract oxygen from water, they undoubtedly played a central role in the oxygenation of the air. "Photosynthetic carbon fixation by phytoplankton [microalgae] is a key component of the global carbon cycle."[4] Algae and higher plants produce sugar and oxygen photo synthetically through a process of carbon-dioxide assimilation. Acre for acre, algae is the most efficient transformers of CO2 and producers of oxygen — much more efficient than trees or other life forms.

"Cyanobacteria account for 20–30% of Earth's photosynthetic productivity and convert solar energy into biomass-stored chemical energy at the rate of approximately 450 terawatts . . . Solar energy reaches the Earth at the rate of the 178,000 terawatts (TW) of which 0.2% to 0.3% is harnessed by cyanobacteria."[5] For comparison sake, all fuel energy use world-wide was estimated at 15 terawatts in 2008.[6] "The amount of energy that passes through cyanobacteria exceeds by more than 25 times the energy demand of human society (15 TW); roughly 1,000 times the energy produced by all the nuclear plants on Earth."[7]

Geoengineering to Increase Oceanic Carbon Uptake

Microalgae's photosynthetic efficiency has led some innovative thinkers to suggest that our current dangerous rise in CO_2, which contributes to the escalating greenhouse effect, might be reversed by creating numerous algae ponds throughout the world, especially in poor soil and hot climate conditions. Furthermore, during glacial periods, low atmospheric carbon dioxide concentrations have been associated with increased oceanic carbon uptake, particularly in the southern oceans.[8]

Phytoplankton or diatom microalgae can draw huge amounts of atmospheric CO2 and transport the carbon molecules to the ocean depths. In the Arctic waters near the poles, the factor limiting phytoplankton algal blooms appears to be a lack of trace minerals, such as iron. It has been proposed that by adding ferric iron to the oceans or stirring up nutrients from the bottom, it might be possible to stimulate enormous ocean microalgae blooms and significantly reduce atmospheric CO2 levels.[9-10]

Climos is one such company working on that idea. According to Dan Whaley, their chief executive, iron readily available as a byproduct of steel making, could be dispersed into strategic areas of the ocean, for a little more than the shipping costs. Iron ore dust is a very common and available material. Australia sells hundreds of millions of tons to China, to make steel. "This technique has already been attempted on a small scale. Last year, an American entrepreneur named Russ George attempted to create an algae bloom that would theoretically suck CO2 out of the atmosphere by dumping iron filings into the Pacific Ocean. Much like fertilizer for plants, the filings provided vital nutrients to the algae and led to the formation of a large bloom."[11]

Even the CIA will partially fund a $630,000 study by the National Academy of Sciences to investigate how techniques using 'geoengineering,' such as seeding the oceans with iron to combat global climate change. "As for why the CIA is involved with projects attempting to alter the Earth's climate, a spokesperson told Mother Jones Magazine, "It's natural that on a subject like climate change the Agency would work with scientists to better understand the phenomenon and its implications on national security."[12]

Another curious ally for the idea is the Haida native people of British Columbia. Due to their concerns over diminishing salmon runs, the tribal council has decided to assist with funding and provide a Haida crew on the ship to support the ocean seeding research project. In defense of the value of the project, late marine biologist, John Martin described how the Galapagos Islands are a massive source of iron, in one of the most iron-depleted oceans, of the southeast Pacific. Here scientists have witnessed massive iron stimulated phytoplankton blooms that go for hundreds of miles off the west coast of the Galapagos Islands, due to iron dust run-off.

Additionally, "in 2008 when 450 million sockeye salmon left the Fraser River, the expectation was that fewer than one million would return. More and more baby salmon go to sea and fewer and fewer adult salmon return. But in August 2008 a volcano dusted ash and the northeastern Pacific Ocean turned into a massive plankton bloom. The plankton bloom was of larger proportion than what we did in the area. Instead of one million, some 40 million fish came home. That offered some hope."[13] Such observations helped to convince the Haida Salmon Restoration Corp. of the potential merits of seeding the ocean with iron to stimulate phytoplankton blooms; as their culture is intimately linked to the fate of salmon.

Another approach would be to support environmental factors to increase the amounts of B vitamin and nitrogen producers, bacteria and cyanobacteria. Low oceanic B12 and nitrogen levels are believed to be primary restricting factors that limit the growth of marine phytoplankton.[14] Altogether cyanobacteria and phytoplankton are prime drives for capturing CO2 and releasing oxygen into the Earth's atmosphere. Only one phytoplankton species, a brown unicellular diatom (*Bacillariophycease*), "is predicted to be comparable to that of all terrestrial rain forests combined," and yet they represent only one third of all marine carbon capture![15] Conversely, anything that adversely affects the health of marine cyanobacteria and their supportive role for phytoplankton would be a major detriment for our biosphere.

Much controversy exists regarding the effectiveness or even the wisdom of trying geo-engineering (planetary medicine) approaches. First, no dependable models exist to predict the influence of microalgae oceanic uptake of atmospheric carbon dioxide (CO_2) as related to global warming. Secondly, initial small pilot projects have yielded mixed results.[16-17] Thirdly, geo-engineering efforts might bring unforeseen dangers to the ocean environment. Still, a few 'geoengineering' pioneers, Russ George and Dan Whaley, who advocate this approach, say the main risk is to do nothing and even larger scale experiments are urgently required. They propose that one side effect of increasing microalgae and phytoplankton blooms would likely be an increase in the whale & fish populations that would feed off phytoplankton. Their ultimate goal is to protect and save the lives of hundreds of millions of people from climate-related disasters.

Conclusion

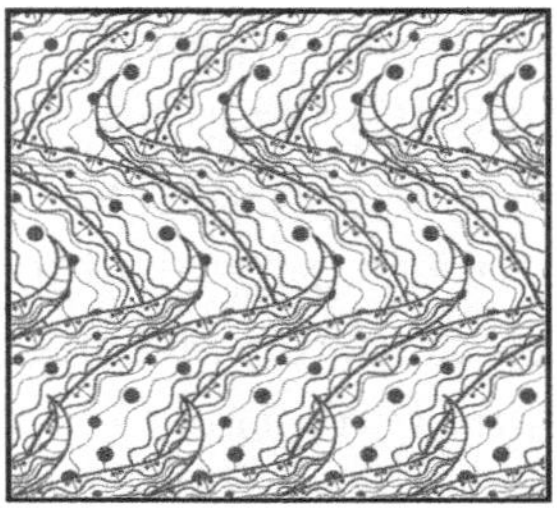

"In blue-green algae...we find three and one-half billion years of life on this planet encoded in their nucleic acids (DNA). At the same time, all microalgae supply that fresh burst of primal essence that manifested when life was in its birthing stages. At a moment in history when the survival of the human species is in jeopardy, many people have begun instinctively to turn to these original life forms for nutritional support."[1]

— *Pitchford P, 1993*

"Algae are the most plentiful plant on earth and produce roughly 40% of the total biomass daily. Most of the algae biomass is eaten because 100 times more organisms consume algae than any other food source. Algae also produce about 70% of the world's oxygen daily, substantially more than all the forests and fields combined."[2]

— *Edwards M, 2012*

Constituting a hefty share of the world's biomass, algae have an enormous effect on the biosphere. It is humbling to consider that many of the challenges humanity is currently struggling with—trying to develop sustainable sources of food and energy, restoring the environment, as well as protecting ourselves from harmful toxins, infections, and cancer—are ones the lowest species on Earth have long since resolved. Microalgae represent a vast, relatively untapped biological resource, an evolutionary genetic encyclopedia, which are readily available. Humans need to take greater stewardship in caring for our precious home planet, Earth. We need to learn life's most ancient lessons of sustainable thriving. We have only begun to uncover the many gifts of microalgae and to learn how we may best align with these extraordinary partners to help restore our health as well as our planetary home.

During the course of hundreds of millions of years, single-celled organisms developed strategies to contend with a hostile environment. Ultraviolet radiation, heat, cold, drought, and attacks from other bacteria or viruses threatened the survival of microalgae lineages. To survive a host of environmental stresses, microalgae developed heat shock proteins, potent antioxidants and other adaptive strategies. Restricted by a limited, mostly passive, range of movement, primitive cyanobacteria evolved sophisticated chemical defenses. Some microalgae species, like mushrooms, even developed deadly toxins. Many of microalgae's remarkable, beneficial, molecular solutions have been widely conserved across many species, including humans.

Twelve years ago, when I created my first microalgae research monograph, I made several predictions worth revisiting. At that time, I wrote "A critical mass of research on microalgae's benefits will tip the scale and draw worldwide scientific and public interest . . . once microalgae's health benefits are well documented, they will be sought. This industry will shift from its current reliance on anecdotal "testimonials" to the use of scientific research to capture the public's interest and demand."[4]

Indeed, according to Robert Henrikson (2011), "In over 40 countries, tablets powder and capsules are widely used supplements. *Spirulina* [the most researched and marketed microalgae] is a featured ingredient in pasta, cookies, snack bars and juice drinks, and in personal care products like skin creams and shampoos . . . Today there may be 1000 tons or more sold in the USA market [more than double the amount of blue green algae sold world-wide only two decades ago] . . . algae are ingredients in thousands of products for food, feed, colors, nutraceuticals, medicinals, cosmetics and personal care, bio-fertilizers and fine chemicals. Even more innovative algae based products are coming."[4] Green super foods have grown tremendously in popularity over the last two decades.

Furthermore, algae biotechnology has matured into a global industry with growing numbers of entrepreneurs beginning to recognize microalgae's remarkable biochemical diversity for a wide array of uses. While hundreds of thousands of microalgae species exist, only a few thousand strains are kept in collections. Fewer microalgae strains are investigated for bio-active molecular content and less than a dozen species are widely cultivated for industrial or nutritive uses. Needless to say, continued investigation and in-depth screening of microalgae is required, to better understand the many uses of algal physiology, biochemistry and genetics.

The capacity of microalgae to produce multiple products, such as oils, proteins and antioxidants, has led to the development of a bio refinery concept for processing microalgae. Akin to the petrochemical industry, where crude oil is processed to yield petroleum and a range of chemicals, microalgae can be employed to yield numerous consumer products. Different industries are able to use various algal products. For instance, the pharmaceutical and nutraceutical industries use high value algal products such as ω-3 fatty acids and carotenoids. The transport industry can use algae produced biofuels. The chemical industry can use products such as glycerin. The cosmetics industry can use environmentally friendly algal sun screen products. The bulk algae biomass can be used for human nutrition or for horticulture as animal feed. The environmental restoration industry can use algae as a remarkably toxic absorbent, as well as a premier fertilizer.

> *Imagine opaque circulating tubes of colorful microalgae integrated into building designs, protected against air-born toxins, adequately temperature controlled, improving indoor air quality, while creating energy and food supplies.*

Undoubtedly, the largest interest in microalgae use lies within biodiesel production. It represents potentially a more sustainable and cleaner alternative than fossil fuels. Microalga production facilities could be located on land ill-suited for traditional agriculture and even incorporated within architectural designs for cities. In a far larger scale than originally proposed for space travel, integrated waste-energy-nutrient systems could be essential components within our municipalities. Imagine opaque circulating tubes of colorful microalgae integrated into building designs, protected against air-born toxins, adequately temperature controlled, and improving indoor air quality, while creating energy and food supplies. Notably, microalgae in comparison to land plants, can produced much more biofuel per acre.

The author is convinced that microalgae are one of the premier antiaging foods. Especially cyanobacteria provide powerful antioxidants, supportive redox-signaling molecules, and may protect cellular mitochondria unlike any other food sources. Microalgae life-extension claims made for any individual is irresponsible. However, the author hypothesizes that a good-sized experimental study between equivalent groups of individuals, comparing lifespan of those who daily consume one or two grams of edible microalgae versus those who do not, would likely discover significant improvements in lifespan and quality of health between experimental and control groups. Microalgae need to be seriously investigated as mitochondrial nutrients.

As "baby boomers" increasingly turn into "senior boomers," microalgae will be highly prized to combat critical problems associated with aging. For example, age-related digestive problems have already created a multibillion-dollar industry out of antacids, which merely mask symptoms and often exacerbate underlying health problems (as well as increase risk for food-borne illness). Algal compounds can stimulate beneficial intestinal flora and support digestion. Algal antioxidants will be increasingly relied upon to prevent cellular breakdown from occurring in the first place. Moreover, microalgae's ability to support both detoxification pathways and to enhance adult stem cell regeneration pathways is unique.

Algae's ability to help protect our body from exposure to many commonly found heavy metals, pesticides, drugs and industrial chemicals is becoming recognized. Research is needed to expand the use of microalgae within substance abuse and addiction programs. Nutritional support can help substance abusers to better detoxify, and to better recover from injuries to the liver, kidney, adrenals, gastrointestinal system, and brain. The mood enhancement benefits of microalgae can assist addicts' recovery, as well.

Especially *AFA* microalgae appear to offer potent brain-protective benefits, as well as enhanced neurotransmitter production. Some of the most promising areas are treatment of mood disorders like depression and anxiety, learning disabilities, problems of attention, brain injuries, and developmental disorders. The neuro-protective and mood enhancement benefits of edible microalgae have been explored by researchers in the United States and in Italy.

A range of human clinical trials need to be conducted in those most promising areas that animal research and retrospective case studies have already identified. Research on microalgae's anticancer, immune-enhancement and antiviral properties will continue to buildup. Additional human microalgae health research is needed for diabetes, hypoglycemia, cardiovascular disease, fibromyalgia, rheumatoid arthritis, chronic fatigue, brain injury and Alzheimer's disease, to name just a few areas.

Alas, human clinical trials take time, expertise, and lots of money, all of which are precious commodities. One of the most promising microalgae health areas is in the prevention and treatment of diabetes. According to Dr. Mark Edwards (2013) professor at Arizona State University, "Research on humans and animals shows algae components offer significant utility in the prevention and control of diabetes. Aligned studies have demonstrated algae's therapeutic value for the diseases common with diabetics; cholesterol management, blood pressure, heart disease and cancers. Algae can moderate chronic inflammation that often proceeds and accompany degenerative diseases. Algae compounds provide therapeutic value for diabetes and fat metabolism . . . Algae pharmaceuticals and medicines are currently in trial phases. These valuable algae therapeutics should move through medical approval relatively quickly since they are natural products and, to date, have shown no allergic reactions."[5]

Because diabetes represents a model for aging – accelerating many of the disease pathologies commonly seen in the elderly – a better understanding of how microalgae compounds support and protect mitochondria could have wider applications in medicine. Better methods of identification, extraction, and standardization of active algal ingredients, like phycocyanin, carotenoids, sulfolipids, and algal oils will result in microalgae products becoming more widely used by the nutraceutical industry and more widely recommended by doctors. This result will be due in part to increased public and professional demand for high-quality standardized nutraceuticals, and greater demand for evidenced-based medicine, with more rigorous clinical outcome studies.

Microalgae are a natural source of valuable compounds and can act as bioreactors for recombinant protein production. Natural high-value microalgae compounds including pigments, long-chain polyunsaturated fatty acids, and polysaccharides, have a wide range of applications in the food, feed, cosmetics, and pharmaceutical industries.

Furthermore, the multi-billion dollar investments, over the next few decades, in the expanding algal biofuel industry will dramatically increase our understanding and ability to harness the wider benefits of microalgae.

Microalgae can act as bio factors for recombinant protein production. Petroleum-based drugs may gradually be over taking by bio [microalgae produced] pharmaceuticals. Certain scientists feel that genetically engineered microalgae will offer nearly unlimited applications.

For example, "transgenic microalgae can be used as bioreactors for the production of therapeutic and industrially relevant recombinant proteins. This technology shows great promise to simplify the production process and significantly decrease the production costs. To date, a variety of recombinant proteins have been produced experimentally from the nuclear or chloroplast genome of transgenic *Chlamydomonas reinhardtii*. These include monoclonal antibodies, vaccines, hormones, pharmaceutical proteins, and others."[6]

Moreover, natural high-value microalgae compounds including pigments, long-chain polyunsaturated fatty acids, and polysaccharides, have a wide range of applications in the food, feed, cosmetics, and pharmaceutical industries.

One traditional use of algae extracts that continues to expand is in cosmetic markets. A recent study found that; nearly 84 percent of macro algae patents are held by Japan, China, and South Korea with the demand for more algal extract cosmetics increasing. Outside of the Asia Pacific region, countries such as USA and France have invested the most resources into algae research and development.[7]

The application of microalgae as a food coloring is especially valued in Japan. *Spirulina*, blue-green algae, is found in numerous food products as a colorant and a natural preservative. A broad rainbow of various microalgae-based natural colorants, antioxidant food preservers, and nutrient additives could be commercially utilized by the food industry.

In conclusion, the remarkable multi-billion year journey of microalgae is far from over. These most primitive of species offer many more volumes of information to unravel and learn from. In microalgae there exists relatively "undiscovered kingdoms" that will yield numerous health benefits. Survival codes preserved in ancient lineages of microalgae offer a biochemical pharmacopoeia that will be increasingly valued and utilized in the twenty-first century. Even the pioneering field of epigenetics, i.e., how the environment can shape genetic expression, will benefit from a deeper exploration of microalgae.

Scientist report microalga "constitutes an excellent model to study the basis of epigenetic reprogramming events that lead to morphological variations in response to external stimuli." [8] Microalgae rely on DNA methylation strategies more extensively than plants, animals and fungi and may add "significantly to our understanding of the evolution of DNA methylation in eukaryotes [multicellular life - like humans]." DNA methylation appears to play critical roles in the expression and regulation of the genome as shaped by the environment.

Likely at this point, many readers are wondering just how far algal genomics and microalgae applications will contribute towards improving our human condition. While the jury is out on that question, as humanity collectively reaches deep into Earth's primal living matter, numerous discoveries and marvelous innovations lie ahead.

Figure 30. Algatron Illumination Chamber

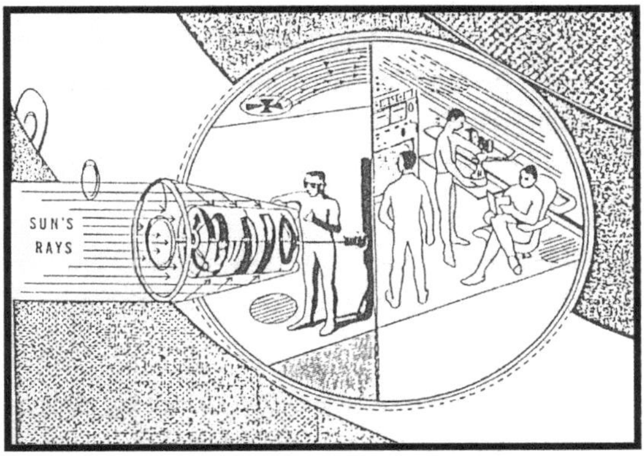

Concept of using microalgae to support space missions, an integrated life support system. Courtesy of Dr. William Oswald (1965) a UC Berkeley professor and Lawrence Berkeley Nation Lab scientist.[9]

Glossary

ABSORPTION of nutrients is the first step in good nutrition. It begins with digestion and takes place mainly through the intestinal mucosa (lining). Simple sugars and mineral ions are able to pass directly into the blood and lymph system. Larger molecules, such as carbohydrates, fats, and proteins, must be broken down into simpler sugars, fatty acids, and amino acids, respectively. This breakdown and the resultant absorption of large molecules are catalyzed by digestive enzymes and a healthy intestine that contains a high population of friendly intestinal bacteria.

ACIDS are molecules that, when added to water (or blood), break down into small positive hydrogen ions and larger negative ions. Acids decrease the pH of water (a pH less than 7 is considered acidic, a pH greater than 7 is considered basic, and a pH of 7 [distilled water] is considered neutral, i.e., neither acidic nor basic). Hydrochloric acid in the stomach has a pH of about 1. Fresh-water alkaline lakes (where blue-green algae flourish) have a pH between 9 and 11.

ADENOSINE TRIPHOSPHATE (ATP) is a complex and profoundly important bio-molecule that stores the body's chemical energy in the mitochondria of each cell.

ADRENAL GLANDS, located above both kidneys, function independently and produce hormones (messenger molecules) from their inner regions (epinephrine and norepinephrine) and outer regions (hydrocortisone).

ADRENALIN. *See* epinephrine.

ALBUMIN is an "escort" protein able to deliver nutrients to the liver by its strong electric affinity. It makes up about half of the protein found in blood serum.

ALANINE is a nonessential amino acid concentrated in muscle tissue. Alanine can be transformed into glucose when needed by muscles (or in the liver) for energy. People with hypoglycemia may have an alanine deficiency. Alanine assimilation is enhanced by the zinc and vitamin B6 found in *AFA* and is known to:

1. Bio stimulates the immune system by increasing the size of the thymus gland and thus lymphocyte production.
2. Be useful for hypoglycemic patients by helping to balance their blood sugar (glucose).
3. Function as a neurotransmitter.

4. Help build the cell walls of friendly bacteria.

ALGINATE is a food additive derived from giant *kelp* algae. It is used as a thickening agent in ice cream, cheese, yogurt, and whipped cream.

ALKALINE ASH is a measure of the mineral content of any food. Since *AFA* is 7 percent alkaline ash by weight, the mineral content of *AFA* algae is consequently extremely rich.

ALLOPHYCOCYANIN from the Greek (allos) meaning "other" and (phykos) meaning "alga," and (kyanos) meaning "blue;" is a pigment-protein complex from the light-harvesting phycobiliprotein family, along with phycocyanin and phycoerthrin. All phycobiliproteins are water soluble and therefore cannot exist within the membrane like carotenoids, but aggregate forming clusters that adhere to the membrane called phycobilisomes. They are accessory pigments to chlorophyll. Allophycocyanin absorbs and emits at longer wavelengths than phycocyanin C or phycocyanin R.

AMINO ACID METABOLISM refers to either the breakdown of individual amino acids for energy purposes or their buildup into larger proteins.

AMINO ACID OXIDASE is an enzyme used in the liver to change "old" amino acids into "new" ones. This enzyme requires riboflavin, a coenzyme, for its bioactivation.

AMINO ACID PROFILES show the proportions of amino acids in a given food compared to healthy human blood. Two foods with amino acid profiles that match most closely are the chicken egg and *AFA*.

AMINO ACIDS are molecules that bond together to form polypeptides or proteins. Dry *AFA* is about two-thirds amino acids.

AMINO SUGARS are sugar molecules that have an amino group of atoms bonded to them. These occur in the cell wall of *AFA*.

AMYLASE is a type of enzyme that digests starch. In the saliva it is called "salivary amylase"; in the pancreas it is called "pancreatic amylase'

ANABOLISM refers to all biochemical reactions that involve the building of larger molecules needed to construct and repair cells. The rich variety of raw materials present in *AFA*, such as amino acids and fatty acids, biostimulate such anabolism processes.

ANEMIA is an unhealthy condition that occurs when the amount of hemoglobin in circulating red blood cells is deficient. "Pernicious anemia" is caused by the inability to absorb vitamin B_{12} through the stomach lining. The extraordinarily high B_{12} content in *AFA* algae often brings relief from the debilitating symptoms of anemia.

ANIONS are ions that have a negative electrostatic charge due to the presence of extra electrons.

ANTIBIOTIC literally means "against life' Antibiotics were once considered "magic bullets" by medicine. Serious doubts about their long-term effectiveness in treating certain diseases have recently been raised because of the high amounts of antibiotics needed for treatment. Typically, antibiotics (e.g., penicillin) will dramatically diminish the population of friendly intestinal micro flora and impair the function of the bowel. Birth control pills, alcohol, caffeine, some food additives, and even chlorinated water can also have similar effects. Taking supplements of friendly intestinal bacteria such as *acidophilus* can often lead to a dramatic normalization of bowel function.

ANTIBODIES are proteins formed in blood serum in response to invading viral or bacterial particles (antigens). Also called immunoglobulins, antibodies function by chemically identifying and marking invaders so that the immune system can then send out white blood cells to destroy them.

ANTIGENS are invading bacterial or viral particles that evoke the formation of antibodies (immunoglobulins) in the blood serum.

ANTIOXIDANTS are cellular biomolecules that protect organisms from the damaging effects of free radicals. *AFA* has an unusually high concentration and wide variety of antioxidants, such as beta carotene, superoxide dismutase, glutathione, and vitamins C and E.

APHANIZOMENON FLOS-AQUAE (*AFA*) is the scientists" name (in Greek, this means "invisible living flower of the water") for the blue-green algae that flourish in Upper Klamath Lake, Oregon, and serve as a nutritional protein-rich anabolic food source.

ARACHIDONIC ACID is a fatty acid found mainly in saturated animal fats. Excessive consumption should be avoided because saturated fatty acids rob the cell membrane of its healthy flexibility.

ARGININE is an amino acid that children and stress-affected people cannot biosynthesize. Dietary arginine is available from nuts, meat, cheese, and *AFA*. Its primary metabolic role is to stimulate the activity of a liver enzyme which initiates the urea cycle, an important bio pathway that metabolizes protein. Arginine, when properly assimilated, can:

1. Biostimulate the thymus gland to enhance the immune system by producing more T-lymphocyte cells, which are effective and active in fighting infection. The activity of natural killer cells is also enhanced to inhibit tumor growth and size. Fifty years of accumulated research evidence attest to this antitumor effect.
2. Biostimulate the pituitary gland to secrete human growth hormone (GH) and produce some muscle-building and fat-burning effects.
3. Build up an "arginine reserve" to protect and detoxify the liver.
4. Increase sperm production and male fertility because human sperm is very rich in arginine.

ARTHRITIS is associated with the inflammation of joints. Poor nutrition, stress, and aging all contribute to the underproduction of natural and protective antioxidants. As a result, the synovial fluids that surround a bone in its joint accumulate excess free radicals that increase swelling and soreness. Fish oils and microalgae oils can reduce arthritic inflammation.

ASPARAGINE is a nonessential amino acid and neurotransmitter that supplies the brain with energy. Asparagine is also an immunostimulant because it biostimulates the thymus gland.

ASPARTIC ACID, a nonessential amino acid, has a higher concentration in the human brain than any other amino acid. It has been useful in the treatment of chronic fatigue syndrome because of its involvement in DNA/RNA synthesis. Aspartic acid has shown usefulness as a biostimulator of the thymus gland and thus as an immune system stimulant.

ASPERGILLUS FLAVUS; ASPERGILLUS NIGER are molds found on grains, corn, and peanuts. These molds are used to produce digestive enzymes.

ATHEROSCLEROSIS is a disease that causes hardening of artery walls and results in a loss of artery elasticity.

ATP (ADENOSINE TRIPHOSPHATE) is an energy-storage molecule used by all cells to drive the biochemical reactions that are needed to sustain their life.

BETACAROTENE (PROVITAMIN A) is a free radical—quenching vitamin that is easily obtained from *AFA* algae, carrots, sweet potatoes, and leafy green vegetables. It is a precursor (provitamin) to vitamin A (retinol), which is essential for night vision. *AFA* is probably the richest beta carotene food source on Earth. Beta carotene offers powerful anticancer, antiaging, and antioxidant properties without the toxic effects of taking fat-soluble vitamin A.

The beta carotene in *AFA* most definitely boosts the immune system by protecting the cell lining of the digestive and respiratory tracts. Apparently, the number and strength of immune cells will dramatically increase so that a variety of infections can be fought off. In general, malignant cell growth appears suppressed by beta carotene. *See* carotenes.

BIFIDOBACTERIUM BIFIDUM is (or should be) the friendly bacteria living mainly in the colon (large intestine). The colons of breast-fed children are inhabited mainly (99 percent) by this bacteria. As aging increases typically the numbers of these bacteria decline. These friendly bacteria produce special acids which prevent the growth of some viruses and bacteria. These friendly microflora even help to prevent nitrite toxicity (present in hot-dogs and other packaged meats).

BILE is a yellow-green bitter fluid continuously secreted by the liver, stored in the gall bladder, and used in the small intestine (duodenum) — along with pancreatic juices — to digest fats. When fats enter the small intestine, a hormone is secreted to signal this process. Bile is actually a mixture of water, pigments from hemoglobin breakdown, bile salts, and cholesterol.

BILIRUBIN is a reddish-yellow pigment found in bile, made by the liver. Excessive bilirubin levels are responsible for the distinctive yellow color that tints the skin and eyes of patients with jaundice. Slightly elevated bilirubin levels have been associated with antioxidant benefits, such as lower risk of diabetes.

BILIVERDIN is a green pigment in bile, an intermediate product of the breakdown of hemoglobin, which normally reduces into bilirubin. Biliverdin is evident in some bruises as a green color. Its breakdown into bilirubin shifts into a yellowish color. Bile pigments like biliverdin have potent antimutagenic and antioxidant properties. Studies have found that people with moderately elevated levels of bilirubin and biliverdin in their bodies have lower risks of cancer, cardiovascular disease and metabolic disorders.

BIOFLAVONOIDS are found in both *AFA* algae and in the pulp of citrus fruits. Once called Vitamin P, bioflavonoids are now considered important phytochemicals that can help boost the immune system. The small amount of bioflavonoids in *AFA* can also help to increase capillary strength so that cellular nutrients and wastes can be more efficiently transported.

BIOSYNTHESIS. *See* Synthesis.

BIOTIN is a water-soluble B vitamin coenzyme that is obtained from a good diet (e.g., brown rice, mushrooms, and *AFA*) and friendly intestinal bacteria. It is used in a variety of important enzyme systems to construct fatty acids by attaching acid functional groups to larger molecules. Because of this, biotin can produce healthy-looking hair. Biotin also improves athletic performance by helping to metabolize branched-chain amino acids.

BLOOD is a life-sustaining fluid with a majority of red blood cells (erythrocytes) and a minority of white blood cells (leukocytes) and platelets. Oxygen is carried by the red blood cells, and larger molecules (e.g., hormones, glucose, lipids) are transported in the blood plasma. The pH of healthy blood is about 7.4.

BLOOD-BRAIN BARRIER refers to the inability of most molecules other than glucose to cross the barrier from the brain's blood vessels into the brain's tissues.

BLOOMS are profuse growths of microalgae that discolor the water in which they live for a short duration. An *AFA* bloom occurs approximately four times per year in Upper Klamath Lake (Oregon).

BLUE-GREEN ALGAE. *See* cyanobacteria.

BMAA β-Methylamino-L-alanine, or BMAA is a nonstandard amino acid produced by species of cyanobacteria. There is currently ongoing research into the role of BMAA as a possible environmental factor in neurodegenerative disease.

BORON is a "bone-strengthening" trace element found in soil. Its concentration in soil varies depending on farming practices. Boron is obtained from apples, pears, leafy greens, and *AFA* and it improves calcium retention. US government-sponsored research shows that boron deficiency in humans may be common, especially because the elderly (and alcoholics) have increasing difficulty absorbing boron from their intestines. It is possible that osteoporosis, arthritis, and hypertension may be lessened with boron-rich foods such as *AFA*. Some researchers even claim that boron may enhance mental clarity. Artery plaque due to calcium deposits (and thus high blood pressure) seems to be lessened in the presence of boron. Citizens of countries with soil containing high quantities of boron (e.g., in Israel) have a lower incidence of arthritis than those living in countries with less rich soil (e.g., in Jamaica).

BOWEL TOXEMIA occurs when the elimination function of the colon is reduced by and impacted with the toxins of unfriendly bacteria, viruses, and parasites. Myriad health problems develop, along with swelling and inflammation of the bowel, which prevent or reduce the absorption of nutrients. Eventually, toxins enter the bloodstream and cause more problems with the circulatory and lymphatic system, even the lungs and kidneys. One way to help insure a properly functioning intestine is to use probiotics regularly. *See* leaky-gut syndrome.

BRANCHED-CHAIN AMINO ACIDS (BCAAs) refer to the essential amino acids valine, leucine, and isoleucine. About one-third of our muscle mass is made of BCCAs, so having a generous and continual supply of them can help prevent muscle loss and even build muscle mass.

BROWN ALGAE, commonly known as "seaweed" or *kelp*, are found floating freely in ocean water.

CALCIUM is the "antiosteoporosis" mineral element that is found in dairy products, leafy green vegetables, tofu, and especially in *AFA*. Swiss cheese contains about 10.0 mg per gram. *AFA* contains 14.0 mg per gram. Because calcium absorption in the intestines diminishes with age, calcium will be slowly lost and cause a degenerative "porous-bone" disorder called osteoporosis. Although as much as 75 percent of dietary calcium does not get absorbed, calcium absorption is enhanced by the beta carotene and chelating amino acids found in *AFA*. Because some enzyme systems involving neurotransmitters are activated by calcium, the calcium in *AFA* has a nerve-calming effect and acts nicely as a bedtime tranquilizer. Muscle cramping is diminished. Calcium also protects against cardiovascular disease by lowering cholesterol and blood pressure.

CANDIDA ALBICANS a common yeast present in the gastrointestinal tract of almost everyone. It can cause a variety of allergic reactions if the intestinal tract is overrun. As *Candida* continues to proliferate, the immune system weakens and problems such as irritable bowel syndrome, premenstrual stress syndrome, food sensitivities, depression, prostate pain, and even chronic carbohydrate cravings begin to develop.

CARBOHYDRATE METABOLISM begins with glucose (blood sugar) being carried to the liver. With the transformative effects of enzymes and the "energy molecule" ATP (adenosine triphosphate), glucose is phosphorylated (i.e., phosphorous and oxygen are added) and thus transformed step-wise into glycogen, a complex carbohydrate that acts as an energy reserve for later combustion.

CARBOHYDRATE DIGESTION depends upon the action of several enzymes. Salivary amylase begins the process whereby complex carbohydrates are broken into polysaccharide fragments. Pancreatic amylase (and intestinal enzymes) further breaks the fragments into individual monosaccharaides.

CARBOHYDRATES may be classified as sugars, starches, or celluloses depending on the number and bonding arrangement of their constituent sugar groups. A somewhat straight chain of about 1000 sugar groups is a starch, whereas a branched-chain of similar length describes glycogen. Such long chains are also called polysaccharides. Shorter chains of two sugar groups are called disaccharides (e.g., lactose, sucrose), and individual sugar groups that are independent of each other are termed monosaccharaides (e.g., glucose and fructose).

CARCINOGENS are chemical substances that initiate tumor formation.

CAROTENES or carotenoids are reddish-yellow plant pigments found abundantly in *AFA* algae, carrots, and leafy green vegetables. When carotene molecules are digested (oxidized and "broken" in half), they are converted from vitamin A precursors to retinol molecules, the liver-storage form of vitamin A. The most active and beneficial type of carotene is the dark-red beta carotene. Alpha carotene is closely related to beta carotene, as are the less well known carotenes, lutein and zeaxanthin.

CATALASE is an enzyme that can absorb and render harmless the dangerous peroxide free radicals created during normal metabolism.

CATALYSTS are substances that speed up a chemical reaction without being consumed. Catalysts made of protein material are called enzymes.

CATECHOLAMINES are an important family of neuropeptides that include dopamine and adrenaline. The tyrosine in *AFA* acts as their precursor.

CELLS are fundamental units of living matter capable of reproduction. Eukaryotic cells have a well-defined central nucleus, whereas prokaryotic cells do not.

CELL MEMBRANES are semifluid lipid and protein coverings that surround the body or cytoplasm of the cell. Billions of years ago, in the ancient oceans of Earth, such lipids gradually formed into primitive cell membranes that were necessary for the maintenance of life and the eventual formation of blue-green algae (cyanobacteria). The fatty acids that form the lipid layers of the cell membrane ultimately affect the membrane's fluidity, depending upon the actual shape and length of each fatty acid. The cell membrane of any cell needs to be "fluid and flexible" if it is to remain healthy. Cholesterol is also important to the cell membrane in that its molecules are imbedded between the fatty acid molecules of the lipid bilayer, helping to hold them together and thus impart integrity and strength to the cell membranes. There is one drawback, however: Too much cholesterol holds the membrane together too rigidly so that the membrane loses its fluidity and thereby its health. The cell membrane of *AFA* algae is very flexible, partly because it does not contain any cholesterol whatsoever.

CELLULOSE is a polysaccharide composed of about 1000 glucose units that are bonded together such that animal enzymes cannot break (digest) them. Cows are able to digest cellulose only because of the unique enzymes produced by the friendly bacteria in their intestines.

CHELATED MINERALS are found in blue-green algae such that the metal atoms of each mineral element (e.g., iron, zinc, magnesium) are naturally bonded to nonmetal atoms inside of larger peptides, protein, or enzyme molecules.

CHELATES are molecules — usually peptides — that can chemically bond to minerals by holding them with their nitrogen or oxygen atoms. The word chelate comes from a Greek word meaning "claw." Iron is chelated within the hemoglobin molecule by being attached to several negatively charged nitrogen atoms.

CHELATION THERAPY is a biochemical process of ridding the body of toxic metals that have been accumulated from a toxic environment. The presence of such metals (e.g., lead, nickel, mercury, cadmium) lowers resistance to disease, causes enzyme dysfunction, and imperils basic physiological functions. Many physicians employ chelation therapy using ethylenediaminetetraacetic acid (EDTA). Although used by the U.S. Navy to help treat lead poisoning, EDTA is a synthetic drug without full FDA approval. Sulfur-containing amino acids (cysteine and methionine) found in *AFA* algae, along with chlorophyll and glutathione (a chelating peptide), naturally and slowly begin to reduce the amount of toxic metals in the body.

CHLORELLA is a green algae used as a food only after its indigestible cell wall is removed.

CHLOROPHYLL is a photosynthetic pigment found abundantly in plants and algae, especially *AFA* algae. *AFA* contains 2 to 3 percent of this wondrous green molecule, placing our blue-green algae among the highest in chlorophyll content on Earth. It is considered by some to be a cell regenerator partially because its central magnesium atom plays an important role in so many (325) different enzyme systems. The chlorophyll in *AFA* may help to inhibit the mutation-promoting effects of some environmental carcinogens. It may do this by acting as an antioxidant and thus protect our DNA during cell division, a very vulnerable time in the life of any cell.

CHLORINE is a poisonous gas, but is found in nature as the chloride ion within sodium chloride (table salt). Chloride is part of our stomach acid in the form of hydrochloric acid and helps our liver to detoxify and maintain the body's acid-base balance.

CHOLESTEROL is a polycyclic molecule found in all animal fats and oils. It can chemically combine with certain proteins to form lipoproteins such as high-density lipoproteins (HDLs) — the "good" cholesterol — or low-density lipoproteins (LDLs) — the "bad" cholesterol. Cholesterol is a precursor molecule of vitamin D and a variety of steroid hormones. However, too much cholesterol can build up and can cause artery blockage.

CHOLINE is a medium-sized molecule, part alcohol and part amine, which occurs as a component of biological tissue in lecithin and acetylcholine. Choline is present in all foods, so deficiencies of choline do not exist.

CHROMIUM is a trace micro mineral with only about 5 milligrams distributed throughout the entire human body. As a cofactor, chromium bioactivates a variety of enzymes which initiates insulin action, controls glucose levels, and stimulates fatty acid synthesis in the liver. In addition, chromium is the "sugar-regulating" trace mineral found in brewer's yeast, black pepper and *AFA* algae. Refined sugars should be avoided because it causes the body to excrete its valuable chromium.

This micronutrient metal is an essential part of a relatively large glucose-regulating tripeptide molecule called the "glucose tolerance factor" (GTF). The exact structure of this molecule is presently unknown. Our mineral-poor topsoil, along with our over-refined foods, has led to an alarming depletion of chromium in the American diet. To make matters worse, chromium is relatively difficult to absorb. Only about 1 percent of ingested chromium even winds up in the bloodstream, the rest being excreted. Absorption may be raised to higher levels when the metal ion is chelated, as it is in *AFA*. There is some evidence that refined white flour should be avoided because it removes chromium from the blood.

CHRONIC describes a disease that lasts a relatively long time. Acute pains go away quickly; chronic complaints might last a lifetime.

CIRCULATING IMMUNE COMPLEXES (GIGS) are incompletely digested proteins in the blood that are marked by antibodies as if they were toxic antigens. This puts a strain on the immune system. Protein-digesting enzymes from *aspergillus* are effective in breaking them up.

CIRRHOSIS is a digestive disease of the liver caused by poor nutrition and/or alcohol or drug abuse. The liver gets "scarred" internally so that its normal function (there are hundreds!) is greatly impaired.

CIS AND TRANS refer to the two different shapes that an unsaturated fatty acid can have. Each has remarkably different biochemical effects on the fluidity of the cell membrane. A "cis" type fatty acid, for example, has a twisted carbon chain and tends to stack up randomly and form flexible cell membranes. A "trans" type fatty acid has a much straighter carbon chain that tends to stack up neatly and form solids or gooey liquids that can block up blood vessels.

CHLORELLA is a green algae that grows in fresh water and is widely used as a nutritional supplement. In its natural form *chlorella* is indigestible and its cell wall needs to be broken to use as a food source. *Chlorella* is spherical in shape and very tiny, about 2 to 10 um in diameter.

COBALAMIN (VITAMIN B12) is the "rejuvenator and energizer" vitamin. It can be obtained from fish, meat, liver, dairy products, and *AFA* algae (the richest known source). Cobalamin is used in a wide variety of enzyme systems. As a result, it can be used to synthesize hemoglobin and thus keep blood oxygenated, to increase energy and promote growth, and to maintain and repair the nervous system. Vegetarians (especially children) need to watch out for this deficiency because vitamin B_{12} is not present in plants. *AFA* is one way the vegetarian can prevent this detrimental deficiency. Cyanocobalamin is one of its more active forms and plays a role in red blood cell synthesis.

COBALT, an important cofactor of vitamin B12, helps to biosynthesize red blood cells and repair nerve tissue. Cobalt is found in *AFA*, oysters, ocean fish, red meat, and liver. The high concentration of cobalt in *AFA* is good news for vegetarians because the U.S. soil content of cobalt is normally very low.

COD LIVER OIL is important fish oil because of its vitamin A and vitamin D contents. Its value, in part, probably derives from the fact that cod fish eat large quantities of algae.

COENZYMES are nonprotein molecules (usually B vitamins) that bond to protein molecules and impart to them the ability to function properly as enzymes.

COENZYME Q10, OR UBIQUINONE, removes electrons from foods (proteins, carbohydrates, and fats) and carries them directly to oxygen molecules, turning them into water. Ubiquinone is the coenzyme that makes the process of food oxidation possible. It occurs mainly in the mitochondria "energy factory" within cells.

COFACTORS are inorganic metals that are required for enzyme activity.

COLITIS is a disease in which the colon becomes irritated and inflamed. The ingestion of probiotics has been shown to have some success in treating this painful disease.

COLLAGEN is a protein that makes up most of the connective tissue between skin cells and between and inside bones.

CONJUGATED PROTEINS are proteins that can be chemically connected to fats (lipoproteins), carbohydrates (glycoproteins), or phosphates (phosphoproteins).

COPPER is an essential trace mineral found in liver, shellfish, nuts, and *AFA* algae. As a micronutrient, it is important along with iron in many enzyme systems as a catalyst for the biosynthesis of hemoglobin. Copper is part of the antioxidant enzyme superoxide dismutase (SOD), which may have an antiarthritic effect in the synovial joints.

COX-2 INHIBITORs are used to treat the pain of inflammatory conditions, like arthritis, menstrual cramps and sports injuries. Aspirin and certain blue-green algae have been shown to act as Cox-2 inhibitors.

C-PHYCOCYANIN is an accessory photosynthetic pigment occurring mainly in cyanobacteria and red algae. It is a phycobiliprotein, in which the pigmented prosthetic group is phycocyanobilin, which give phycocyanin and thus blue-green algae its blue color.

CYANOBACTERIA are blue-green prokaryotic algae such as *Aphanizomenon flos-aquae* that carry out photosynthesis. They are the simplest of the algae in that they do not have a nuclear membrane. Although there are thousands of different species, only *AFA*, *Spirulina,* and a few dozen other species are known to be edible.

CYANOPHYCIN STORAGE GRANULES are found inside the *AFA* cell and are composed of a polymer of aspartic acid and arginine, two amino acids that stimulate the immune system.

CYSTINE is an important sulfur-containing nonessential amino acid that is synthesized from cysteine in the liver and then used in a variety of interesting and powerful ways. The shapes of large and complex protein molecules depend on the placement of cystine. By use of its sulfur atoms it can hold the otherwise straight chain of amino acids in more complex patterns. Cystine is sometimes considered an "antiaging" nutrient because of its detoxifying properties. Cystine helps to detoxify carcinogens by using its sulfur atoms to absorb the damaging effects of free radicals.

CYTOPLASM refers to the cellular contents between the cell membrane and the nucleus.

DERMATITIS is a skin disorder caused by irritants in the environment and aggravated by stress and fatigue. Eczema, with its flaky and itchy blisters, is caused in part by a deficiency of B vitamins and/or amino acids. *AFA* algae, being rich in both, are a good preventative for dermatitis.

DOCOSAHEXAENOIC ACID (DHA) is an omega-3 fatty acid that is a chief structural component of the human brain, cerebral cortex, skin, sperm, testicles, and retina. DHA is distributed throughout the body. DHA levels have been reported to decrease the risk of dementia and cardiovascular disease as well as to improve mood.

DIGESTION involves the use of the alimentary canal (mouth, stomach, intestines) and the gastrointestinal (GI) tract, supported by digestive organs (salivary and intestinal glands, pancreas, liver, gall bladder) where enzymes and hydrochloric acid break down and convert (1) fats to fatty acids, (2) proteins to amino acids, and (3) carbohydrates to sugars.

DIGESTIVE ENZYMES are used to break down food. They are either endogenous or exogenous, depending on their origin. When food is cooked, inherent and exogenous digestive enzymes are destroyed (denatured). This forces our digestive system to compensate by producing more endogenous digestive enzymes.

DNA (DEOXYRIBONUCLEIC ACID) is found in the central region of all cells. DNA biochemically transfers genetic information to successive cellular generations by making a copy of itself (replicating). The DNA molecule itself is a "double helix" of molecular groups (sugar, phosphate, and nitrogen bases) that are arranged into chains from which the proteins (and enzymes) needed for the life of a cell may be constructed. These chains, sometimes called genes, are also grouped together into larger units called chromosomes. The DNA in *AFA* algae consists of a single circular molecule.

DOPAMINE is an important neurotransmitter that helps to promote memory and improve alertness. It is derived from the amino acid tyrosine.

DUNIELLA salina is a type of green micro-algae especially found in salty water environments. Too salty to eat as a pure food source, Duniella is harvested because of its orange-colored antioxidant carotenoids. These are used in cosmetics and dietary supplements.

E. COLI / (ESCHERICHIA COLI) BACTERIA are rod-shaped bacteria that live in human intestines. They are considered "friendly" in that they provide us with vitamin K and some B vitamins. They could, however, be considered "unfriendly" at times because they can produce infection if introduced into open wounds.

EICOSAPENTAENOIC ACID (EPA) is a semi essential fatty acid present in cold-water fish such as cod, mackerel, salmon, and *AFA*. Greenland Eskimos have greatly reduced cardiovascular disease because they eat such EPA-containing fish, which are known to dine on large quantities of blue-green algae.

ELECTROLYTES are substances that, when added to water, break up or ionize into positively and negatively charged electrical particles called ions.

ENDOCRINE GLANDS are among those six ductless glands (e.g., thyroid, adrenals, and pancreas) that secrete hormones carried throughout the bloodstream. These hormones are incredibly powerful even though their amounts are measured in micrograms.

ENDORPHINS are morphine-like neuropeptides. They may be formed in small quantities following the digestion of low molecular weight peptides found in *AFA*.

ENDOSYMBIOSIS is a mutually beneficial arrangement where one organism lives inside the other; the two evolve into behaving as a single organism. Over time, genes become exchanged between the two organisms, so they become reliant upon each other. Examples include chloroplasts and mitochondria, from cyanobacteria and bacteria, which transformed into organelles inside their cellular hosts.

ENDOTOXINS are powerful lipopolysaccharide (LPS) substances found within the cell walls of some bacteria and blue-green algae (cyanobacteria). LPS endotoxins often biostimulate the immune system by increasing the disease-fighting activity of white blood cells.

ENZYMES are biochemical protein catalysts that speed up reactions in the body without themselves being consumed or chemically altered. Most enzymes also require a vitamin coenzyme and/or a metal ion cofactor to be bioactive and functional. Each cell requires several thousand different enzymes. Typically, enzymes are named by adding "ase" to the name of the substrate molecule. For example, maltase catalyzes the hydrolysis (breakdown) of maltose, lactase hydrolyzes lactose, and lipase digests lipids.

EPIGENETICS is the study of how genes may express themselves differently especially when interacting with the environment. Some have compared genetics as to a book and epigenetics more like the different interpretations one can derive from reading a given book. More formally, epigenetics is the study of heritable changes in gene activity that are not caused by changes in the DNA sequence.

EPINEPHRINE (OR ADRENALIN) is an enzyme activator secreted from the adrenal glands (one of six endocrine glands) that causes glucose to be quickly released from the liver into the bloodstream, allowing for an emergency "fight or flight" situation. Each adrenalin molecule is responsible for the release of roughly 30,000 glucose molecules. The essential amino acid tyrosine (with vitamin C) is needed to biosynthesize adrenalin.

ERYTHROCYTES. *See* red blood cells.

ESSENTIAL refers to those molecules that are necessary for life processes but impossible for the body to produce. As a result, essential molecules (e.g., essential amino acids, essential fatty acids) must be supplied by the diet.

ESSENTIAL AMINO ACIDS are amino acids that humans cannot biosynthesize at the rate needed. There are eight of them from among the twenty amino acids needed for protein synthesis. The quality of any protein source is determined by the amount and proportion ("amino acid profile") of these eight essential amino acids.

ESSENTIAL FATTY ACIDS (EFAS) are fatty acid molecules which cannot be biosynthesized by humans (hence the word "essential"). They must be procured from food. Linoleic and linolenic fatty acids — once called vitamin F — are essential to survival. EFAs provide energy, insulation, and structure while maintaining fluid and flexible cellular membranes. They are precursors of prostaglandins, which are required for hormone-like purposes. The World Health Organization recommends that 3 to 5 percent of the diet include these essential fatty acids. *AFA* algae are rich in such compounds.

EUKARYOTIC cells have a true nucleus and a variety of organelles, all of which are surrounded by membranes. Eukaryotic cells evolved from the more basic prokaryotic cells, exemplified by cyanobacteria, the blue-green algae.

FATS are made up of glycerol (a sweet polyalcohol) bonded to one, two, or three fatty acids. The resulting fats — called oils in the liquid form — are digested mostly in the small intestine, where they are broken down by pancreatic and intestinal enzymes. After being delivered by the blood to muscle and liver tissue, fats are either oxidized to supply 70 percent of the body's energy needs or stored in fat tissue for later use.

FATTY ACID ENZYMES are used in the biosynthesis of nonessential fatty acids. A family of seven such enzymes — collectively called "fatty acid synthetase" — is used.

FATTY ACIDS typically have a long hydrocarbon chain that always ends with an organic *acid* group of atoms. The molecule may be monounsaturated or polyunsaturated. Linoleic and linolenic acids are both abundant polyunsaturated fatty acids. They are called "essential" when they cannot be biosynthesized by humans.

FIBER refers to indigestible toxin-absorbing polysaccharides, such as cellulose, hemicellulose, and pectin, found within cell walls. *AFA* contains about 5 percent dietary fiber, ranking it high in those polysaccharides needed by the colon to eliminate toxic wastes.

FIXATION refers to the chemical incorporation of carbon dioxide and/or nitrogen from the air into cellular material by plants or algae.

FLOW CYTOMETERY is a technique for counting and to examine microscopic particles, such as cells and chromosomes, by suspending them in a stream of fluid and passing them by an electronic detection device. The auto-fluorescent properties of certain photosynthetic microalga and bacteria can be used by flow cytometery in molecular biology and medical research.

FLUORINE, like chlorine, is a poisonous gas. However, as the negatively charged fluoride *ion* it is known to be potent against tooth decay and possibly to help slow down osteoporosis. Fluoride ions are not yet considered essential to health. They are found naturally in most food, water, and soil. Apparently, fluoride in our water (or our algae) is protective against osteoporosis because of the biostimulation of new bone growth and the concurrent hardening of the bone itself.

FOLIC ACID is the "DNA synthesizer" vitamin molecule that is obtained from a good diet of *AFA* algae and fresh leafy green foliage. Deficiency of this vitamin may lead to anemia, weakness, and irritability. Alcoholics and elderly people are at risk because of intestinal malabsorption problems. Because of this, the ingestion of a combination of *AFA*, plant enzymes, and probiotics is recommended. Folic acid also appears to improve mental function (possible IQ increase) in children.

FOOD ADDITIVES range from coloring agents to antimicrobials and number around 2000. Red Dye Number 3 is probably neurotoxic because it may interfere with neuropeptide function; Blue Dyes 1 and 2 may be carcinogenic. Some people in state and city governments are taking notice of the potential harm of these food additives.

FOOD-ENZYME STOMACH REGION refers to the upper portion of the stomach where predigestion occurs for about one hour. Predigestion is followed by gastric acid secretions, which then deactivate the endogenous digestive enzymes at work.

FREE RADICALS are molecules (or ions) that are highly reactive chemical cousins of molecular oxygen and are often generated by otherwise harmless biochemical processes. They are disruptive to all living organisms because they do not have all of their electrons paired up as do other more stable molecules. Thus, free radicals will steal back their missing electrons from such important biomolecules as DNA or membrane lipids, causing disease and speeding up the aging process. Many antioxidant substances in *AFA* algae are able to absorb and disarm such free radicals.

FRIENDLY BACTERIA are those symbiotic microorganisms that live in the gastrointestinal tract of humans (and animals) and contribute to good health. The 400 or so species present in humans weigh roughly 4 pounds and number in the trillions, equaling or exceeding the total number of cells in the body. Typically, one-third of dry fecal matter is actually dead bacteria, friendly or otherwise. Antibiotics such as penicillin and steroids (e.g., prednisone, cortisone, and birth control pills) greatly reduce the population of these important micro bacterial floras.

GAMMA-LINOLENIC ACID (GLA) is a semi essential fatty acid that is critical to the maintenance of a healthy and fluid cell membrane. GLA is enzymatically biosynthesized from linoleic acid — an essential fatty acid — and is present to varying extents in evening primrose oil, borage oil, mother's milk, and most abundantly in *AFA* algae. GLA effectively lowers high cholesterol levels over 100 times more efficiently than does linoleic acid.

GENES are working subunits of DNA, which provide instructions to make molecules called proteins. A gene is a basic physical and functional unit of heredity. Each gene contains a particular set of instructions, usually coding for a particular protein or for a particular function. DNA is coiled and supercoiled (a double helix) to form chromosomes. There are numerous genes within each chromosome.

GERMANIUM is an important micro mineral with a variety of positive healthful benefits. Sometimes it is useful in controlling the virus in chronic Epstein-Barr disease. Organic germanium seems to stimulate the immune system and have anticancer activity.

GLUCAGON is a hormone that increases the blood sugar (glucose) level by stimulating the liver to break down its polysaccharide glycogen store into glucose.

GLUCOSE is the blood sugar and monosaccharide unit used in building glycogen, the "animal starch" stored in the liver as a source of energy. Glucose supplies energy to the brain because it so easily crosses the "blood-brain barrier!'

GLUCOSE TOLERANCE FACTOR (GTF) refers to a glucose-regulating molecule that is bioactivated by chromium. GTF helps to transfer glucose from the blood to individual cells. The chromium, niacin, and amino acids in *AFA* may be useful for the construction of GTE

GLUTAMIC ACID is a nonessential amino acid and "brain fuel" readily obtainable from animal, vegetable, or algae protein. It is concentrated in the human brain's memory center (hippocampus) as a "stimulant" neurotransmitter. It is for this reason that glutamic acid can improve mental functioning. Glutamic acid also helps to reduce alcohol and sugar cravings.

GLUTAMINE is a nonessential amino acid derived from glutamic acid. It plays a major role in DNA synthesis. Glutamine, like glutamic acid, is a neurotransmitter "brain fuel." Its concentration in blood is typically much higher than any other amino acid.

GLUTATHIONE is a tripeptide coenzyme that is made from glutamic acid, cysteine, and glycine — all amino acids present in *AFA* algae. It can function as a "biological electrical conductor" by transferring electrons from one molecule to another. Glutathione is also considered an antioxidant and is able to protect the body from attack by dangerous free radicals.

GLYCEROL OR GLYCERIN is a sweet, syrupy, polyalcohol liquid.

GLYCINE is a nonessential amino acid that is important in brain chemistry and the synthesis of DNA, hemoglobin, and collagen. Found in protein foods, it is the simplest amino acid, tastes like glucose (hence its name), and has a calming effect. Glycine receptor sites are found throughout the central nervous system. Glycine has a variety of nutritional uses:

1. Can lower cholesterol and triglyceride fats in the blood.
2. Detoxifies benzoic acid, a common food additive.
3. Stimulates the production of glutathione, probably the most important endogenous antioxidant in all living organisms.
4. Stimulates the secretion of glycogen, and therefore balances blood sugar levels.
5. Heals wounds because skin collagen requires high amounts of glycine.

GLYCOGEN is the "animal starch" stored in the liver. It serves as our carbohydrate source of energy. Glycogen can be broken down into glucose or energy by a variety of life-sustaining enzymes.

GLYCOGEN GRANULES are naturally found inside the *AFA* cell and are composed of polymers of glucose that serve as a carbohydrate energy reserve.

GLYCOLIPIDS are fatty acids that are chemically bonded to sugar molecules and found mainly in brain tissues. *AFA* contains these important brain chemicals.

GLYCOPROTEINS consist of amino acids and sugar molecules chemically bonded together. They are important in the structure of skin, blood vessels, and cell membranes. Glycoproteins "actively transport" metals in and out of cells on the surface of the cell membrane.

GREEN ALGAE are mostly fresh-water inhabitants that contain a high concentration of chlorophyll. *Chlorella* algae are inedible unless their cellulose cell walls are removed.

HEALTHY COLON MUCOSA is achieved when the cell surface lining (mucosa) efficiently secretes necessary hormones and lubricants, aggressively prevents toxin absorption, and systematically allows usable nutrients into the bloodstream. Healthy intestinal peristalsis that produces about one contracting movement every four seconds is imperative for regular bowel movement and a properly functioning colon. A regular regimen of probiotics is strongly recommended. Some of the many symptoms that can be corrected with probiotics are:

- Backache and headache

- Fatigue and concentration loss
- Constipation, gas, bloating and indigestion
- Bad breath and coated tongue

HEAT SHOCK PROTEINS (HSPs) are nearly universal in cells under normal conditions, but are expressed at greater levels when cells are exposed to a sudden temperature change or under other forms of stress. HSPs play protective and restorative roles in a wide variety of organisms. The discovery of heat shock proteins came from the observation that brief exposure of cells to high temperature can provide protection from subsequent larger increases in temperature, even at temperature levels which are normally lethal. HSPs play multiple roles from transporting antigens to the immune system to aiding protein folding and shape.

HEAVY METALS such as lead, mercury, cobalt, and cadmium are environmental toxins that cause a variety of physical, emotional, and even mental problems. The powerful short chain peptides in *AFA* may help to gradually remove these contaminants by the process of chelation.

HEME is the deep red iron pigment component of hemoglobin. Heme is essential for the maintenance of cellular homeostasis by sensing or using oxygen. Plant heme (chlorophyll) is the green magnesium pigment associated with the chloroplast. Both play vital roles in electron transfer. The heme ring is nearly identical to chlorophyll, except iron is replaced by magnesium. Altogether a wider family of hemoprotein shares many vital and diverse biological roles. Some scientists propose the original evolutionary function of hemoprotein was electron transfer in primitive sulfur-based photosynthesis pathways in ancestral cyanobacteria-like organisms before the appearance of molecular oxygen.

HEME OXYGENASE (HO-1) is an essential enzyme, within the hemoprotein family, that cleaves heme to form biliverdin. The ability of heme oxygenase to split heme, thus producing carbon monoxide, iron and bile pigments, contributes to anti-inflammatory effects. HO-1 has potent anti-inflammatory, antioxidant and antiproliferative effects, is up-regulated by multiple stimuli and provides protection against oxidative stress.

HEMOGLOBIN consists of a complex protein molecule (globin) bonded to a nonprotein portion (heme) which contains iron that is capable of bonding to oxygen long enough to deliver it to the animal cells that require it. This ability facilitates the process of delivering oxygen to and removing carbon dioxide from each cell. There are 574 amino acid molecules required to biosynthesize hemoglobin.

HETEROCYSTS are specialized blue-green algae cells within which nitrogen fixation takes place so that amino acids may be enzymatically constructed.

HISTIDINE is a conditionally essential amino acid needed by children and stress-affected people. Dietary histidine is available from most animal and plant protein, but not very much is really known about it other than its role in boosting the immune system and in producing red and white blood cells. Histidine is also a mineral chelator and thus enhances intestinal absorption of minerals — especially zinc and copper.

HOMEOSTASIS refers to that natural and healthy tendency associated with the internal chemistry of the body to continually maintain a balanced, resilient, dynamic, and fluid internal stability. This state occurs when there is a coordinated response of specific parts of the body — organs, tissues, hormones, enzymes — to intelligently respond to the stresses of outside stimuli so that normal function is not disturbed or "dis-ease."

HORMONES are regulatory molecules secreted by specialized cells, such as those of the endocrine glands, throughout the body. Hormones dramatically affect such bodily functions as digestion, growth, and sexual development.

HUMUS is a dark brown or black colloidal mass of partially decomposed but stable organic matter in the soil. Humus enhances most plant growth and aids water retention in soils. Humus (rich) soil is considered to be one of the most fertile types of soil in the world.

HYDROCHLORIC ACID (HCI) is a digestive acid in the stomach. It is secreted by the gastric cells of the stomach wall and converted to pepsin, a protein-digesting enzyme. Since HC1 is strongly acidic (pH = 1), invading organisms can be destroyed. The HC1 also helps enzymes to break the bonds within carbohydrates and proteins.

HYDROLYSIS refers to the chemical process by which chemical bonds between small molecular units (e.g., amino acids, sugars) in a polymer are enzymatically broken by the action of water.

HYDROPHILICITY comes from the Greek (hydros), meaning water, and (philia), meaning love or a molecule or molecular compound that has a strong affinity for water.

HYPERVITAMINOSIS, OR VITAMIN TOXICITY, is a disorder brought on by ingesting too many vitamin supplements containing excessive fat-soluble vitamins such as A and E.

IMMUNOSTIMULAT ION, or IMMUNE SYSTEM ENHANCEMENT, occurs when there is an increase in white blood cell activity. Healthy probiotics with a strong population of friendly *acidophilus* bacteria is a prerequisite. Pathogenic (harmful) bacteria such as *Candida* and *Clostridium,* which can impair the immune system, are destroyed by the lactic and acetic acids produced by such friendly bacteria.

INSULIN is a protein secreted by the pancreas (endocrine gland) in response to the rise in blood sugar (glucose) that follows carbohydrate assimilation. Insulin is the only hormone that initiates the conversion, in the liver, of monosaccharide glucose to polysaccharide glycogen.

INTERFERON is a protein, produced by white blood cells, that provides protection from invading viruses.

IODINE is a trace mineral needed by the thyroid (endocrine) gland to biosynthesize a variety of hormones which control energy expenditure and how fast beta carotene is converted to vitamin A. The best source of dietary iodine is *kelp* and ocean fish. *AFA* provides a small daily amount. Unfortunately, fast food restaurants use too much "iodized" salt in their food. Such high doses of iodine often cause acne.

IRON is the "rosy cheek" trace mineral absolutely essential for human survival. Fatigue, irritability, depression, and low enthusiasm are all typical symptoms of iron deficiency. Iron has been used to bioactivate enzymes, which have been essential in DNA synthesis by ancient microorganisms like blue-green algae for 3-4 billion years. Iron also biostimulates the immune system by generating free radical "bullets" that can help to destroy invading bacteria.

Iron is available in red meat, salmon, brewer's yeast, green vegetables, and *AFA.* Iron absorption takes place via the intestinal walls over a period of roughly three hours. Vitamin C increases iron absorption, as does a small amount of copper, cobalt, or manganese, all of which are present in *AFA.* Hemoglobin molecules give blood its red color and contain iron ions that are used to transport oxygen to every cell of our body.

ISOLEUCINE is an essential branched-chain amino acid (BCAA) that is available in meat, fish, and *AFA.* Concentrated mostly in our muscle tissue, it is biochemically used in energy production and muscle building by those who have been ill for a long time and need to restore lost muscle mass. It seems to reduce muscle twitching and tremors by stabilizing blood sugar. Isoleucine is also useful in treating liver damage from alcoholism and neurological side-effects from an impaired liver.

JAUNDICE refers to the yellow tint imparted to the skin and eyes by an excess of yellow bile pigment (bilirubin), the natural breakdown product of red hemoglobin. This can occur when red blood cells are dying faster than the liver can process them.

KREBS CYCLE, or citric acid cycle, is a series of important biochemical events wherein specific enzymes oxidize proteins, carbohydrates, and fatty acids for energy production. The cycle may be thought of as a "wheel" that turns faster (and thereby produces more energy) as the amount of enzyme is increased. The energy produced in this cycle supplies us with 90 percent of our needs.

LACTASE is a digestive enzyme secreted by gastrointestinal mucous cells that "breaks" (hydrolyzes) lactose (milk sugar) into glucose and galactose. Some people have permanently lost this enzyme and so have a condition called "lactose intolerance:'

LACTIC ACID is an acid produced by the bacterial fermentation of lactose (milk sugar) and during muscle contraction.

LACTOBACILLUS ACIDOPHILUS is a type of friendly bacteria that primarily inhabits the small intestine in humans and animals and manufactures the enzyme lactase, which digests sugars such as lactose (milk sugar). *Acidophilus* bacteria also produce lactic acid and natural antibiotics, which kills off *Candida* yeast as well as the invading population of 27 different bacteria, including *salmonella.* They even help lower cholesterol. Unfortunately, the use of certain medicines and drugs can drastically reduce the population of these friendly bacteria in the intestine.

LACTOBACILLUS BULGARICUS may be used to make yogurt, along with *Streptococcus thermophilus,* another friendly intestinal bacterium. Both types produce antibiotics and lactic acid, which kill harmful bacteria, thus creating a better intestinal environment for the proliferation of other friendly bacteria, such as *Lactobacillus acidophilus* and *Bifidobacterium bifidum.*

LACTOSE is milk sugar, a disaccharide of glucose and galactose.

LACTOSE INTOLERANCE is the inability to digest lactose due to the absence of the enzyme lactase. Because lactase diminishes with age, "lactose intolerance" can develop in many adults and lead to stomach cramps, nausea, and even diarrhea. Most non-Caucasian adults have this condition and should avoid or minimize milk product ingestion.

LAKE CHAD is an alkaline African lake in Chad, Africa, where the local Kanembu tribal people gather blue-green algae and dry it into protein-rich cakes called "dihe."

LEAKY-GUT SYNDROME occurs when toxins, bacteria, and/or undigested food molecules pass into the bloodstream due to complications from bowel toxemia. Such molecules cause an immune reaction when antibodies are formed against them. Unfortunately, healthy tissue is also attacked. This condition may contribute to degenerative diseases such as Rheumatoid Arthritis and thyroid disease. Since these toxins can damage and deform the DNA structure in healthy cells, cancer may develop as well.

LECITHIN is a phospholipid made up of a glycerol molecule chemically bonded to two fatty acids and one phosphate group. As a result, lecithin acts as an emulsifier and may help to dissolve cholesterol deposits. The lecithin in *AFA* helps to create a "healthy cell membrane" Lecithin deficiency may even lead to Alzheimer's disease. The lecithin molecule is technically named phosphatidylcholine because it can be broken down into glycerin, two fatty acids, phosphorous and choline. The choline portion keeps the cell membrane fluid and flexible, allowing for the passage of nutrients into (and wastes out of) the cell.

Some British researchers believe that lecithin repairs the cell membranes of unhealthy liver cells and thus has some antiaging properties. Lecithin may possibly protect against heart disease because it helps to elevate "good cholesterol" It may also improve short-term memory because the choline in the lecithin is used in the biosynthesis of acetylcholine, a neurotransmitter required by many nerve cells.

LEUCINE, an essential branched-chain amino acid, is available in animal protein, wheat germ, and *AFA* algae. Similar to isoleucine, it is concentrated in muscle tissue and used to produce energy. It stabilizes blood sugar and may be particularly helpful in reducing some of the uncomfortable symptoms of hypoglycemic conditions.

LEUKOCYTES. *See* white blood cells.

LINOLEIC ACID, an essential fatty acid, must be obtained from a good diet. It cannot be biosynthesized by humans and must be continuously supplied to sustain life. Linoleic acid is a precursor for gamma-linolenic acid (GLA), which is required for needed prostaglandin synthesis. However, various factors "block" linoleic acid from being converted to GLA. Saturated fats, cholesterol, alcohol, viral infections, and zinc deficiency act as blocking agents. Thus, even though we may get linoleic acid from our diet, we often do not have the ability to metabolize it to GLA. Consuming *AFA* algae on a daily basis offers a direct source of GLA.

LIPASE is a digestive enzyme that breaks down lipids, usually in the form of triglycerides, in the small intestine. The enzyme originates in the pancreas and is called pancreatic lipase.

LIPIDS are fats, waxes, and fatty acids found in all plants and animals. They are insoluble in water. Lipids are usually found in the form of triglycerides, a chemical combination of the "sweetener" glycerol with three attached fatty acids that vary in their similarity to each other. Fat digestion takes place in the intestine with the help of the liver's bile (an emulsifier) and pancreatic enzymes. They are broken into fatty acids (and glycerol), which are later resynthesized back into lipid triglycerides. Since lipids make up about 60 percent of the brain, their dietary source ought to be as good as those found in *AFA*.

LIPID PEROXIDATION refers to the oxidative weakening of lipids. A reaction occurs whereby oxygen atoms are altered into peroxides. Peroxides act as free radicals that "steal" electrons from lipid-rich membranes, which results in cell damage. Exposure to toxins or infections in the body can stimulate lipid peroxidation. In addition, end products of lipid peroxidation may be mutagenic and carcinogenic.

LIPOPROTEINS are complex proteins that are chemically bonded to lipids (fat). Low density lipoproteins ("bad" cholesterol) contain about 35 percent cholesterol, whereas high-density lipoproteins ("good" cholesterol) contain about half that much (17 percent).

When white blood cells encounter an LDL molecule damaged by free radicals, they engulf them, become engorged, and then stick to the artery walls as "foam cells" This blocks arteries and causes high blood pressure. *AFA* may help to reverse this effect.

LIVER, the largest organ in the body, has an extraordinary wide variety of functions. The liver secretes bile which breaks up fat particles in the small intestine. This organ also stores glycogen for later use, along with iron and various B vitamins. All blood returning from the stomach, intestines, spleen, and pancreas is "detoxified" by the liver such that most poisons, drugs, and medicines can be broken down and rendered harmless.

LYMPH is a fluid that flows parallel to the blood circulatory system and primarily contains lymphocytes. The Latin word *lympha* means water.

LYMPHOCYTES are white blood cells that fight infection. When stimulated by an antigen such as foreign proteins or polysaccharides, lymphocytes produce antibody proteins that bind to their surface. They number about a trillion in our bodies and pass through the thymus gland, a walnut-sized organ at the base of the neck. When "fighter cells" called T-lymphocytes (T for thymus) encounter an invading virus, they surround the human cell within which the invading virus has hidden. These T-cells, sometimes called "natural killer" cells (NKs), release toxic molecules that open up these cells, releasing the foreign virus (antigen). The T-cells then release their antibodies which bind to the surface of the protein-coated virus. Fever, swollen lymph glands, and muscle aches usually accompany this process. Now that the virus is "marked" by the bonded antibody, white blood cells called macrophages ("big eater" in Greek) engulf, digest, and destroy the virus. Vitamins B_6 and B_{12} along with amino acids help the T-cells to produce the antibodies that mark invaders. *AFA* is very rich in all of these immune-stimulating ingredients.

LYSINE is an essential amino acid that is available in wheat germ, pork, fish, dairy products, and *AFA*. Concentrated in muscle tissue, lysine, along with vitamin B6, helps in the intestinal absorption of calcium. This helps to prevent osteoporosis. Lysine may also help to reduce lead toxicity by chelation. Along with vitamin E and iron, lysine is used in collagen (skin protein) formation.

LYSOZYME, an enzyme in teardrops, protects the eyes by chemically destroying the cell walls of invading bacteria, rendering them harmless.

MAGNESIUM, an important "antistress" mineral, plays a role in about 325 different enzyme systems that control almost all vital biochemical processes. A magnesium deficiency can bring on migraine headaches and high blood pressure; getting enough magnesium will contribute toward a healthy and vibrant feeling. Unfortunately, magnesium deficiency is much too common. The USDA tells us that the average American gets only 25 percent of the RDA of magnesium. Long-term deficiency includes fatigue, twitching, and confusion. The magnesium in *AFA* may also be helpful in reducing anxiety, irritation, mood swings, and fatigue. The best dietary sources of magnesium are *AFA*, seafood, greens, nuts, and grains.

The magnesium in *AFA* is chelated inside of the chlorophyll molecule. As such, it is effective as a natural tranquilizer because it relaxes skeletal muscles and the heart and also reduces fatigue. Since hard water is high in both calcium and magnesium, in geographical areas where the water is hard, people have a lower incidence of cardiovascular problems. The magnesium in *AFA* may also be helpful in reducing premenstrual stress syndrome symptoms. Magnesium absorption is low partly because sugar and dairy products seems to lessen its absorption. The phosphate in all soft drinks will leech valuable magnesium from the body. Diabetics, the elderly, alcoholics, and pregnant women would all probably benefit from the high concentration of magnesium in *AFA*.

MALABSORPTION SYNDROME is a general disease in which nutrients cannot be properly absorbed and utilized. Malnutrition is the resulting condition.

MANGANESE is a little-understood trace mineral that is important in the formation of bone and in the metabolism of glucose and protein. Although the human body contains only a "pinch" (15-20 milligrams) of this metal, it is enough to catalyze the biosynthesis of dopamine, an important neurotransmitter. Manganese is found in small amounts in greens, nuts, seeds, whole grains, and *AFA*. Unfortunately, high-tech farming depletes manganese such that most people are deficient in this mineral. Refinement of wheat to white flour depletes 90 percent of its manganese.

Manganese probably helps to decrease bone joint inflammation by increasing the production of soothing mucopolysaccharides. This is partly related to the role manganese plays in a variety of antioxidant enzyme systems. It may be helpful in relieving fatigue, nervousness, and poor memory. Since cancerous tumors are known to be exceptionally low in manganese, this mysterious mineral may play an unknown anticancer role.

MANNA was a food referred to in the Bible and eaten by the wandering Hebrews in Israel. Some archaeologists speculate that it was a blue-green algae, which dried on rocks and turned into a sweet white polysaccharide.

METABOLISM refers to all biochemical reactions which involves either the synthesis (anabolism) or breakdown (catabolism) of molecules inside any cell.

METHIONINE is an essential amino acid found in nuts, sunflower seeds, rice, corn, liver, eggs, fish, and *AFA* algae. It contains sulfur (like cysteine) and is therefore important for skin and nail growth, fat reduction, and preventing fatigue, depression, and some allergies. When bio converted to cystine, methionine also serves as an antioxidant. Some studies suggest that "excessive" alcohol destroys or otherwise greatly reduces the amount of methionine.

Methionine biostimulates specific enzymes with the help of vitamins B_{12} and B6. It can then increase the amount of memory-enhancing neurotransmitters such as dopamine, and mood-elevating polypeptides such as endorphins. These polypeptides can properly be absorbed from *AFA* directly into the bloodstream, thus positively affecting physiological behavior directly. Methionine was probably one of the first amino acids available in earth's ancient primordial seas, billions of years ago. This amino acid was (and is still) used by blue-green algae to biosynthesize glutathione, Earth's first antioxidant. Methionine has also been shown to help humans detoxify lead and copper in the blood. Studies have even shown radiation protection as well. Some of our friendly bacteria (if we have healthy intestinal flora) can synthesize this needed methionine from the more accessible amino acid, aspartic acid.

MICRO GRAM is a unit of weight equal to one-millionth of a gram. It is abbreviated as *mcg.*

MICRONUTRIENTS are profoundly necessary physiological substances such as trace minerals, vitamins, certain essential amino acids, coenzymes, antioxidants, bioflavonoids, and essential fatty acids present in and required by the body with a concentration typically measured in micrograms or milligrams.

MICROVESICLES sometimes called exosomes are small membrane bound fragments of between 50 and 1,000 nanometers (nm) in diameter and are found in many types of body fluids, such as blood. Microvesicles are released by most cell types in physiological conditions.

MINERAL ASH (OR ALKALINE ASH) refers to the residue of a food after its complete combustion or burning with oxygen gas. Mineral compounds such as magnesium oxide and calcium oxide are left over and weighed as a measure of alkalinity.

MINERALS are rocks and ores that contain metallic elements (e.g., iron, calcium, and zinc) and nonmetal elements (e.g., chloride and phosphate) that are absolutely essential for the existence of all life forms — plant, animal, bacteria, or algae. Minerals such as calcium phosphate are used by the body in the form of bones and teeth; minerals such as iron or zinc are essential components for the glorious workings of enzymes.

MITOCHONDRIA are regions within eukaryotic cells where energy is stored as adenosine triphosphate, or ATP. Each mitochondrial "organelle" is a magnificent factory of "oxidative enzymes" that slowly release enough energy so that the cell can survive. Ninety percent of the molecular oxygen used by the cell is used up in the mitochondria.

MOLECULE is the smallest unit of a chemical compound capable of existing independently while retaining properties of the original substance.

MOLECULAR SHIELDS are protective antioxidant molecules that can disarm dangerous forms of oxygen called free radicals, which are generated during many stages of cellular metabolism. The most important antioxidant molecular shields in *AFA* are beta carotene, superoxide dismutase, glutathione, vitamin C, and vitamin E.

MOLYBDENUM is a very rare mineral and important micronutrient. It is available in *AFA* algae and can bioactivate several human enzymes associated with longevity enhancement, free-radical absorption, and the ability to neutralize carcinogenic nitrosamine molecules, which are linked to colon cancer.

MONOCYTES are large leukocytes (white blood cells) that can attack and engulf great numbers of invading bacteria, destroying them with the aid of powerful enzymes.

MONOSACCHARIDES are simple sugars composed of single molecules. *AFA* contains glucose, galactose, mannose, and ribose monosaccharaides.

NEURONS are specialized nerve cells that can communicate with each other by long cellular extensions that conduct electrical impulses. The flexibility of neurons is enhanced by high-quality essential fatty acids, such as those found in *AFA*.

NEUROTRANSMITTERS are molecules that transmit messages through nerve tissue. They are usually composed of several amino acids and play a key role in mental states such as anxiety and depression. The interaction of neuropeptides and receptor sites are where images turn into chemistry, chemistry into images.

NIACIN (VITAMIN B3) is a "stress reducer" B vitamin water-soluble molecule. It is readily obtained from *AFA*, meat, fish, and green vegetables and is used in enzymes to repair cells and convert food into energy. Niacin actually refers to either nicotinic acid or nicotinamide. A lot of research points to a marked cholesterol-lowering effect by niacin, even in relatively small amounts. In the form of nicotinic acid, it may be able to reverse some atherosclerosis. Blood triglycerides have been lowered by as much as one-half in some human studies, followed by blood pressure lowering. Niacin often detoxifies damaged liver cells in alcoholics and heroin addicts and thus lowers cravings for these substances.

NICKEL is thought to be an essential trace mineral and is involved in bioactivating certain enzymes that break down glucose. Deficiency of this mineral may impair the liver. Nickel is found in soybeans, green peas, and *AFA* algae.

NITROGEN FIXATION is carried out in the heterocyst cells of blue-green algae. With the help of nitrogenase enzymes, nitrogen gas can be transformed into ammonia for the eventual manufacture of amino acids.

NOSTOC is a genus of blue-green algae with cells arranged in beadlike chains that are grouped together into a gelatinous mass. Some species of *Nostoc* have been prized as a food source in Asia for over a thousand years

NUCLEIC ACIDS are polymers of ribose sugar and phosphate, alternating with purine and pyrimidine molecules (nitrogenous organic bases), that are found in all cells except human red blood cells. The two major types of nucleic acids are DNA and RNA. DNA (found in the nucleus) and RNA (outside the nucleus) work together to store genetic information that dictates how all proteins and enzymes can be correctly synthesized.

NUTRIENTS are biochemical substances needed for the fundamental survival of life. Nutrients are made available to us as proteins, carbohydrates, lipids (fats), vitamins, minerals, and water.

NUTRITION is a constant and daily process of being nourished. This can refer to food, water, or air, as well as all else that enters our body and mind, such as our daily thoughts, sights, and psychological experiences. Good nutrition also means ingesting those nutrients which — by optimum digestion and nutrition — positively affect the health of every cell of our body. This makes nutrition an important aspect of preventive medicine as well as an often overlooked component in corrective medicine.

OLIGOSACCHARIDES are monosaccharide polymers found within the cell walls of blue-green algae.

ORGANELLES are substructures or "little organs" within the cell such as chloroplasts, mitochondria, and ribosomes. Most evolutionary biologists believe that organelles were once independent cells. The photosynthetic chloroplasts that give plants their green color have been hypothesized to be linked to primitive, yet independent forms of blue-green algae.

ORGANIC is a word with two meanings. To the chemist it simply means molecules that contain carbon (other than CO_2 and metal carbonates). To the consummate organic farmer, it refers to farming practices that do not employ the use of pesticides and herbicides.

ORGANIC ACIDS are molecules that contain one small cluster of carbon, oxygen, and hydrogen atoms (called a functional group) usually placed at one end of the molecule. When placed in water, the hydrogen atom is pulled off as an ion, and the pH is lowered. Organic fatty acids such as linoleic or linolenic acids are well-known examples of long-chain acids; acetic and lactic acids are examples of short-chain acids.

ORTHOMOLECULAR MEDICINE is a medical treatment that uses the body's own naturally occurring substances (e.g., amino acids) to promote nutritional balance. The word "ortho" simply means normal or endogenous to human physiology. According to the late Dr. Linus Pauling — who originated the term — a natural micronutrient rich diet is essential for good health. *AFA* provides just such a vital spectrum of orthomolecular nutrients.

OSTEOPOROSIS is a disease whereby bones become porous due to insufficient bone protein production.

OXIDATION refers to a general type of chemical reaction, which involves the addition of oxygen atoms, the removal of hydrogen atoms, or both. For example, when glucose is oxidized by molecular oxygen, glucose's carbon and hydrogen atoms both break away and become enriched with oxygen. When a saturated fatty acid is oxidized, unsaturation develops as hydrogen atoms are removed. In general (and from a more modern perspective), carbon in food undergoes oxidation by *losing electrons* during the Krebs cycle within the cell's mitochondria.

PANCREAS is an organ (behind the stomach) that secretes pancreatic enzymes, which digest a variety of starches, fats, and proteins. In addition, the hormones insulin and glucagon are produced to regulate glucose levels.

PANCREATIN is a secretion from the pancreas that contains digestive enzymes (e.g., amylase, trypsin, and lipase).

PANCREOZYMIN, a hormone produced by the wall of the small intestine, stimulates the production of pancreatic "juices!' These juices contain various types of digestive enzymes such as proteases (to digest proteins), lipases (to digest fats), amylases (to digest carbohydrates), and nucleases (to digest nucleic acids).

PANTOTHENIC ACID (VITAMIN B5), the "antiaging" vitamin, plays a variety of roles in enzymatic reactions throughout the body. It is obtained from liver, eggs, *AFA* algae, and whole grains. This vitamin is especially important for how your body uses amino acids to form proteins and manufacture hormones. Vitamin B_5 is also an antioxidant that helps to protect cells from the damage of free radicals. It alleviates some morning arthritis pain and joint disorders because of such antioxidant properties. Vitamin B5 biostimulates the immune system by activating certain types of white blood cells. One of the B_5 enzymes may even speed up the detoxification of alcohol toxins (acetaldehyde) in the blood.

PASTEURIZATION is a food process that uses mild heat to reduce microbial levels in substances, such as milk, which are sensitive to heat. Endogenous enzymes are usually destroyed in this process.

PHAGOCYTES refer to specialized immune cells (macrophages) in the blood and lymph fluid that are capable of traveling freely in body fluids to ingest bacterial invaders. Since their ability to do so depend partly on the flexibility of their cell membranes, ingesting essential fatty acids — such as those in *AFA* — is helpful.

PHARMACOLOGY is the study of the action of drugs on the body. Drugs will only increase or decrease a specific action already present in a cell. Drugs will *not* introduce a new action in the body as will the biostimulating effects of nutrient-dense foods such as wheat germ, alfalfa sprouts, and *AFA*.

PHENOTYPE is the composite an organism's observable traits or characteristics. Whereas a genotype is the sum total of inheritance, only a small number of traits are outwardly visible or functionally active (phenotypical expression), most genetic traits lie silent, unexpressed.

PHOSPHOROUS is a nonmetal mineral element that is present in every cell, especially in bones and teeth. It is vital to energy production and protein synthesis and is readily available from most protein foods such as meat and *AFA* algae.

PEPTIDE BONDS are those which hold one amino acid to another in a peptide or protein. Such as bond is sometimes called an "amide linkage and digestion is needed, with an appropriate enzyme, to break it.

PEPTIDOGLYCAN is the rigid layer of a cyanobacterial (blue-green algae) cell wall that is composed of several types of complex sugars and short polypeptides.

PESTICIDES are chemicals used to destroy insects and animals that impact agriculture. Currently the Environmental Protection Agency allows more than 400 varieties of pesticide, equaling 2.6 billion pounds, to be used per year in the United States. This comes out to approximately 10 pounds per year for every man, woman, and child in America. When laboratory animals are given three pesticides (instead of an equal weight of one), death often results. This phenomenon is called "multiple exposure syndrome.

pH is a measure of the acidity (or basicity) on a scale ranging from 0 to 14. The letter "p" stands for "power" and the letter "H" stands for hydrogen ion concentration. A pH value less than 7 is acidic, more than 7 is basic, and equal to 7 is neutral. The pH of blood is basic 7.4.

PHENYLALANINE is an essential amino acid found in milk, meat, wheat germ, and *AFA*. This interesting molecule has a direct effect on brain chemistry because of its relatively rare ability to very quickly cross the "blood-brain barrier:' In the liver, phenylalanine is easily bioconverted to such important neurotransmitters as adrenaline and dopamine.

With the help of B vitamins, phenylalanine enhances alertness, concentration, and learning abilities. Phenylalanine seems to diminish addictive cravings for alcohol and sugar. By slowing down the breakdown of morphine-like enkephalin and endorphin peptides, back pain, arthritic pain, and even depression are reduced. The phenylalanine in *AFA* is also useful in a wide variety of brain and intestinal peptides. Long term use of smaller amounts (as in *AFA*) has not yet been carefully studied. Unfortunately, drinking diet sodas that contain the artificial sweetener aspartame, a dipeptide of phenylalanine and aspartic acid, can cause excessive absorption by the brain and diminish some of the beneficial effects of *AFA*.

PHOSPHOLIPIDS are triglyceride fat molecules that help to form flexible cell membranes.

PHOSPHATIDYLCHOLINE is the most common phospholipid present in and important to the structure and flexibility of all cellular membranes — plant, animal, or algae. It consists of phosphate, choline, and a B vitamin.

PHOTOSYNTHESIS is a process whereby plants and algae utilize the energy of the sun and the catalytic presence of chlorophyll to combine CO_2 and H_2O to synthesize glucose. Carried out by a complex chain of chemical events, all life in Earth's biosphere ultimately depends upon this transformation. Actually about half of all photosynthesis on Earth is carried out by blue-green algae, one of the earliest pioneers of this life-giving process.

AFA carries on photosynthesis within ancient bio factories called thylakoid sacs, which are complex beta carotene- and chlorophyll-containing factories that assimilate sunlight. Their photosynthetic efficiency is reflected by the fact that no other life form has such generous amounts of these transformative pigments.

PHYCOBILIPROTEINS are part of a larger light harvesting antennae protein complexes called phycobilisomes. Phycobiliproteins are water-soluble, so they cannot exist within the membrane like carotenoids can. Instead, phycobiliproteins form clusters that adhere to the membrane called phycobilisomes. Phycobiliproteins have fluorescent properties that are used in immunoassay kits for labeling antibodies, among other applications. Each phycobiliprotein has a specific absorption and fluorescence emission maximum in the visible range of light.

PHYCOBILISOMES are light-harvesting "antennae" structures, which use groups of photosynthetic accessory pigments to channel a wider spectrum of light to chlorophyll, thereby increasing energy for photosynthesis. This is especially useful under water conditions, where longer wavelengths of light become inaccessible to chlorophyll. Phycobilisomes are located on the surface of thylakoid membranes in all cyanobacteria and red algae.

PHYCOCYANIN is from the Greek *phyco* meaning "algae" and *cyanin* is from the English word "cyan", which refers to a shade of blue-green (close to "aqua"). The product phycocyanin, produced by *Aphanizomenon flos-aquae* and *Spirulina*, is a potent antioxidant pigment. Also, phycocyanin is used by the food and beverage industry as a natural coloring agent 'Lina Blue' and is found in sweets and ice cream.

PHYCOERYTHRIN is an antioxidant pigment that can give blue-green algae a red coloration. Phycoerythrin is a pigment-protein complex from the light-harvesting phycobiliprotein family, along with allophycocyanin and phycocyanin. They are accessory pigments to chlorophyll. Also, phycoerythrins are among the brightest fluorescent dyes ever identified.

PHYCOLOGY is the study of algae.

PHYTOCHROME is a light sensitive pigment, a photoreceptor, which plants and algae use to detect light. Phytochromic pigments change their "color" upon light absorption. They are especially sensitive to light in the red and far-red region of the visible spectrum. A phytochrome is a protein with a bilin chromophore.

PHYTOPLANKTONS are suspended algae that drift freely with the water's current. The great blue whale subsists solely on such algae.

PIGMENTS are colored molecules. Important examples are green chlorophyll, orange-red carotenoids, and blue-green phycocyanin.

PLAQUE found on artery walls mostly contains cholesterol and triglycerides. It contributes to artery "hardening" and partial artery blockage. This typically results in high blood pressure or even the complete blockage of arteries.

PLASMA is the liquid that carries blood cells, nutrients, gases, wastes, hormones, and antibodies. When blood cells are removed from plasma, blood serum remains.

PLATELETS are a type of white blood cell that causes blood clotting and, after liberating serotonin, causes small blood vessels to severely contract to prevent further blood loss.

POLYPHOSPHATE GRANULES, found inside the *AFA* cell, are composed of phosphate polymer chains of phosphorous and oxygen, and are used to construct *AFA*s cell membrane.

POLYSACCHARIDES are polymer chains of simple sugars that form starch, cellulose, and glycogen.

POLYPEPTIDES are made up of several peptides. Peptides are chains of at least two or more amino acids. Polypeptides perform many functions in the body. Generally very short polypeptides are called peptides, while typically long ones with more 100 amino acids, are called proteins.

POLYUNSATURATED refers to organic molecules that have one or more double bonds between one or more pairs of bonded carbon atoms. Polyunsaturated fatty acids in *AFA* algae help to keep cell membranes fluid and flexible.

PREBIOTIC SOUP refers to the physical and chemical conditions on planet Earth right before the beginning of life and the proliferation of one-celled alga organisms. PRECURSORS are biomolecules that can be synthetically converted to another useful form. For example, the beta carotene of *AFA* is a precursor to vitamin A.

PROKARYOTIC cells, such as bacteria and blue-green algae (cyanobacteria), do not have a nucleus or nuclear membrane.

PROSTAGLANDINS are powerful hormone-like molecules that affect inflammation, muscle contraction, cholesterol levels, and blood pressure. The essential fatty acids of *AFA* help to synthesize those prostaglandin hormones that are beneficial in raising HDL ("good" cholesterol).

PROBIOTICS is a Greek word meaning "for life" and refers to the friendly bacteria (about 400 different species) that are symbiotically needed in the gastrointestinal tract to maintain good health. They improve digestion and profoundly increase proper bowel function. Typically, they change the acidity of their surroundings and thus kill invading, unfriendly toxic bacteria.

PROLINE, a nonessential amino acid, is a component of collagen and thus helpful in tendon and cartilage formation. Readily found in meat, dairy, wheat germ, and *AFA*, pro-line peptides also contribute to enhanced learning abilities.

PROSTAGLANDINS are extremely potent fatty acids that act like regulator hormones to produce neurochemical effects in very small amounts. They are vital enzyme regulators and hormones that control every cell in every organ.

PROTEASE is a protein-digesting enzyme, which includes pepsin, trypsin, chymotrypsin, and peptidases.

PROTEINS are polypeptide macromolecules made up of a linear arrangement of twenty participating amino acids. The sequence and number of amino acids affect the shape and function of the resulting protein. Some are catalysts (enzymes), and others are structural agents inside and outside the cell. *AFA* algae are about 60 percent protein with all eight of its essential amino acids in perfect balance. Beef, by comparison, is only about 20 to 25 percent protein.

PUTREFACTION refers to the decomposition of proteins in the small intestine, which gives off a foul smell or bad breath.

PYRIDOXINE (VITAMIN B_6) is the "immune system booster" that is obtained primarily from whole grains, brewer's yeast, and *AFA*. Because this vitamin bio activates more than sixty enzyme catalysts, it plays a key role in a wide variety of biochemical reactions such as the production of immune and red blood cells as well as the biosynthesis of neurotransmitters.

RDA means "recommended daily allowance" in reference to vitamins, minerals, and other micronutrients. Because of poor diets, most people do not reach these allowances. Older Americans, for example, may only receive about two-thirds of their B vitamins and even less of the minerals they need. The RDA of nutrients is based on the standards established by the Food and Nutrition Board of the National Research Council.

"The RDA for a vitamin is not the allowance that leads to the best of health for most people. It is, instead, only the estimated amount that for most people would prevent death or serious illness from overt vitamin deficiency."

— Linus Pauling (two-time Nobel Prize winner)

RECEPTOR SITES are regions on the surface of cell membrane proteins that allow messenger molecules (e.g., hormones, neurotransmitters) to temporarily connect so that important chemical reactions may commence.

RED ALGAE, or rhodophyta (from the Greek *rhodos*, red), are a type of "seaweed" that carry on photosynthesis deep within the sea. They are most abundant in warm tropical coastal waters and have been found at depths of 800 feet near the Bahamas. They have a limited use as a harvested food source because of their cellulose cell wall.

RED BLOOD CELLS (ERYTHROCYTES) number about 25 billion in the body. They live about 4 months while being bent, squeezed and deformed as they pass through our tiny capillaries. Creation of a new red blood cell takes about a week in the bone marrow with the help of vitamin B_{12}. Red blood cells do not undergo cell division. Instead, about 1.2 billion red blood cells die every hour and need replacement. *AFA* is unsurpassed as a source of vitamin B_{12}.

REDOX (reduction-oxidation) reactions including all chemical reactions in which atoms have their oxidation state changed. REDOX reactions involve the transfer of electrons between chemical species.

REDOX Signaling is cellular communication using short-lived reactive oxidative species (ROS), reactive nitrogen species (RNS) or reactive sulfur species (RSS), or through changes in the electrochemical potential of REDOX couples within biological systems.

REDUCTION is the gain of electrons or decrease in oxidation state by a molecule, atom, or ion.

RHAMNOSE SUGARS are complex natural plant sugars found in algae.

RIBOFLAVIN (VITAMIN B 2) is an "antioxidant" water-soluble, B-vitamin molecule that helps to biosynthesize glutathione, an important antioxidant that protects against free radical damage. Riboflavin is obtained from *AFA*, milk, and green vegetables. Riboflavin deficiencies are very common, especially in the elderly and alcoholics. Cracks on lips or in the corners of the mouth show this deficiency.

RIBOSOMES are cellular particles in all cells made up of proteins and ribonucleic acid (RNA). Their function is to accurately transcribe the genetic code of DNA and manufacture peptides and proteins from amino acids.

RNA, OR RIBONUCLEIC ACID, is a polymer of ribose sugar, phosphate, and, most critically, the nucleic acids that are intimately involved with DNA in the synthesis of proteins. *Aphanizomenon flos-aquae* is rich in RNA, the nucleic acids of which help regenerate the "mind-brain" function and reportedly enhance memory, concentration, and mental clarity.

SATURATED FATTY ACIDS are derived from animal and vegetable fats. Dietary sources of these molecules should be avoided because they impart rigidity to the cell membrane. They are composed of long-chain organic acids, the carbon atoms of which have the maximum (saturated) number of bonded hydrogens.

SELENIUM is an antioxidant and an immune-stimulating mineral that also provides valuable protection against certain forms of cancer. It is found in wheat germ, liver, and *AFA*. Selenium concentration varies dramatically with the soil from which it is derived. For this reason, garlic and onions vary widely in selenium content.

SERINE is a nonessential amino acid used primarily in brain proteins and nerve coverings. It is used in DNA and RNA synthesis and is especially important in the formation of the phospholipids needed for flexible cell membranes. Found in meat, dairy, and *AFA*, serine is good for the skin and has often been used as a natural moisturizer in skin creams.

SEROTONIN is a powerful neurotransmitter. It is biosynthesized from the amino acid tryptophan found in *AFA*. Serotonin is a brain hormone that can reduce depression, elevate mood, constrict blood vessels, regulate sleep, trigger intestinal peristalsis, and change into melatonin within the pineal gland.

SPECIES are a collection of closely related strains.

SPIRULINA is a most common species of blue-green algae that is sold worldwide as a food supplement, for cosmetic use, and as a natural food coloring agent. *Spirulina* is the most widely researched edible microalgae and one of the easiest to cultivate and harvest.

SPLEEN is an "organ" that is actually a collection of lymph tissue and serves as a reservoir for red blood cells.

STANDARD AMERICAN DIET (S.A.D.) usually consists of processed fats, refined flour and sugar, and too many food additives and pesticides. Most of the U.S. population receives only about three-quarters of their needed vitamins and minerals. The reason for this is that most of the U.S. food supply is grown in nutritionally depleted soil. This leads to an alarming variety of diseases such as heart disease, diabetes, schizophrenia, arthritis, and cancer. Early warning signs of chronic vitamin and mineral deficiency, such as fatigue, mood swings, confusion, anemia, and insomnia, affect the quality of life, and immune deficiencies affect our very life span. Sensible prison officials have discovered that by simply removing refined flour and sugar from the meals served to their prisoners, there was a dramatic reduction in the amount of violent behavior.

STARCH is a glucose polymer found and stored in plant seeds and roots.

STEM CELLS are unique unspecialized cells that have the potential to develop into specialized cells or divide to produce more stem cells. Stem cells are found in multicellular organisms. In a developing embryo, stem stems can turn into all the various specialized cells necessary for the body. In adult organisms adult stem cells act as a repair system for the body.

STRAINS are a population of cells descended from a single cell of a particular species. There are at least seven different Aphanizomenon species and ten different strains of *AFA*.

STRESS is a general term that describes a biochemical tension — whether psychological, environmental, or physiological — to which the body is subjected. Since micronutrients are expended in dealing with stress, micronutrients need to be continually replenished to avoid the weakening of the entire immune system.

STROMATOLITES are among the world's oldest fossils, some of which are estimated at 3.5 billion years old. Stromatolites appear as a rocky column or rounded mass of layered sediment formed by the growth of cyanobacteria. Shark's Bay Australia and off islands in the Caribbean one can view stromalites, which appear similar to fossils billions of years old.

SUBSTRATE MOLECULES are nutrients that are targeted and transformed into cellular components by the catalytic action of biochemical enzymes.

SUPEROXIDE DISMUTASE (SOD) is a significant enzyme and a ubiquitous antioxidant that protects important biomolecules such as DNA from being damaged by naturally generated superoxide free radicals. All organisms contain SOD to varying extents, *AFA* more than most.

SUBSTANCE P is a powerful neurotransmitter composed of interconnected amino acids, which are available in *AFA*. The amazing effect of this neuropeptide is that it "sharpens" the mind by stimulating brain cells to grow more dendritic spines that can enhance learning ability and elevate our sense of self-worth.

SYNTHESIS is the combining of constituents into a unified whole. In biochemistry it refers to the combining of atoms, ions, and/or molecules for the purpose of creating or manufacturing new substances. When new substances are synthesized in the body, enzymes are almost always used.

SYNERGY is the energetic interaction of various nutrients so that their combined effect is greater than the sum of their individual effects.

T-CELLS are a type of white blood cells called lymphocytes that play an important role in the immune system such as helping the body fight diseases or harmful substances.

TETRAPYRROLES are molecules with a chemical structure in which four pyrrole rings are joined in straight chains, as in phycobilins, or as joined rings, as in a chlorophyll or heme. Tetrapyrrole pigments are found in most living organisms and most are light-sensitive.

THIAMINE (VITAMIN B1) is the "nerve and energy" B-vitamin molecule (water-soluble) that is involved in numerous nerve, heart, and muscle tissue reactions. It is obtained from *AFA* algae, brown rice, and grains, and helps to convert blood glucose into energy by activating the necessary enzymes. Deficiency of vitamin B_1 produces fatigue and mental confusion (beriberi disease). Alcohol can produce this deficiency by chemically impairing and reducing this or any B vitamin's absorption in the intestines. Heavy tea and coffee drinkers and the elderly may also have this problem. Vitamin B_1 seems to prevent lead — a ubiquitous pollutant- from depositing in nerve and brain tissue. The biologically active and chelated form of vitamin B_1 (thiamin carboxylase) is easily absorbed and assimilated. The vitamin B_1 in *AFA* is chelated in a similar manner, allowing even a small amount to have beneficial effects. Vitamin B_1 is helpful toward alleviating nervous system disorders. It seems to improve mental clarity.

THIAMINASE is an enzyme present in some fish and thought to be produced by some of the friendly intestinal bacteria of humans.

THREONINE, often called the "immune booster," is an essential amino acid found in dairy, meat, wheat germs, nuts, seeds, and *AFA*. It is needed especially for skin proteins such as collagen and elastin and plays a minor role in reducing fat buildup in the liver. There is some evidence that threonine can promote healing by bio stimulating the thymus gland to produce T-cells with more flexible cell membranes.

THROMBIN is an enzyme found in blood that helps bring about blood clotting. Vitamin K, produced by friendly bacteria, is used in the biosynthesis of this enzyme.

THYLAKOIDS are membrane-bound compartments inside chloroplasts and cyanobacteria (blue-green algae). Thylakoids are the site for light-dependent reactions of photosynthesis. The word thylakoid comes from the Greek thylakos meaning "sac" or "pouch."

THYROID GLAND is an endocrine gland located in the front of the neck, behind the larynx. It produces thyroxin hormone, an important blood-circulating molecule that regulates cellular energy production. Thyroxin also controls glucose absorption in the intestine and glucose metabolism in each cell. The cells in this "master" gland need iodine to biosynthesize this hormone.

TOXIC BOWEL is a common condition in which toxins from bowel bacteria contribute to ulcerative colitis, thyroid conditions, and assimilation and digestive problems. Fiber present in *AFA* tends to bind to bacterial toxins in the bowel and get excreted. Along with a good probiotic regimen that includes *acidophilus* bacteria, *AFA* algae's assortment of complex carbohydrates, balanced profile of amino acids, and wide range of minerals and vitamins (especially beta carotene) are therapeutically significant to reverse such symptoms.

TRACE ELEMENTS are micronutrients such as selenium and chromium that are required by humans in very small quantities. These micronutrient trace elements are typically utilized as enzyme activators. *AFA* offers the widest range of such trace elements — *more than any known food.*

TRANS-FATTY ACIDS are unsaturated chains of carbon atoms, ending in an organic acid group. The "trans" variation usually contributes to an inflexible cell membrane. This negative effect is due to the relatively straight shape of this molecule.

TRANSAMINASE enzymes are those that allow for the chemical transfer of an amino group from one amino acid molecule to another. Vitamin B6 is required as a coenzyme for this transamination process Amino acids are termed "essential" when there are no transaminase enzymes available to make them.

AFA has all the transaminase enzymes it needs to make the amino acids that humans require.

TRYPSIN, a protein-digesting enzyme, disassembles partially digested proteins in the small intestine. It is so specific that it only breaks the bond between lysine and arginine.

TRYPTOPHAN, an essential amino acid, may be considered an essential vitamin because it is used in the biosynthesis of niacin, or vitamin B_3. Tryptophan is a precursor for the very important neurotransmitter serotonin, a brain hormone and general messenger molecule needed by the brain to bring on sleep and affect mood patterns. Tryptophan is found in small amounts in *AFA*, turkey, and milk, as well as some nuts and seeds.

TYROSINE is a nonessential amino acid that is concentrated mostly in muscle tissue and a little in the brain. It is biosynthesized from and biochemically similar to phenylalanine, an essential amino acid. Known as the "antidepressant" amino acid because of its importance to brain nutrition, it promotes mental alertness, learning, and memory.

UPPER KLAMATH LAKE is the largest lake in Oregon and the home of *Aphanizomenon flos aquae* — *AFA* blue-green algae. Luckily, the lake averages only eight feet deep, thousands of years of mineral runoffs and about 300 days of sunshine per year help to create the ideal conditions needed for our precious algae blooms.

VALINE is an essential branched-chain amino acid. Charitably enough, it is available in most foods and builds muscle tissue. A valine deficiency has produced "aimless circling" in laboratory animals. This could be caused if too much isoleucine and leucine go to the brain, blocking transport of valine. Amazingly, the proportions of all three of these interrelated BCAAs in *AFA* are nearly identical to the proportions recommended by the National Academy of Sciences.

VANADIUM is an essential trace metal slightly present in good soil. It is found in unsaturated olive oil, carrots, radishes, cabbage, and *AFA*. It is especially concentrated in black pepper and dill seed. Vanadium is involved in enzyme systems that regulate blood sugar, bone formation, fat metabolism, and red blood cell production. There was a time when vanadium was considered a cure-all by turn-of-the-century French physicians. Still today there is much that is unknown and more research is needed to verify its biochemical functions.

VIRUSES are living bundles of protein and nucleic acids capable of destroying the host cells within which they reproduce, especially in weakened immune systems.

VITAMIN A is an antioxidant vitamin that is stored in animal liver and available from plants and blue-green algae in the form of beta carotene.

VITAMIN B1. *See* thiamine. VITAMIN B2. *See* riboflavin. VITAMIN B5. *See* pantothenic acid. VITAMIN B3. *See* niacin. VITAMIN B6. *See* pyridoxine. VITAMIN 12. *See* cobalamin.

VITAMIN C, OR ASCORBIC ACID, is an antioxidant and an immune-stimulating vitamin that helps produce collagen and aids in the absorption of iron. It is somewhat effective in reducing the duration of the common cold by about 30 to 50 percent, deters gum bleeding, maintains good vision, and may be helpful in preventing diabetes. Research evidence shows that a daily intake of this vitamin may reduce the incidence of stomach and esophagus cancer.

VITAMIN E (TOCOPHEROL) is a "nerve healing" vitamin that is essential to animals and is an important antioxidant (free radical protector), as is beta carotene. It is widely distributed and concentrated in vegetable oils, whole grains, eggs, and *AFA*. Unfortunately, half the amount of ingested vitamin E is excreted. Tocopherols *(tocopherol is* a Greek word meaning "body bearer") help to prevent the oxidation of other fat-soluble vitamins. This vitamin, by strengthening cell membranes, keeps invading viruses from penetrating into a cell. Also, red blood cells appear healthier and have prolonged lives. It is now an established fact that vitamin E will slow down or even reverses a variety of neurological disorders caused by nerve damage.

VITAMIN F. *See* essential fatty acids.

VITAMIN K, OR PHYLOQUINONE, is a fat-soluble micronutrient produced by friendly intestinal bacteria and associated with the chlorophyll of green plants and algae. It is sometimes called the "antihemorrhage" vitamin. In fact, the letter K stands for the German word *koagulation.* Deficiency is uncommon so long as friendly intestinal bacteria continue to manufacture this mysterious tumor-inhibiting substance. People with dysfunctional bowel syndrome have a dangerously-low probiotic population and risk vitamin K deficiency. Bruising easily and developing black-and-blue marks is a warning of this deficiency. Taking *acidophilus* to nourish the microflora is strongly recommended.

VITAMINS are biomolecules that are required for health and survival. Because English scientist, Casimir Funk, showed in 1906 that these molecules were "vital; and thought they belonged to the "amine" family, he gave vitamins their name. Although they are vital, they are now known to extend well beyond the amine family. Sometimes essential fatty acids are collectively referred to as vitamin E

WHITE BLOOD CELLS (LEUKOCYTES) include many different cell types with varying functions and purposes. Roughly, for every 1000 red blood cells there is one white blood cell. White blood cells live approximately 12 days, while red blood cells live from 100 to 120 days. There is a rise in the white blood cell population after an infection. White blood cells then destroy invading viral and bacterial particles with an "amoebic-like" action. Even oxidized fats left over from partially digested meals are engulfed. White blood cells use their supple and graceful motion to actually pass through capillary walls and enter infected tissues, as needed.

ZINC is a profoundly important micronutrient mineral because of the role it plays in more than 100 different enzymes. It can build healthy cell membranes, maintain healthy skin, bio stimulate the immune system, and help in the synthesis of DNA and RNA. One zinc-containing enzyme (carboxypeptidase) improves protein digestion. Zinc is found readily chelated in *AFA* algae, oysters, liver, and pumpkin seeds. It can help to reverse the gradual breakdown of an aging immune system and also prevent and treat the common cold. Stored in the pancreas, zinc reduces some of the symptoms of diabetes and improves the health of the prostate gland. Zinc can also act as an antioxidant and anticancer mineral when associated with SOD (superoxide dismutase). Americans get only a fraction of the zinc they need because much of U.S. farmland has been de-zinc defiled by the agro-industry.

References

Foreword

1. Bruno, J. (2000). *Edible Microalgae: A Review of the Health Research.*
2. Haydarizadeh P., et al. (2013) Plastids of Marine Phytoplankton Produce Bioactive Pigments and Lipids. *Mar. Drugs.* 11. 3425-3471.
3. Tara L. Walker, Chris Collet, Saul Purton. (2005) Algal Transgenics in the Genomic Era1. *Journal of Phycology.* 41(6). 1077–1093.
4. Zarzycki J, Axen SD, Kinney JN, Kerfeld CA. (2012, Oct 23). Cyanobacterial-based approaches to improving photosynthesis in plants. *J Exp Bot.*
5. Koksharova O. Genetic tools for cyanobacteria. (2002, Feb). *Appl Microbiol Biotechnol.* 58(2). 123-37.
6. Bjerga A. USDA 'Plant Hardiness' Map Shifts Temperature Zones North. 1/25/2012; Retrieved from <http://www.bloomberg.com/news/2012-01-25/usda-plant-hardiness-map-shifts-temperature-zones-north-1-.html>
7. Telford W, Moss M, Morseman J, Allnutt F. (2001, May 1). Cryptomonad algal phycobiliproteins as fluorochromes for extracellular and intracellular antigen detection by flow cytometry. *Cytometry.* 44(1). 16-23.
8. Shalay E., Shanab, S., Singh V. (2010). Salt stress enhancement of antioxidant and antiviral efficiency of *Spirulina platensis. J Med Plants Res.* 4. 2622-2632.

Introduction

1. Waterbury J. (2004). Little Things Matter A Lot Overlooked in the ocean until the 1970s, cyanobacteria are among Earth's most important organisms. *Oceanus Magazine.* 43(2). Retrieved from <http://www.whoi.edu/oceanus/viewArticle.do?id=3808>
2. Waterbury J. (2004). Little Things Matter A Lot Overlooked in the ocean until the 1970s, cyanobacteria are among Earth's most important organisms. *Oceanus Magazine.* 43(2). Retrieved from <http://www.whoi.edu/oceanus/viewArticle.do?id=3808>
3. Pope John Paul II. (1996, Oct 22). Message to the Pontifical Academy of Sciences: On Evolution. Retrieved from <http://www.ewtn.com/library/PAPALDOC/JP961022.htm>

Part I Algae to the Rescue: Earth's First Food

Chapter 1: Drawing from the Past to Create Our Future

1. Henrikson, R. and Edwards, M. (2012). *Imagine Our Algae Future: Visionary Algae Architecture and Landscape Designs.* Ronore Enterprises, Inc.

2. Deng, R., Chow, T. (2010, Aug 28) Hypolipidemic, antioxidant, and antiinflammatory activities of microalgae *Spirulina. Cardiovasc Ther.* 4. e33-45.
3. Srinivasakumar K., Rajashekhar M. (2010) Drugs from marine microalgae: investigating medicinal properties of few marine microalgae. Lap Lambert Academic Publ.
4. Texas A&M AgriLife Communications (2011, July 7). Microalgae could be Texas' next big cash crop. *ScienceDaily.* Retrieved October 28, 2013, from <http://www.sciencedaily.com /releases/2011/07/110706144622.htm>
5. Muramoto T., et al. (1999) The Arabidopsis photomorphogenic mutant hy1 is deficient in phytochrome chromophore biosynthesis as a result of a mutation in a plastid heme oxygenase. *The Plant Cell,* 11, 335–348.
6. Bjorn LO, Wang T. (2000, Jan). Vitamin D in an ecological context. *Int J Circumpolar Health.* 59(1). 26-32.
7. Study suggests nutrient decline in garden crops over past 50 years. (2004, Dec. 1). *Austin News Archive,* University of Texas. Retrieved from <http://www.utexas.edu/news/2004/12/01/nr_chemistry/>
8. Relating Soil Depletion and Nutrition Loss. (2011, Apr 23). *E-The Environmental Magazine.* Retrieved from <http://blastmagazine.com/culturefashion/food-and-drink/relating-soil-depletion-and-nutrition-loss/#sthash.0zjQM1kb.dpuf>
9. Daniel Marano. (2014) Rich Dirt, Poor Dirt. *Psychology Today.* Jan./Feb. 39.
10. The New IMA List of Minerals – A Work in Progress. Retrieved October 2013 from <http://pubsites.uws.edu.au/ima-cnmnc/IMA_Master_List_(2013-10).pdf>
11. Soetan K, Olaiya C, Oyewole O. (2010, May). The importance of mineral elements for humans, domestic animals and plants: A Review. *African Journal of Food Science.* 4(5). 200-222.
12. McMaster University. "'Knowing it in your gut': Cross-talk between human gut bacteria and brain." ScienceDaily. ScienceDaily, 24 March 2011. <www.sciencedaily.com/releases/2011/03/110323140247.htm>.
13. Clarke G, Grenham S, et al. (2012, June). The microbiome-gut-brain axis during early life regulates the hippocampal serotonergic system in a sex-dependent manner. Mol Psychiatry. <www.sciencedaily.com/releases/2012/06/120612115812.htm>.
14. Bravo J, Forsythe P, et al. (2011, Sept. 20). Ingestion of Lactobacillus strain regulates emotional behavior and central GABA receptor expression in a mouse via the vagus nerve**.** *Proceedings of the National Academy of Sciences.* 108(38): 1650-16055.
15. Kay B. (1992) Cell Tech August Celebration Lecture.
16. Rasmussen H, Martinez I, Lee J, Walter J. (2009). Alteration of the Gastrointestinal Microbiota of Mice by Edible Blue-green Algae. *Journal of Applied Microbiology.* 107: 1108-1118.
17. Feichl R, Pollan M. (2013, June). The American Table. *Smithsonian.* 82.

Chapter 2: An Evolutionary Perspective on Amino Acids

1. Mullen L. (2001, May 9). Life's Baby Steps - Origin & Evolution of Life. Retrieved from <http://www.astrobio.net/exclusive/73/lifes-baby-steps>
2. Gutiérrez-Preciado, et al. (2010). An Evolutionary Perspective on Amino Acids. *Nature Education*. 3(9). 29. Retrieved from <http://www.nature.com/scitable/topicpage/an-evolutionary-perspective-on-amino-acids-14568445>
3. Moskowitz C. (2011, Aug 18). Origin & Evolution of Life. *Astrobiology Magazine*. Retrieved from <http://www.astrobio.net/exclusive/4161/amino-acid-alphabet-soup>
4. Olson J. (2006). Photosynthesis in the Archean era. *Photosyn. Res.* 88(2).109–17.
5. Imura K, Okada A. (1998). Amino acid metabolism in pediatric patients. *Nutrition* 14(1). 143–8.
6. Webel D, & Baker D. (1999). Cysteine is the first-limiting amino acid for the utilization of endogenous amino acids in chicks fed a protein-free diet. *Nutr. Res.* 19. 569–577.
7. Parsons C, Hashimoto K, et al. (1992). Effect of overprocessing on availability of amino acids and energy in soybean meal. *Poult. Sci.* 71. 133–140.
8. Parsons C. (1992, Jan). Effect of Overprocessing on Availability of Amino Acids and Energy in Soybean Meal. *Poult. Sci.* 71(1). 133-140.
9. Kaufman D. (2006, Aug). Temperature sensitivity of aspartic and glutamic acid racemization in the foraminifera Pulleniatina. *Quaternary Geochronology;* 1(3). 188–207.
10. Fürst P, Stehle P. (2004, June 1). What are the essential elements needed for the determination of amino acid requirements in humans? *Jour of Nutr.* 134 (6 Suppl): 1558S–1565S.
11. Reeds P. (2000, July 1). Dispensable and indispensable amino acids for humans. *J. Nutr.* 130 (7). 1835S–40S.
12. Braverman E, Pfeiffer C. (1987). *The Healing Nutrients Within: Facts, Findings and New Research on Amino Acids.* Keats Publishing, New Canaan, CT.

Chapter 3: Remarkable Nutritive Qualities of Microalgae

1. Abrams K. (1996). Algae to the Rescue. Logan House. 31.
2. Braverman R, Pfeiffer C. (1987). *The Healing Nutrients Within: Facts, Findings and New Research on Amino Acids.* Keats Publishing.
3. Pirt S. Biochemical Society Trans. 1980; 8. 479 cited by Aaronson and Dubinsky. Mass production of microalgae. *Experientia.* 1982; 43(suppl). 42.
4. Ku C, Yang Y, Park Y, Lee J. (2013). Health Benefits of Blue-Green Algae: Prevention of Cardiovascular Disease and Nonalcoholic Fatty Liver Disease. *Jour of Med Food.* 16(2). 103-111.

5. Patent January 3, 2008, WO 2008/000430 A2. Banfi P, et al. Alphanizomenon Flos Aquae Preparation, Extracts and Purified Components thereof for the treatment of neurological, neurodegenerative and mood disorders.
6. Abrams K. (1996). Algae to the Rescue. Logan House. 31.
7. Shimizu Y. (1996). Microalgal metabolites: a new perspective. *Annual Review of Microbiology*. 50. 442.
8. Cohen A, Norman H, Heimer Y. (1995). Microalgae as a source of omega 3 fatty acids. *World Review of Nutrition and Diet*. 77. 1-31.
9. Simopoulou A. (1991). Omega 3 fatty acids in health and disease in growth of development. *American Journal of Clinical Nutrition*. 54. 438-63.
10. Retrieved from <http://spirulinamedicalresearch.blogspot.com/>

Chapter 4: Cyanobacteria Captured Sunlight and Forever Changed our Planet

1. Awramik SM. (1992). The oldest records of photosynthesis. *Photosynth Res*. 33. 75-89.
2. Lane N. (2002). *Oxygen the Molecule that Made the World*. Oxford University Press.
3. Blankenship R. (2010, Oct). Early Evolution of Photosynthesis. *Plant Physiology*. 154(2). 434
4. Margulis L. (1992) *Symbiosis in Cell Evolution: Microbial Communities in the Archean and Proterozoic Eons*. W.H. Freeman.
5. Arias J, Aller M, Arias J. (2009). Surgical inflammation a pathophysiological rainbow. *Journal of Translational Medicine*. 7. 19.
6. Patent January 3, 2008, WO 2008/000430 A2. Banfi P, et al. Alphanizomenon Flos Aquae Preparation, Extracts and Purified Components thereof for the treatment of neurological, neurodegenerative and mood disorders.
7. Patent January 3, 2008, WO 2008/000430 A2. Banfi P, et al. Alphanizomenon Flos Aquae Preparation, Extracts and Purified Components thereof for the treatment of neurological, neurodegenerative and mood disorders.

Chapter 5: A Brief History of Heme Oxygenase (HO)

1. Morse D, Choi A. (2005). Heme Oxygenase-1: From Bench to Bedside. *Am J Respir Crit Care Med*. 172. 660-670.
2. Blankenship R. (2010, Oct). Early Evolution of Photosynthesis. *Plant Physiology*. 154(2). 435.
3. Shekhawat G, Verma K. (2010). Heme oxygenase (HO): an overlooked enzyme of plant metabolism and Defense. *J. Exp. Botany*. 61. 2255-2270.
4. Shekhawat G, et al., (2011). Heme Oxygenase: Enzyme with Functional Diversity. *Journal of Stress Physiology and Biochemistry*. 7(1).

5. Hardison R. (1999). The Evolution of Hemoglobin: Studies of a very ancient protein suggest that changes in gene regulation are an important part of the evolutionary story. *American Scientist.* 87(2). 126.
6. University of Leicester (2009, June 12). Heme Proteins: ubiquitous and essential for every organism. *ScienceDaily.* Retrieved October 20, 2013; <http://www.sciencedaily.com /releases/2009/06/090612123914.htm>
7. University of Maryland (2013, Feb 5). Protein that allows safe recycling of iron from old red blood cells identified. *ScienceDaily.* Retrieved October 20, 2013; <http://www.sciencedaily.com /releases/2013/02/130205131752.htm>
8. Troxler R, Dokos J. (1973, Jan). Formation of Carbon Monoxide and Bile Pigment in Red and Blue-Green Algae. *Plant Physiol.* 51(1). 72–75.
9. Kikuchi G, Yoshida T, Noguchi M. (2005, Dec). Heme oxygenase and heme degradation. *Biochem. Biophys. Res. Commun.* 338(1). 558–67.
10. Gambetta G. and Lagarias J. (2001, Sept 11). Genetic engineering of phytochrome biosynthesis in bacteria. *PNAS.* 98(19).10566-10571.
11. Gambetta G, Lagarias J. (2001, Sept 11). Genetic engineering of phytochrome biosynthesis in bacteria. *PNAS.* 98(19).10566-10571
12. Morse, D., and Choi A. (2005). Heme Oxygenase-1: From Bench to Bedside. *Am J Respir Crit Care Med.* 172. 660-670.
13. Bugl H, Fauman E, et al. (2000). RNA Methylation under Heat Shock Control. Molecular Cell. 6. 349-360
14. Retrieved December 15 2013; Learn Genetics: The University of Utah Website; <http://learn.genetics.utah.edu/content/epigenetics/>
15. Retrieved December 15 2013; Learn Genetics: The University of Utah Website; <http://learn.genetics.utah.edu/content/epigenetics/>
16. Retrieved December 15 2013; Learn Genetics: The University of Utah Website; <http://learn.genetics.utah.edu/content/epigenetics/>
17. Henderson B, Calderwood S, Coates A, et al. (2010). Caught with their PAMPs down? The extracellular signaling actions of molecular chaperones are not due to microbial contaminants. *Cell Stress Chaperones.* 15. 123-141.
18. Sakthivel K, Watanabe T, Nakamoto H. (2009, Apr). A small heat-shock protein confers stress tolerance and stabilizes thylakoid membrane proteins in cyanobacteria under oxidative stress. Arch Microbiol. 191(4). 319-28.
19. Maridonneau-parini I, Clerc J, Polla B. (1988, July 15). Heat shock inhibits NADPH oxidase in human neutrophils. *Biochemical and Biophysical Research Communications.* 154(1).179–186.
20. Borchers A, Belay A, Keen L, Gershwin E. Spirulina and Immunity. Chpt. 8. In *Spirulina in Human Nutrition and Health.* (2008). Edited by M.E. Gershwin and Amha Belay. CRC Press. 190
21. Smith R. (1991). Where is the wisdom...! *British Medical Journal.* 303. 798-9.

Part II: Algae to the Rescue: Microalgae and Human Health

Chapter 6: Masters of Oxygen, Antioxidants & Inflammation

1. Robbins J. (1987). *Diet for a New America: How Your Food Choices Affect Your Health, Happiness and the Future of Life on Earth*. HJ Kramer
2. Weil A. (2011, Nov. 4). The Depression-Inflammation Connection. The BLOG. Retrieved from <http://www.huffingtonpost.com/andrew-weil-md/depression-and-inflammation_b_1071714.html>
3. Zhang K. and Kaufman R. (2008, July 24). From endoplasmic-reticulum stress to the inflammatory response. *Nature*. 454(7203). 455–462.
4. Cohen S, Janicki-Deverts D, et al. (2012, April 17). Chronic stress, glucocorticoid receptor resistance, inflammation, and disease risk. *PNAS*. 109(16). 5995-5999.
5. Black PH. (2002, Dec). Stress and the inflammatory response: a review of neurogenic inflammation. *Brain Behav Immun*. 16(6). 622-53.
6. Cowles M, Miller A. (2009). Stress, Cytokines and Depressive Illness. *Encyclopedia of Neuroscience*. Academic Press. 519–527.
7. Zhang K, Kaufman R. (2008, Jul 24). From endoplasmic-reticulum stress to the inflammatory response. *Nature*. 454(7203). 455-62.
8. Arias J-I, Aller M, Arias J. (2009). Surgical inflammation: a pathophysiological rainbow. *J Transl Med*. 7. 19.
9. Arias J-I, Aller M, Arias J. (2009). Surgical inflammation: a pathophysiological rainbow. *J Transl Med*. 7. 19.
10. Arias J-I, Aller M, Arias J. (2009). Surgical inflammation: a pathophysiological rainbow. *J Transl Med*. 7. 19.
11. Eriksen N. (2008, Aug). Production of phycocyanin--a pigment with applications in biology, biotechnology, foods and medicine. *Appl Microbiol Biotechnol*. 80(1). 1-14.
12. Lane N. (2002). *Oxygen the Molecule that Made the World*. Oxford University Press. 127.
13. Tomitani A, et al. (1999, July). Chlorophyll b and phycobilins in the common ancestor of cyanobacteria and chloroplasts. *Nature*. 400(6740). 159-62.
14. Johnson E, Schroeder W. (1996). Microbial carotenoids. *Advances in Biochemical Engineering and Biotechnology*. 53. 119-78.
15. American Academy of Dermatology (2011, Feb 6). Incidence of skin cancer rising at alarming rate. *ScienceDaily*. Retrieved October 5, 2013, from <http://www.sciencedaily.com /releases/2011/02/110205140318.htm>
16. American Chemical Society (2010, April 14). Fats for health and beauty: Giving soybean oil a new role in serving society. *ScienceDaily*. Retrieved October 5, 2013, from http://www.sciencedaily.com /releases/2010/03/100323133039.htm

17. Missouri University of Science and Technology (2012, May 7). Sunscreen ingredient may increase skin cancer risk. *ScienceDaily*. Retrieved October 5, 2013, from http://www.sciencedaily.com /releases/2012/05/120507131951.htm
18. Matsui K, Nazifi E, et al. (2012). The cyanobacterial UV-absorbing pigment scytonemin displays radical-scavenging activity. *J Gen Appl Microbiol.* 58(2).137-44.
19. Gao Q, Garcia-Pichel F. (2011, Oct 3). Microbial ultraviolet sunscreens. *Nat Rev Microbiol.* 9(11). 791-802.
20. Sunscreen In A Pill: Dermatologists Discover Sun Protection Under The Sea. Retrieved from <http://www.*sciencedaily*.com/videos/2007/1108-sunscreen_in_a_pill.htm>
21. Wu L, Lin Y, Yang S, Weng Y, Tsai Y. (2011, Oct). Antimelanogenic effect of c-phycocyanin through modulation of tyrosinase expression by upregulation of ERK and downregulation of p38 MAPK signaling pathways. *J Biomed Sci.* 11. 18:74.
22. Matsui M, Muizzuddin N, Arad S, Marenus K. (2003, Jan). Sulfated polysaccharides from red microalgae have antiinflammatory properties in vitro and in vivo. *Appl Biochem Biotechnol.* 104(1). 13-22.
23. D'Souza D, et al. Intimal hyperplasia. Retrieved December 13, 2013 from <http://radiopaedia.org/articles/intimal-hyperplasia>
24. Sheu M, Cheng H, et al. (2010, Aug.). Molecular mechanism of green microalgae, *Dunaliella* salina, involved in attenuating balloon injury-induced neointimal formation. *Br J Nutr.* 104(3). 326-35.
25. Manoj G, Venkataraman L, Srinivas L. (1992). Antioxidant properties of Spirulina *(Spirulina platensis).* In: Seshadri and Bai. Spirulina MCRC. 48-154.
26. Bhat V, Madyastha K. (2000). C-Phycocyanin: a potent peroxyl radical scavenger in vivo and in vitro. *Biochem Biophys Res Commun.* 275. 20-25.
27. Benedetti S, Benvenuti F, Pagliarani S, Francogli S, Scoglio S, Canestrari F. (2004, Sep. 24). Antioxidant properties of a novel phycocyanin extract from the blue-green alga *Aphanizomenon flos-aquae. Life Sci.* 75(19). 2353-62.
28. Benedetti S, Benvenuti F, Scoglio S, and Canestrari. (2010). Oxygen Radical Absorbance Capacity of Phycocyanin and Phycocyanobilin from the Food Supplement *Aphanizomenon flos-aquae. Jour of Med Food.* 13(1). 223-227.
29. Benedetti S, Benvenuti F, Scoglio S, and Canestrari. (2010). Oxygen Radical Absorbance Capacity of Phycocyanin and Phycocyanobilin from the Food Supplement *Aphanizomenon flos-aquae. Jour of Med Food.* 13(1). 223-227.
30. Lee W. (2004). Acetaminophen and the U.S. Acute Liver Failure Study Group: lowering the risks of hepatic failure. *Hepatology.* 40. 6–9.
31. Kuriakose GC, Kurup MG. (2010). Antioxidant and hepatoprotective activity of *Aphanizomenon flos-aquae* Linn against paracetamol intoxication in rats. *Indian Journal of Experimental Biology.* 48. 1123-1130.
32. Kim S, Pangestuti R. (2011). Biological activities and potential health benefits of fucoxanthin derived from marine brown algae. *Adv Food Nutr Res.* 64. 111-28.

33. Kelman D, Posner E, McDermid K, et al. (2012, Feb.). Antioxidant activity of Hawaiian marine algae. *Mar Drugs.* 10(2). 403-16.
34. Ben-Amotz A, Mokady S, Edelstein S, Avron M. (1989, July). Bioavailability of a natural isomer mixture as compared with synthetic all-trans-beta-carotene in rats and chicks. *Journal of Nutrition.* 119(7). 1013-9.
35. Ben-Amotz A, Levy Y. (1996, May). Bioavailability of a natural isomer mixture compared with synthetic all-trans beta-carotene in human serum. *American Journal of Clinical Nutrition.* 63(5). 729-34.
36. Levin G, Mokady S. (1994, July). Antioxidant activity of 9-cis compared to all-trans beta-carotene in vitro. *Free Radical Biological Medicine.* 17(1). 77-82.
37. Dartsch PC. (2008, May). Antioxidant potential of selected *Spirulina platensis* preparations. *Phytother Res.* 22(5). 627-33.
38. Balz Frei (2000, Jan 9) cited in: The Wall Street Journal - Interactive Edition, Green and slimy and two billion years old? Eat it.
39. Kelly F. (1998, Mar). Use of antioxidants in the prevention and treatment of disease. *Journal of the International Federation of Clinical Chemistry.* 10(1). 21-3.
40. Frei B. (1994, Sept). Reactive oxygen species and antioxidant vitamins: mechanism of action. *American Journal of Medicine.* 97(3A). 5S-13S. 22S-28S.
41. Kehrer, J.P. and Smith, C.V. (1994). Free radical in biology: sources, reactivities, and roles in the etiology of human disease. In *Natural Antioxidants in Human Health and Disease,* Frei, B., (Ed.). Academic Press. 25.
42. Romay C, Gonzalez R, Ledon N, Remirez D, Rimbau V. (2003, June). C-phycocyanin: a biliprotein with antioxidant, anti-inflammatory and neuroprotective effects. *Curr Protein Pept Sci.* 4(3). 207-16.
43. Romay C, Gonzalez R. (2000, April). Phycocyanin is an antioxidant protector of human erythrocytes against lysis by peroxyl radicals. *Journal of Pharmacy and Pharmacology.* 52(4). 367-8.
44. Gonzalez R et al. (1999). Anti-inflammatory activity of phycocyanin extract in acetic acid-induced colitis in rats. *Pharmacology Research.* 39(1). 55-9.
45. Romay C, Ledon N, Gonzalez R. (1999, May). Phycocyanin extract reduces leukotriene B4 levels in arachidonic acid-induced mouse-ear inflammation test. *Journal of Pharmacy and Pharmacology.* 51 (5). 641 -2.
46. Chopra K., Bishnoi M. (2008). Antioxidant Profile of Spirulina: A Blue-Green Microalga. In: *Spirulina in Human Nutrition and Health.* M.E. Gershwin and Amha Belay, (Ed.). Boca Raton, FL: CRC Press.
47. Chopra K., Bishnoi M. (2008). Antioxidant Profile of Spirulina: A Blue-Green Microalga. In: *Spirulina in Human Nutrition and Health.* M.E. Gershwin and Amha Belay, (Ed.). Boca Raton, FL: CRC Press. 107.
48. Chopra K., Bishnoi M. (2008). Antioxidant Profile of Spirulina: A Blue-Green Microalga. In: *Spirulina in Human Nutrition and Health.* M.E. Gershwin and Amha Belay, (Ed.). Boca Raton, FL: CRC Press. 107.

49. Romay C, Ledon N, Gonzalez R. (1998, Aug). Further studies on anti-inflammatory activity of phycocyanin in some animal models of inflammation. *Inflammation Research.* 47(8). 334-8.
50. Romay C, et al. (1998). Antioxidant and anti-inflammatory properties of C-phycocyanin from blue-green algae. *Inflammation Research.* 47(1). 36-41.
51. Chopra K, Bishnoi M. (2008). Antioxidant Profile of Spirulina: A Blue-Green Microalga. In: *Spirulina in Human Nutrition and Health.* M.E. Gershwin and Amha Belay, (Ed.). Boca Raton, FL: CRC Press. 108.
52. Chopra K, Bishnoi M. (2008). Antioxidant Profile of Spirulina: A Blue-Green Microalga. In: *Spirulina in Human Nutrition and Health.* M.E. Gershwin and Amha Belay, (Ed.). Boca Raton, FL: CRC Press. 109.
53. Baker J. (1983, Aug). Modern Drug Research: The Potential and the Problems of Marine Natural Products. In: *Natural Products and Drug Development.* Krogsgaard-Larsen, Brogger, Kofod, Eds. Muksgaard, Copenhagen. 145-63.
54. Okai Y, Higashi-Okai K. (1997, June). Potent anti-inflammatory activity of pheophytin A derived from edible green algae, Enteromorpha profilera (Sujiao-nori). *International Journal of Immunopharmacology.* 19(6). 355-8.
55. Jensen G. (2000, Feb). Written communication.
56. Vijayavel K, Anbuselvam, Balasubramanian M. (2007). Antioxidant effect of the marine algae *Chlorella vulgaris* against naphthalene-induced oxidative stress in the albino rats. *Molecular and Cellular Biochemistry.* 303(1). 39-44.
57. Young Ae Son, Jee Ae Shim, et al. (2009). Intake of *Chlorella vulgaris* Improves Antioxidative Capacity in Rats Oxidatively Stressed with Dietary Cadmium. *Annals of Nutrition and Metabolism.* 54(1). 7-14.
58. Wu L, Ho J, Shieh M, Lu I. (2005, May 18). Antioxidant and antiproliferative activities of Spirulina and Chlorella water extracts. *J Agric Food Chem.* 53(10). 4207-12.
59. Mahendra J, Mahendra L, Muthu J, John L, Romanos G. (2013 Oct). Clinical effects of subgingivally delivered spirulina gel in chronic periodontitis cases: a placebo controlled clinical trial. *J Clin Diagn Res.* 7(10). 2330-3.
60. Grether-Beck S, Mühlberg K, et al. (2008, Sep). Bioactive molecules from the Blue Lagoon: in vitro and in vivo assessment of silica mud and microalgae extracts for their effects on skin barrier function and prevention of skin aging. *Exp Dermatol.* 17(9). 771-9.
61. Tolomio C et al. (1999). Colonization by diatoms and antirheumatic activity of thermal mud. Cell Biochemistry and Function. 17. 29-33.
62. Ceccettin M, Bellometti S et al. (1995). Serum interleukin-1 changes in arthrosic patients after mud-pack treatment. *Phys Rehab Kur Med.* 5. 92-3.
63. Weil A. (2011, Nov 4). The Depression-Inflammation Connection. The BLOG. Retrieved from <http://www.huffingtonpost.com/andrew-weil-md/depression-and-inflammation_b_1071714.html>
64. New Study Links Inflammatory Cytokines With Anger Problems. (2013, Dec. 23). Retrieved from <http://alert.psychiatricnews.org/2013/12/new-study-links-inflammatory-

cytokines.html?utm_source=feedburner&utm_medium=feed&utm_campaign=Feed%3A+PsychiatricNewsAlert+%28Psychiatric+News+Alert%29>

65. Kremer J, Lawrence D, Jubiz W. (1989). Different doses of fish-oil fatty acid ingestion in active rheumatoid arthritis: a prospective study of clinical and immunological parameters. In: Dietary Ω3 and Ω6 Fatty Acids: Biological Effects and Nutritional Essentiality. Galli C, Simopoulos AP, eds. Plenum Publishing. 343-50.
66. Sukenik A, Takahashi H, Mokady S. (1994). Dietary lipids from marine unicellular algae enhance the amount of liver and blood omega-3 fatty acids in rats. *Annals of Nutrition and Metabolism*. 38(2). 85-96.
67. Krylov V, Drapeau C, et al. (2001). Retrospective epidemiological study using medical records to determine which diseases are improved by *Aphanizomenon flos-aquae* supplements. Submitted for publication.

Chapter 7: All Life Begins and Stays Alive Because of Cells

1. Aposhyan S. (1999). *Natural Intelligence: Mind-Body Integration and Human Development.* William and Wilkins. 150.
2. Baker M. (2012, Dec 10). Brain Cells Made from Urine. Scientific America. Retrieved from <http://www.scientificamerican.com/article.cfm?id=brain-cells-made-from-urine>
3. Poole A. My Name is LUCA – The Last Universal Common Ancestor. Retrieved from <http://www.actionbioscience.org/evolution/poolepaper.html>
4. Pennisi E. (1994, Mar 12). *Science News.*
5. Grúz P, Shimizu M. (2010, Apr-Jun). Origins of age-related DNA damage and dietary strategies for its reduction. *Rejuvenation Res.* 13(2-3). 285-7.
6. Tomitani A, et al. (1999, July). Chlorophyll b and phycobilins in the common ancestor of cyanobacteria and chloroplasts. *Nature.* 400(6740). 159-62.
7. Rothschild L, Cockell C. (1990). Radiation: microbial evolution, ecology, and relevance to Mars missions. *Mutation Research.* 430. 281.
8. Kornberg, A. (1989). *For the love of enzymes.* Harvard University Press. Cambridge, MA. 65.
9. Reddy, H. (2008). Mitochondrial Medicine for Aging and Neurodegenerative Diseases. *Neuromolecular Med.* 10(4). 291-315.
10. Rugarli E, Langer T. (2012, Mar 21). Mitochondrial quality control: a matter of life and death for neurons. *EMBO J.* 31(6). 1336–1349.
11. Rugarli E, and Langer T. (2012, Mar 21). Mitochondrial quality control: a matter of life and death for neurons. *EMBO J.* 31(6). 1336–1349.
12. Wallace D. (2005). A Mitochondrial Paradigm of Metabolic and Degenerative Diseases, Aging, and Cancer: A Dawn for Evolutionary Medicine. *Annu Rev Genet.* 39. 359.
13. Harman, D. (1972). The biological clock: the mitochondria? *J Am Geriatr Soc.* 20. 145-47.

14. Wallace D. (2005). A Mitochondrial Paradigm of Metabolic and Degenerative Diseases, Aging, and Cancer: A Dawn for Evolutionary Medicine. *Annu Rev Genet.* 39. 359.
15. Reddy P. (2007). Mitochondrial dysfunction in aging and Alzheimer's disease: Strategies to protect neurons. *Antioxidants Redox Signaling.* 9. 1647–1658.
16. Brown W, George M, Wilson A. (1979). Rapid evolution of animal mitochondrial dna. *PNAS,* 76(4).
17. Patel C. Ouroboros: research in the biology of aging. Retrieved from <http://ouroboros.wordpress.com/2008/12/11/calorie-restriction-speeds-mitochondrial-turnover/>
18. Mitochondrial DNA mutations can cause degenerative heart and muscle disease. (2008, Feb 14). Retrieved from <http://phys.org/news122221913.html#jCp>
19. Choo-Kang A, Lynn S, et al. (2002, July). Defining the Importance of Mitochondrial Gene Defects in Maternally Inherited Diabetes by Sequencing the Entire Mitochondrial Genome Diabetes. 51(7). 2317-2320
20. Naidu, A. (2013). *REDOX Life.* Bio-REp Network Media.
21. Bachstetter A., et al. (2010, May). Spirulina promotes stem cell genesis and protects against LPS induced declines in neural stem cell proliferation. *PLoS One.* 5;5(5). e10496.
22. Bickford P, Tan J, Shytle R, et al. (2006, Feb). Nutraceuticals synergistically promote proliferation of human stem cells. *Stem Cells Dev.* 15(1).118-23.
23. Shytle D, Tan J, et al. (2010, Jan.). Effects of blue-green algae extracts on the proliferation of human adult stem cells in vitro: a preliminary study. *Med Sci Monit.* 16(1). BR1-5.
24. Bachstetter A, Jernberg J, Schlunk A, et al. (2010, May 5). Spirulina promotes stem cell genesis and protects against LPS induced declines in neural stem cell proliferation. *PLoS One.* 5(5). e10496
25. Das, S, Basu, A. (2008, May 1). Inflammation: a new candidate in modulating adult neurogenesis. *J Neurosci Res.* 86(6). 1199-208.
26. Gemma C, et al. (2002, July 15). Diets enriched in foods with high antioxidant activity reverse age-induced decreases in cerebellar beta-adrenergic function and increases in proinflammatory cytokines. *J Neuroscience.* 22(14). 6114-20.
27. Yasuhara T, Hara K, Maki M, Masuda T, Sanberg CD, et al. (2008). Dietary supplementation exerts neuroprotective effects in ischemic stroke model. *Rejuvenation Res.* 11. 201–214
28. Acosta S, Jernberg J, Sanberg C, Sanberg P, Small B, Gemma C, Bickford P. (2010 Oct.). NT-020, a natural therapeutic approach to optimize spatial memory performance and increase neural progenitor cell proliferation and decrease inflammation in the aged rat. *Rejuvenation Res.* 13(5). 581-8.
29. Stromberg I, Gemma C, Vila J, Bickford P. (2005). Blueberry- and spirulina-enriched diets enhance striatal dopamine recovery and induce a rapid transient microglia activation after injury of the rat nigrostriatal dopamine system. *Exp Neurol.* 196. 298–307.

30. Gemma C, Mesches M, Sepesi B, Choo K, Holmes D, et al. (2002). Diets enriched in foods with high antioxidant activity reverse age-induced decreases in cerebellar beta-adrenergic function and increases in proinflammatory cytokines. *J Neurosci.* 22. 6114–6120.
31. Wang Y, Chang CF, Chou J, Chen HL, Deng X, et al. (2005). Dietary supplementation with blueberries, spinach, or spirulina reduces ischemic brain damage. Exp Neurol. 193. 75–84.
32. Bachstetter et al. Spirulina Promotes Stem Cell Genesis and Protects against LPS Induced Declines in Neural Stem Cell Proliferation. (2010). *PLoS One.* 2010; 5(5). e10496.
33. Acosta S, Jernberg J, Sanberg C, Sanberg P, Small B, Gemma C, Bickford P. (2010, Oct.). NT-020, a natural therapeutic approach to optimize spatial memory performance and increase neural progenitor cell proliferation and decrease inflammation in the aged rat. *Rejuvenation Res.* 13(5). 581-8.
34. Pabon M, Jernberg J, Morganti J, Contreras J, Hudson CE, Klein R, Bickford P. (2012). A spirulina-enhanced diet provides neuroprotection in an α-synuclein model of Parkinson's disease. *PLoS One.* 7(9). e45256.
35. Yasuhara T, Hara K, Maki M, Masuda T, Sanberg CD, et al. (2008). Dietary supplementation exerts neuroprotective effects in ischemic stroke model. *Rejuvenation Res.* 11. 201–214
36. Drapeau C, Antarr D, Ma H, Yang Z, Tang L, Hoffman RM, Schaeffer DJ. (2010, May). Mobilization of bone marrow stem cells with StemEnhance improves muscle regeneration in cardiotoxin-induced muscle injury. *Cell Cycle.* 9(9). 1819-23.
37. Tipton K, Ferrando A, Phillips S, Doyle D, Wolfe R. (1999). Postexercise net protein synthesis in human muscle from orally administered amino acids. *Am J Physiol Endocrinol Metab.* 276. E628-E634.
38. The secret of 'muscular' old age (2005, Dec 11). Retrieved from http://news.bbc.co.uk/1/hi/health/4502862.stm
39. Norton LE, Layman DK. (2006, Feb.). Leucine regulates translation initiation of protein synthesis in skeletal muscle after exercise. *J Nutr.* 136(2). 533S-537S.
40. Valerio A, D'Antona G, Nisoli E. (2011, May). Branched-chain amino acids, mitochondrial biogenesis, and healthspan: an evolutionary perspective. Aging, 3(5). Retrieved from www.impactaging.com>
41. Anderson R et al. (2000, July). Reduction in free-radical-induced DNA strand breaks and base damage through fast chemical repair by flavonoids. *Free Radical Research.* 33(1). 91-103.
42. Ma G, Xue K, Wu J, Yuan S, Qin H. (1998, May). Antimutagenic effects of beta-carotene from Dunaliella salina (Chinese). *Chung Kuo Yao Li Hsueh Pao.* 19(3). 282-4.
43. Gagne S. (1990). *Energetics of Food.* Spiral Sciences Inc. 273.
44. Kumamoto S. (1984, Aug 28) U.S. Patent # 4,468,460 for Method of Human Cell Culture.
45. Apsley J. (1996). *The Regeneration Effect.* Genesis Communications.

46. Krasnikova N. (1973, Oct). Proliferation of the epithelium surrounding a skin wound in hairless mice exposed to sodium chlorophyllin. *Biull Eksp Biol Med* (Russian). 75(10). 99-102.
47. Krasnikova N. (1972, Sep-Oct). Experimental use of an aqueous solution of sodium chlorophyllin to speed the healing of skin wounds. *Eksp Khir Anesteziol* (Russian). 17(5). 19-22.
48. Hughes J, Latner L. (1936). Chlorophyll and hemoglobin regeneration after hemorrhage. *Journal of Physiology*. 86. 388.
49. Patek A. (1936). Chlorophyll and regeneration of blood. *Archives of Internal Medicine*. 57. 76.
50. Adams M, Ovokaitys. (2010). *Dietary Nucleic Acids: A Discussion with Dr. Todd Ovokaitys*. Truth Publishing.
51. Cummins J. (2001, Mar 3). The fate of food genes and the DNA CpG motif and its impact. Toxicology Symposium, University of Guelph. Retrieved from http://www.psrast.org/jcfateofgen.htm
52. Zhang, et al. (2011, 20 Sept). Exogenous plant MIR168a specifically targets mammalian LDLRAP1: evidence of cross-kingdom regulation by microRNA. *Cell Research*. doi:10.1038/cr.2011.158
53. Greenwood V. (2011, Sept 21). What You Eat Affects Your Genes: RNA from Rice Can Survive Digestion and Alter Gene Expression. *Discover Magazine*. Retrieved 2013 from <http://blogs.discovermagazine.com/80beats/2011/09/21/what-you-eat-affects-your-genes-rna-from-rice-can-survive-digestion-and-alter-gene-expression/?utm_source=feedburner&utm_medium=feed&utm_campaign=Feed%3A+80beats+%2880beats%29>
54. Greife H, Molnar S. (1978). Studies of nucleic acid metabolism in rats by use of carbon-14-labeled purine and pyrimidine bases and nucleic acids: Anabolic pathways of nucleic acid derivatives. *Tierphysiologie* (German). 40(5). 248-56.
55. Dr. Todd Ovokaitys, *Dietary Nucleic Acids*, www.TruthPublishing.com, 2009.
56. Berthold H et al. (1995, Oct.). Evidence for incorporation of intact dietary pyrimidine (but not purine) nucleosides into hepatic RNA. *Proceedings of the National Academy of Science USA*. 92(22). 10123-7
57. Devi M. (1983). The effect of algal protein diets on the regeneration of serum and liver proteins in protein depleted rats. *Plant Foods for Human Nutrition*. 33. 287.
58. Devi M et al. (1979). Serum protein regeneration studies on rats on algal diets. *Nutrition Research International*. 19. 785.
59. Voltarelli FA, de Mello M. (2008). Spirulina enhanced the skeletal muscle protein in growing rats. *Eur J Nutr*. 47. 393–400.
60. Arasaki S, Arasaki T. (1983). Low Calorie, High Nutrition Vegetables From the Sea: To Help You Look and Feel Better. Japan Publications. 60.

Chapter 8: Radiation-Protection for a Modern World

1. H. Needleman and P. Landrigan. (2000). Raising Children Toxic Free, Avon Books.
2. Weiss J, Landauer M. (2000). Radioprotection by antioxidants. *Annals of the New York Academy of Science.* 899. 44-60.
3. Lane N. (2002). Oxygen: The Molecule that Made the World. Oxford University Press. 128-30
4. Lane N. (2002). Oxygen: The Molecule that Made the World. Oxford University Press. 124-25
5. Retrieved from <http://nutritionmedicinedisease.typepad.com/nutrition-medicine-and-d/2011/03/vitamins-and-nutrients-that-protect-against-radiation.html>
6. Rae D. (2011, April 24). Protect your family from radiation exposure with superfoods. Retrieved from <http://www.naturalnews.com/032162_superfoods_radiation_exposure.html#ixzz2KWUPZtI1>
7. Bill Bodri (2004). How to Help Support the Body's Healing after Intense Radioactive or radiation exposure. Top Shape Publishing. Retrieved from www.RadiationDetox.com
8. von Hippel, F. (2011, Sept/Oct). The radiological and psychological consequences of the Fukushima Daiichi accident. *Bulletin of the Atomic Scientists.* 67(5). 27–36.
9. Loseva L, Dardynskaya I. (1993, Sep). Spirulina- natural sorbent of radionucleides. Research Institute of Radiation Medicine, Minsk, Belarus. *6th Int'l Congress of Applied Algology,* Czech Republic. Belarus.
10. Ben-Amotz A et al. (1988, Oct). Effect of natural beta-carotene supplementation in children exposed to radiation from the Chernobyl accident. *Radiation Environmental Biophysics.* 37(3). 187-93.
11. Drapeau C. (1998, Aug), Cell Tech former research director. Personal communication.
12. Rangsayatorn N, Pokethitiyook P, Upatham ES, Lanza GR. (2004, Mar.). Cadmium biosorption by cells of *Spirulina platensis* TISTR 8217 immobilized in alginate and silica gel. *Environ Int.* 30(1). 57-63.
13. Life Extension Magazine (2010, Aug). Retrieved from http://www.lef.org/magazine/mag2010/aug2010_Protect-Your-DNA-from-CT-Scans-X-rays_01.htm
14. American College of Radiology (2010, Feb 2). Epigenetics could help researchers determine any risks associated with low-dose radiation. *ScienceDaily.* Retrieved September 21, 2013, from http://www.sciencedaily.com /releases/2010/02/100201091628.htm
15. Ben-Amotz A et al. (1996, Nov). Natural beta-carotene and whole body irradiation in rats. *Radiation and Environmental Biophysics.* 35(4). 285-8.

16. Umegaki K, Takeuchi N, Ikegami S, Ichikawa T. (1994). Effect of beta-carotene on spontaneous and X-ray-induced chromosomal damage in bone marrow cells of mice. *Nutrition and Cancer*. 22(3). 277-84.
17. Bewick E et al. (1984). *Chlorella: The Emerald Food*. Ronin Publishing. 20.
18. Kumar S et al. (1999, Mar). Inhibition of radiation-induced DNA damage in plasmid pBR322 by chlorophyllin and possible mechanism(s) of action. *Mutation Research*. 425(1). 71-9.
19. Karpov L et al. (2000, May-Jun). The postradiation use of vitamin-containing complexes and a phycocyanin extract in a radiation lesion in rats. *Radiats Biol Radioecol* (Russian). 40(3). 310-4.
20. Sarma L, Tiku A, Kesavan P, Ogaki M. (1993, Dec). Evaluation of radioprotective action of a mutant (E25) form of *Chlorella vulgaris* in mice. *Journal of Radiation Research* (Tokyo). 34(4). 277-84.
21. Vacek A, Rotkovska D, Bartonickova A. (1990). Radioprotection of hemopoiesis conferred by aqueous extract from chlorococcal algae (*Ivastimul*) administered to mice before irradiation. *Experimental Hematology*. 18. 234-37.
22. Rotkovska D, Vacek A, Bartonickova A. (1989, Nov). The radioprotective effects of aqueous extract from chlorococcal freshwater algae (*Chlorella kessieri*) in mice and rats. *Strahlenther Onkologie*. 165(11). 813-6.
23. Rotkovska D, Vatsek A, Bartonichikova A. (1989, Sep-Oct). Increase in the radiation resistance of mice using *Ivastimul*. *Radiobiologiia* (Russian). 29(5). 652-4.
24. Qishen P, Guo B, Kolman A. (1989, Aug). Radioprotective effect of extract from *Spirulina platensis* in mouse bone marrow cells studied by using the micronucleus test. *Toxicology Letters*. 48(2). 165-9.
25. Singh S, Tiku A, Kesavan P. (1995, Aug). Post-exposure radioprotection by Chlorella vulgaris (E-25) in mice. *Indian Journal of Experimental Biology*. 33(8). 612-5.
26. Apsley J. (1996). *The Regeneration Effect*. Genesis Communications. 75.
27. Abrams K. (1996). *Algae to the Rescue*. Logan House. 32.
28. Qishen, P. et al. (1988). Enhancement of endonuclease activity and repair DNA synthesis by polysaccharide of spirulina. Chinese Genetics Journal. 15(5). 374-381. China.
29. Lahitova, N, Doupovcova M, Zvonar J, Chandoga J, Hocman G. (1994). Antimutagenic properties of fresh-water blue-green algae. *Folia Microbiologica*. 39(4). 301-3.

Chapter 9: Eat Light and Feel Bright: Ultimate Brain Food

1. University of California - Davis Health System (2010, Nov 30). Children with autism have mitochondrial dysfunction, study finds. *ScienceDaily*. Retrieved September 18, 2013, from http://www.sciencedaily.com /releases/2010/11/101130161521.htm
2. Leonard, W. (2003, Aug. 24). Food for Thought. Scientific America. 65.

3. Sun et al. (2013, July 25). Motile Axonal Mitochondria Contribute to the Variability of Presynaptic Strength. *Cell Reports*. DOI: 10.1016/j.celrep.2013.06.040
4. VIB (2013, May 16). Faulty energy production in brain cells leads to disorders ranging from Parkinson's to intellectual disability. *ScienceDaily*. Retrieved September 18, 2013, from http://www.sciencedaily.com /releases/2013/05/130516123804.htm
5. University of California - Davis Health System (2010, Nov 30). Children with autism have mitochondrial dysfunction, study finds. *ScienceDaily*. Retrieved September 18, 2013, from http://www.sciencedaily.com /releases/2010/11/101130161521.htm.
6. Raine C. (1999). Characteristics of the Neuron. *Basic Neurochemistry: Molecular, Cellular and Medical Aspects*. 6th edition.
7. May M. (2000, June). Disturbing Behavior. *Environmental Health Perspectives*. 108(6). A262-67.
8. Retrieved from <http://www.boston.com/lifestyle/health/articles/2011/11/29/teens_still_not_eating_enough_fruits_and_vegetables/#sthash.rMfrxKEo.dpuf>
9. Serdula M et al. (1995). Fruit and Vegetable Intake among Adults in 16 States: Results of a Brief Telephone Survey. *American Journal of Public Health*. 85(2). 236-39.
10. The Centers for Disease Control and Prevention, 2012, 2nd National Report on Biochemical Indicators of Diet and Nutrition in the U.S. Population. Retrieved from http://www.cdc.gov/nutritionreport/pdf/ExeSummary
11. Gordon et al. (2004). Prevalence of Vitamin D Deficiency among Healthy Adolescents. *Arch Pediatr Adoles Med*. 158. 531-537.
12. Munoz K, Krebs-Smith S, et al. (1997, Sep 1). Food Intakes of US children and Adolexcents Compared with Recommendations. *Pediatrics*. 100(3). 323-29.
13. James SJ, Melnyk S, et al. (2006). Metabolic endophenotype and related genotypes are associated with oxidative stress in children with autism. *Am J Med Genet B: Neuropsychiatr Genet*. 141. 947–56.
14. National Institute of Mental Health. (2001, Jan 1). The Invisible Disease: Depression. NIH Publication No. 01-4591. Retrieved November 6, 2001 from the World Wide Web: http://www.nimh.nih.gov/publicat/invisible.cfm.
15. Nesse R, Williams G. (1999, Jan). *Why We Get Sick: The New Science of Darwinian Medicine*. A Division of Random House Inc., New York.
16. World Health Organization. (2000). WHO/OMS, 2000. Mental Health and Brain Disorders. Depression. Retrieved November 8, 2001 from the World Wide Web: <http://www.who.int/mental_health/Topic_Depression/depression1.htm>
17. Pill-Popping Pre-Schoolers, Published in Psychiatric Services, April 3, 2004; Retrieved from <http://www.upliftprogram.com/h_depression.html#h77>
18. Gray P. (2010, Jan. 26). The Decline of Play and Rise in Children's Mental Disorders. Retrieved from <http://www.psychologytoday.com/blog/freedom-learn/201001/the-decline-play-and-rise-in-childrens-mental-disorders.>

19. National Health Interview Surveys. Retrieved from <http://www.cdc.gov/nchs/nhis.htm; www.childtrendsdatabank.org.>
20. National Survey of Children's Health, United States, 2003, 2007 and 2010. National Parental Survey of Children's Health (U.S.A. Center for Disease Control). Retrieved from <http://www.cdc.gov/mmwr/preview/mmwrhtml/mm5944a3.htm?s_cid=mm5944a3_w#tab2>
21. Williams F. (2010, Jan. 26). Generation Toxic. *OnEarth Magazine*. Retrieved from <http://issuu.com/onearth/docs/onearth_14wtr>
22. Retrieved from <http://www.childtrendsdatabank.org/sites/default/files/109_Children_with_Special_Health_Care_Needs.pdf>
23. Crawford M. (1990). The early development and evolution of the human brain. *Upsala Journal of Medical Sciences. Supplement.* 48. 43-78.
24. Crawford M et al. (1999). Evidence for the unique function of docosahexaenoic acid during the evolution of the hominid brain. *Lipids.* 34(suppl). S39-S47.
25. Broahurst C, Cunnane S, Crawford M. (1998, Jan). Rift Valley lake fish and shellfish provided brain-specific nutrition for early Homo. *British Journal of Nutrition.* 79(1). 3-21.
26. Cunnane, S., and Stewart K. (2010) *Human Brain Evolution: The Influence of Freshwater and Marine Food Resources.* John Wiley & Sons Inc.
27. Michael Crawford, chp. 2: Cunnane, S., and Stewart K. (2010) *Human Brain Evolution: The Influence of Freshwater and Marine Food Resources.* John Wiley & Sons Inc.
28. Cunnane S. (2005). *Survival of the Fattest: the Key to Human Brain Evolution.* World Scientific Pub Co.
29. *Proceedings of the National Academy of Sciences, May 22, 2001* http://www.fi.edu/learn/brain/fats.html
30. Milton K. (2000, Dec). Reply to SC Cunnane. *Am J Clin Nutr.* 72(6). 1586-1588
31. Slagle P. (1987).*The Way Up from Down.* Random House. 197-211.
32. DeCava J. (1997). *The Real Truth about Vitamins & Antioxidants.* Price-Pottenger Foundation. 121.
33. Rose W, Haines W, Warner D. (1951). The amino acid requirements of man. III. The role of isoleucine; additional evidence concerning histidine. *J Biol Chem* 193(2). 605-612.
34. Cheraskin E, Ringsdorf W, Brecher A. (1974). *Psychodietetics: Food as the Key to Emotional Health.* Stein and Day Publishers, New York.
35. Simopoulos A, Leaf A, Salem N. (2000). Workshop statement on the essentiality of and recommended dietary intakes for Omega-6 and Omega-3 fatty acids. *Prostaglandins Leukot Essent Fatty Acids.* 63(3). 119-121.
36. Kris-Etherton P, Harris W, Appel L. (2002). Fish consumption, fish oil, omega-3 fatty acids, and cardiovascular disease. *Circulation.* 106(21). 2747-2757.
37. Hibbeln J, Salem N. (1995). Dietary polyunsaturated fats and depression: when cholesterol does not satisfy. *Am J Clin Nutr.* 61(1). 1-9.

38. Adams P, Lawson S, Sanigorski A, et al. (1996). Arachidonic acid to eicosapentaenoic acid ratio in blood correlates positively with clinical symptoms of depression. *Lipids.* 31. S157-61.
39. Stevens L, Zentall S, Abate M, et al. (1996). Omega-3 fatty acids in boys with behavior, learning, and health problems. *Physiol Behav.* 1(59). 915-20.
40. Kalmijn S, van Boxtel M, Ocke M, et al. (2004). Dietary intake of fatty acids and fish in relation to cognitive performance at middle age. *Neurology.* 62. 275-80.
41. Kalmijn S, Launer L, Ott A, et al. (1997). Dietary fat intake and the risk of incident dementia in the Rotterdam Study. *Ann Neurol.* 42. 776-82.
42. Morris M, Evans D, Bienias J, et al. (2003). Consumption of fish and n-3 fatty acids and risk of incident Alzheimer disease. *Arch Neurol.* 60. 940-6.
43. Lukiw W, Bazan N. (2008). Docosahexaenoic acid and the aging brain. *J Nutr.* 138. 2510-4.
44. Schaefer E, Bongard V, Beiser A, et al. (2006). Plasma phosphatidylcholine docosahexaenoic acid content and risk of dementia and Alzheimer disease: the Framingham Heart Study. *Arch Neurol.* 63. 1545-50.
45. Fedorova I, et al. (2009, Dec). Deficit in prepulse inhibition in mice caused by dietary n-3 fatty acid deficiency. *Behav Neurosci.* 123(6). 1218-25. Retrieved from <http://www.sciencedaily.com/releases/2009/12/091216130718.htm>
46. Colter AL et al. (2008, Feb.). Fatty acid status and behavioural symptoms of attention deficit hyperactivity disorder in adolescents: a case-control study. *Nutr J.* 14(7). 8
47. Richardson AJ, Puri BK. (2002). A randomized double-blind, placebo-controlled study of the effects of supplementation with highly unsaturated fatty acids on ADHD-related symptoms in children with specific learning difficulties. *Prog Neuropsychopharmacol Biol Psychiatry.* 26(2). 233-239.
48. Valerio, A., D'Antona, G., and Nisoli, E. (2011, May). Branched-chain amino acids, mitochondrial biogenesis, and healthspan: an evolutionary perspective. *Aging.* 3(5). 464–478.
49. Cole J, et al. (2010, Jan 5). Dietary branched chain amino acids ameliorate injury-induced cognitive impairment. *Proc Natl Acad Sci USA.* 107(1): 366–371.
50. Aquilani R, et., al. (2005). Branched-chain amino acids enhance the cognitive recovery of patients with severe traumatic brain injury. *Arch Phys Med Rehabil.* 86. 1729-1735.
51. Aquilani R, et., al. (2008). Branched-chain amino acids may improve recovery from a vegetative or minimally conscious state in patients with traumatic brain injury: a pilot study. *Arch Phys Med Rehabil.* 89. 1642-1647.
52. Aquilani, R. et al. (2011, June). Nutrition for brain recovery after ischemic stroke: an added value to rehabilitation. *Nutr Clin Pract.* 26(3). 339-45.
53. Laguna M, Villar R, Calleja J, Cadavid I. (1993, April). Effects of *Chlorella stigmatophora* extract on the central nervous system. *Planta Medica.* 59(2). 125-30.
54. Villar R, Laguna R, Calleja J, Cadavid I. (1992, Oct). Effects of *Skelettonema costatum* extracts on the central nervous system. *Planta Medica.* 58(5)398-404.

55. Villar R, Laguna M, Calleja J, Cadavid I. (1992). Effects of *Dunaliella tertiolecta* extracts on the central nervous system. *Planta Medica*. October. 58(5). 404-9.
56. Benedetti S, Benvenuti F, Scoglio S, and Canestrari. (2010). Oxygen Radical Absorbance Capacity of Phycocyanin and Phycocyanobilin from the Food Supplement *Aphanizomenon flos-aquae*. *Journal of Medicinal Food*. 13 (1). 223-227.
57. Scogio S, Canestrari F, Benedetti S, Benedetti Y, Delgado-Esteban M. (2008, Jan. 3). Alphanizomenon Flos Aquae Preparation, Extracts and Purified Components Thereof for the Treatment of Neurological, Neurodegenerative and Mood Disorders. Internation Patent Application. Number WO 2008/000430 A2
58. Liebowitz, Michael, R. (1983). *The Chemistry of Love*. Boston: Little, Brown, & Co.
59. Sabelli H, Borison RL, Diamond BI, Havdala HS, Narasimhachari N. (1978). Phenylethylamine and brain function. Biochem Pharmacol. 27. 1707–1711.
60. Grimsby J, Toth M, et al. (1997). Increased stress response and β-phenylethylamine in MAOB-deficient mice. *Nat Genetics*. 17. 206–210.
61. Sabelli H, Fink P, Fawcett J, Tom C. (1996). Sustained antidepressant effect of PEA replacement. J *Neuropsychiatry Clin Neurosci*. 8. 168–171.
62. Baker G, et al. (1991). Phenylethylaminergic mechanisms in attention-deficit disorder. *Biological Psychiatry* 29 (1). 15–22.
63. Sabelli H, Fink P, Fawcett J, Tom C. (1996). Sustained antidepressant effect of PEA replacement. J *Neuropsychiatry Clin Neurosci*. 8. 168–171.
64. Nakamura M, Ishii A, Nakahara D. (1998). Characterization of β-phenylethylamine-induced monamine release in rat nucleus accumbens: a microdialysis study. *Eur J Pharmacol*. 349. 163–169.
65. Yang HY, Neff NH. (1973, Nov). "Beta-phenylethylamine: a specific substrate for type B monoamine oxidase of brain". *The Journal of Pharmacology and Experimental Therapeutics* 187(2). 365–71.
66. Suzuki O, Katsumata Y, Oya M. (1981, Mar). Oxidation of beta-phenylethylamine by both types of monoamine oxidase: examination of enzymes in brain and liver mitochondria of eight species. *The Journal of Neurochemistry* 36(3). 1298–301.
67. Scogio S, Canestrari F, Benedetti S, Benedetti Y, Delgado-Esteban M. (2008, Jan. 3). Alphanizomenon Flos Aquae Preparation, Extracts and Purified Components Thereof for the Treatment of Neurological, Neurodegenerative and Mood Disorders. Internation Patent Application. Number WO 2008/000430 A2
68. Cousens G. (1985). Report of treatment of Alzheimer's disease with Alphanae Klamathonmenon flos-aqua [sic]. *Orthomedicine*. Winter/Spring. 8(1). 2.
69. Cousens G. (1995, April). Microalgae: First and finest superfood. *Body Mind Spirit*. 14(3). 12-18.
70. Dr. Cousens (2006, Oct). Interview, *Focus Magazine*.
71. Cousens G. (1995, May). *Body Mind Spirit Magazine*.
72. Cousens G. (1986). *Spiritual Nutrition and the Rainbow Diet*. Cassandra Press.

73. Deisseroth K. (2010, Oct. 20). Optogenetics: Controlling the Brain with Light. *Scientific America*. Retrieved from http://www.scientificamerican.com/article.cfm?id=optogenetics-controlling
74. Deisseroth K. (Oct. 20, 2010). Optogenetics: Controlling the Brain with Light. *Scientific America*. Retrieved from http://www.scientificamerican.com/article.cfm?id=optogenetics-controlling
75. Takenaka H et al. (1993, Oct). Protective effect of *Dunaliella* bardawil on water-immersion-induced stress in rats. *Planta Medica*. 59(5). 421-4.
76. Tanaka K et al. (1997, Oct). Oral administration of a unicellular green algae, *Chlorella vulgaris*, prevents stress-induced ulcer. *Planta Medica*. 63(5). 465-6.
77. Challem J. (1981). *Spirulina*. Keats Publishing. 16.
78. Shinpo K et al. (1982). *36th Conference of Japanese Society of Nutrition and Food Science*. 102.
79. Jarratt C, Jewett M, Peters S, Tragash E. (1995).*The Children and Algae Report*. The Center for Family Wellness, Harvard, Mass.
80. Bruno J, Gittelman J, Tuchfeld B. (2001) Lowered aluminum, with better detoxification and improved cognitive, behavioral ratings by children eating *Aphanizomenon flos-aquae*. Submitted for publication.
81. Bitterman N, Melamed Y, Ben-Amotz A. (1994, Mar). Beta-carotene and CNS oxygen toxicity in rats. *Journal of Applied Physiology*. 76(3). 1073-6.
82. Komatsu M, Hiramatsu M. (2000, Aug). The efficacy of an antioxidant cocktail on lipid peroxide level and superoxide dismutase activity in aged rat brain and DNA damage in iron-induced epileptogenic foci. *Toxicology*. 148(2- 3). 143-8.
83. Vatassery G. (1998, Sept). Vitamin E and other endogenous antioxidants in the central nervous system. *Geriatrics*. 53(Suppl 1); S25-S7.
84. Shohami E et al. (1999, May). Closed head injury in the rat induces whole body oxidative stress: overall reducing antioxidant profile. *Journal of Neurotrauma*. 16(5). 365-76.
85. Annapurna A, et al. (1991). Bioavailability of blue-green algae carotenes in preschool children. *Journal of Clinical Biochemistry and Nutrition*. 10. 145-51.
86. Jama J et al. (1996). Dietary antioxidants and cognitive function in a population-based sample of older persons. The Rotterdam Study. *American Journal of Epidemiology*. 144(3). 275-80.
87. Brunett L. (2000, summer). New ways to explain the brain. *Wisconsin Alumni Journal*. 36.
88. Stromberg I, Gemma C, Vila J, Bickford P. (2005). Blueberry- and spirulina-enriched diets enhance striatal dopamine recovery and induce a rapid, transient microglia activation after injury of the rat nigrostriatal dopamine system. *Exp Neurol*. 196. 298–307.
89. Wang Y, Chang CF, et al. (2005). Dietary supplementation with blueberries, spinach, or spirulina reduces ischemic brain damage. *Exp Neurol*. 193. 75–84.
90. Bachstetter et al. (2010). Spirulina Promotes Stem Cell Genesis and Protects against LPS Induced Declines in Neural Stem Cell Proliferation. *PLoS One*. 2010; 5(5). e10496.

91. Yasuhara T, Hara K, Maki M, Masuda T, Sanberg CD, et al. (2008). Dietary supplementation exerts neuroprotective effects in ischemic stroke model. *Rejuvenation Res.* 11. 201–214
92. Mild Traumatic Brain Injury in the United States: Steps to Prevent a Serious Public Health Problem. (2003, Sept). Centers for Disease Control and Prevention. Retrieved from <http://www.cdc.gov/ncipc/pub-res/mtbi/mtbireport.pdf.>
93. Alvarez L. (August 25, 2008). War Veterans' Concussions Are Often Overlooked. The New York Times. Retrieved form <http://www.nytimes.com/2008/08/26/us/26tbi.html>
94. Liu CY. Combined therapies: National Institute of Neurological Disorders and Stroke funding opportunity in traumatic brain injury research. *Neurosurgery.* 2008; 63. N12.
95. Yu S, Kaneko Y, Bae E, et al. (2009) Severity of controlled cortical impact traumatic brain injury in rats and mice dictates degree of behavioral deficits. *Brain Res* 1287. 157–163.
96. Acosta, S., Tajiri, N. and Borlongan, C. (2013, Aug 9). Long-Term Upregulation of Inflammation and Suppression of Cell Proliferation in the Brain of Adult Rats Exposed to Traumatic Brain Injury Using the Controlled Cortical Impact Model. *PLoS One.* 8(8).
97. S. Acosta, J. Jernberg, and Paula C. Bickford. (2010, October). NT-020, a Natural Therapeutic Approach to Optimize Spatial Memory Performance and Increase Neural Progenitor Cell Proliferation and Decrease Inflammation in the Aged Rat. *Rejuvenation Research.* 13(5). 581-588.
98. Kaneko Y, Cortes L, et al. Dietary Supplementaitons as Neuroprotective Therapies: Focus on NT-020 Diet Benefits in a Rat Model of Stoke. Retrieved from <http://www.ncbi.nlm.nih.gov/pmc/articles/PMC3397535/>
99. Valencia A, Walker J. (1999, June). A multi-axial treatment paradigm for mild traumatic brain injury to achieve reparative functional metaplasticity. Paper presented at the *Third World Congress on Brain Injury.* Quebec City, Canada.
100. Sevilla I, Aguiree N. (1995). Study on the Effects of Super Blue-Green Algae on the Nutritional Status and School Performance of First-, Second-, and Third-Grade Children Attending the Monsenor Velez School in Nandaime, Nicaragua [dissertation, in Spanish]. Universidad CentroAmericana, Nicaragua.
101. Jarratt C, Jewett M, Peters S, Tragash E. (1995). *The Children and Algae Report.* The Center for Family Wellness, Harvard Mass.
102. Jarratt C. (1998, Dec). Personal communication.
103. Bruno J, Gittelman J, Tuchfeld B. (2001). Lowered aluminum, with better detoxification, and improved cognitive, behavioral ratings by children eating Aphanizomenon flos-aquae. Unpublished manuscript.
104. Foldoe M. (1999). A Case Study of The Perceived Effects of Eating Aphanizomenon Flos-Aquae, a Blue-Green Algae, On the Attention, Academic, and Social Behavior of Three Elementary School-Aged Children diagnosed with

Attention Deficit/Hyperactivity Disorder (AD/HD) [thesis]. Sonoma State University.

105. Krylov V, et al. (2000). Retrospective epidemiological study using medical records to determine which diseases are improved by *Aphanizomenon flos-aquae* supplements. Submitted for publication.
106. Drapeau C. (1998, Aug 8). Results of Survey of AFA consumers. Presented at Klamath Falls, Oregon.
107. Buletsa B, Ihnatovych I, Lupych P, Pulyk O. (1996, Oct-Dec). The prevalence, structure and clinical problems of multiple sclerosis in the Transcarpathian area based on epidemiological study data. *Lik Sprava.* (10-12). 163-5.
108. Bourre J et al. (1993). Function of dietary polyunsaturated fatty acids in the nervous system. *Prostaglandins Leukotrienes and Essential Fatty Acids.* 48. 5-15.
109. Kushak R et al. (2000). Favorable effects of blue-green algae *Aphanizomenon flos-aquae* on rat plasma lipids. *Journal of the American Nutraceutical Association.* 2(3). 59-65.
110. Jacobson M. and Schardt D. (1999). Diet, ADHD & Behavior: A quarter-century review. *Center for Science in the Public Interest,* Washington, D.C.
111. Arnold LE, Bozzolo H, et al. (2005, Aug). Serum zinc correlates with parent- and teacher- rated inattention in children with attention-deficit/hyperactivity disorder. *J Child Adolesc Psychopharmacol.* 15(4). 628-36.
112. Searight R, Robertson K, et al. (2012). Complementary and Alternative Therapies for Pediatric Attention Deficit Hyperactivity Disorder: A Descriptive Review. *International Scholarly Research Network. ISRN Psychiatry.* 804127.
113. Kozielec T, Starobrat-Hermelin B. (1997, June). Assessment of magnesium levels in children with Attention Deficit Hyperactivity Disorder (ADHD). *Magnesium Research.* 10(2). 143-8.
114. Arnold L, Bozzolo H, et al. (2005, Aug). Serum zinc correlates with parent- and teacher- rated inattention in children with attention-deficit/hyperactivity disorder. *J Child Adolesc Psychopharmacol.* 15(4). 628-36.
115. Colquhoun I, Bunday S. (1981, May). A lack of essential fatty acids as a possible cause of hyperactivity in children. *Medical Hypotheses.* 7(5). 673-9.
116. Duvner T. (1994). Gamma-linolenic acid as a treatment in AD/HD. Paper presented at the Sixth European Conference of Neurodevelopmental Delay, Stockholm, reported in *Hyperactive Children Support Group Journal.* 48.
117. Murray C, Lopez A. (1997, May). Alternative projections of mortality and disability by cause 1990 to 2020: Global Burden of Disease Study. *Lancet.* 349(9064). 1498-504.
118. University of Chicago Medical Center (2010, Oct 4). Children with ADHD at increased risk for depression and suicidal thoughts as adolescents; *ScienceDaily.* Retrieved February 8, 2013, from http://www.sciencedaily.com /releases/2010/10/101004162834.htm
119. Christensen L. (1996). *Diet-Behavior Relationships.* American Psychological Association Publications. 168.

120. Taylor J. (1998, Feb.). The algae AD/HD connection: Can blue-green algae be of help with attention deficit/hyperactivity disorder? *Network of Hope Newsletter.*
121. Cited in: Carroll L, Tober J. (1999). *The Indigo Children: The New Kids Have Arrived.* Hay House Inc. 188.

Chapter 10: Microalgae Solutions for the Diabetes Epidemic

1. Mark Edwards. (2013). Algae Diabetes Solutions; http://www.algaeindustrymagazine.com/algae-medical-solutions-part-5/
2. Brownlee, M. (June 2005). The Pathobiology of Diabetic Complications: A Unifying Mechanism. Diabetes. 54.
3. Giacco F, Brownlee M. (2010, Oct 29). Oxidative Stress and Diabetic Complications. *Circ Res.* 107(9). 1058-70.
4. Mark Bittman (2013). It's the Sugar, Folks. Retrieved from <http://opinionator.blogs.nytimes.com/2013/02/27/its-the-sugar-folks/>
5. Retrieved from <http://wgntv.com/2013/03/11/the-soaring-costs-of-diabetes-in-america/>
6. Reinberg, S. (2012, Nov 15). U.S. Diabetes Ratings Soaring. *HealthDay.* Retrieved from <http://health.usnews.com/health-news/news/articles/2012/11/15/us-diabetes-rates-soaring-cdc>
7. Basu S, Yoffe P, Hills N, Lustig R. (2013). The Relationship of Sugar to Population-Level Diabetes Prevalence: An Econometric Analysis of Repeated Cross-Sectional Data. *PLoS ONE* 8(2). e57873.
8. Last A, Wilson S. (2006, Jun 1). Low-Carbohydrate Diets. *Am Fam Physician.* 73(11). 1942-8.
9. Surgeon General's Report (*1988*) on Nutrition and Health *Public Health Service Washington, DC. USDHHS Publication. 88. 50210.*
10. Last A, Wilson S. (2006, Jun 1). Low-carbohydrate diets. Am Fam Physician. 73(11). 1942-8.
11. Furukawa S, Fujita T, et al. (2004). Increased oxidative stress in obesity and its impact on metabolic syndrome. *J Clin Invest.* 114. 1752–1761;
12. Urakawa H, Katsuki A, Sumida Y, Gabazza EC, Murashima S, Morioka K. (2003). Oxidative stress is associated with adipocity and insulin resistance in men. *J Clin Endocrinol Metab.* 88. 4673–4676.;
13. Mantena S, King A, Andringa K, Eccleston H, Bailey S. (2008). Mitochondrial dysfunction and oxidative stress in the pathogenesis of alcohol- and obesity-induced fatty liver diseases. *Free Radic Biol Med.* 44. 1259–1272.
14. Diniz Y, Rocha K, Souza G, Galhardi C, Ebaid G, Rodrigues H. (2006). Effects of N-acetylcysteine on sucrose-rich diet-induced hyperglycaemia, dyslipidemia and oxidative stress in rats. *Eur J Pharmacol.* 543. 151–157.
15. Fridyand L, Philipson L. (2006). Reactive species, cellular repair and risk factors in the onset of type 2 diabetes mellitus: review and hypothesis. *Curr Diabetes Rev.* 2. 241–259.

16. Lund University (2013, Mar 19). New clues in hunt for heredity in type 2 diabetes. *ScienceDaily*. Retrieved September 20, 2013, from http://www.sciencedaily.com /releases/2013/03/130319091144.htm
17. Lund University (2013, July 3). Epigenetic changes to fat cells following exercise. *ScienceDaily*. Retrieved September 20, 2013, from http://www.sciencedaily.com /releases/2013/07/130703101344.htm
18. Lund University (2013, Mar 19). New clues in hunt for heredity in type 2 diabetes. *ScienceDaily*. Retrieved September 20, 2013, from http://www.sciencedaily.com /releases/2013/03/130319091144.htm
19. Gicacco, F., Brownlee, M. (2010). Oxidative Stress and Diabetic Complications, *Circ Res.* 107. 158-1070.
20. Alberts B, Johnson A, Lewis J, et al. (2002). The Evolution of Electron-Transport Chains. Molecular Biology of the Cell. 4th edition. New York: Garland Science.
21. Cousens, G. (2008). *There is a Cure for Diabetes: The Tree of Life 21-Day+ Program.* North Atlantic Books.
22. Halliwell B. (1999, Oct). Antioxidant defense mechanisms: from the beginning to the end (of the beginning). *Free Radic Res.* 31(4). 261-72.
23. Benzie, I (2003). Evolution of dietary antioxidants. *Comparative Biochemistry and Physiology.* 136 (1). 113–26.
24. Brownlee, M. (2005 Jun). The Pathobiology of Diabetic Complications: A Unifying Mechanism. *Diabetes.* 54(6). 1615-1625.
25. Kim J, Wei Y, Sowers J. (2008). Role of Mitochondrial Dysfunction in Insulin Resistance. *Circulation Research.* 102. 401-414
26. Giacco F, Brownlee M. (2010). Oxidative Stress and Diabetic Complications. *Circulation Research.* 107. 1058-1070.
27. Baynes J. (1991). Role of oxidative stress in development of complications in diabetes. *Diabetes.* 40. 405–412.
28. Oberley L. (1988). Free radicals and diabetes. *Free Radic Biol Med.* 5. 113–124.
29. Kakimoto M, Inoguchi T, et al. (2002). Accumulation of 8-hydroxy-2'-deoxyguanosine and mitochondrial DNA deletion in kidney of diabetic rats. *Diabetes.* 51. 1588–1595.
30. Boveris A, Chance B. (1973, July). The mitochondrial generation of hydrogen peroxide. General properties and effect of hyperbaric oxygen. *Biochem. J.* 134,707–716.
31. Nishikawa T, Edelstein D, et al. (2000). Normalizing mitochondrial superoxide production blocks three pathways of hyperglycaemic damage. *Nature.* 404. 787–790.
32. Inoguchi T, Li P, Umeda F, et al. (2000). High glucose level and free fatty acid stimulate reactive oxygen species production through protein kinase C: dependent activation of NAD(P)H oxidase in cultured vascular cells. *Diabetes.* 49. 1939–1945.
33. Bruno R, et al., (2013). Current status of NADPH oxidase research in cardiovascular pharmacology. *Vasc Health Risk Manag.* 9. 401–428.

34. Gao L, Mann G. (2009, April 1). Vascular NAD(P)H oxidase activation in diabetes: a double-edged sword in redox signalling. *Cardiovasc Res.* 82(1). 9-20.
35. Cheng X, Siow RC, Mann GE. (2011, Feb 1). Impaired redox signaling and antioxidant gene expression in endothelial cells in diabetes: a role for mitochondria and the nuclear factor-E2-related factor 2-Kelch-like ECH-associated protein 1 defense pathway. *Antioxid Redox Signal.* 14(3). 469-87
36. Shuji Sasaki and Toyoshi Inoguchi. (2012, Aug). The Role of Oxidative Stress in the Pathogenesis of Diabetic Vascular Complications. *Diabetes Metab J.* 36(4). 255–261
37. Stocker R, Yamamoto Y, et al. (1987). Bilirubin is an antioxidant of possible physiological importance. *Science.* 235. 1043–1046.
38. Lanone S, Bloc S, et al. (2005). Bilirubin decreases nos2 expression via inhibition of NAD(P)H oxidase: implications for protection against endotoxic shock in rats. *FASEB J.* 19. 1890–1892.
39. Zheng J, et al. (2013, Jan 15). Phycocyanin and phycocyanobilin from *Spirulina platensis* protect against diabetic nephropathy by inhibiting oxidative stress. *Am J Physiol Regul Integr Comp Physiol.* 304(2). R110-20.
40. Beale S, Cornejo J. (1984, Dec). Enzymatic heme oxygenase activity in soluble extracts of the unicellular red alga, Cyanidium caldarium. *Arch Biochem Biophy.* 235(2). 371-84.
41. Borchers A, Belay A, Keen L, Gershwin, E. (2008). Spirulina and Immunity. Chpt. 8. In Spirulina in Human Nutrition and Health. Edited by M.E. Gershwin and Amha Belay. CRC Press.
42. Libor Vítek. (2012). The role of bilirubin in diabetes, metabolic syndrome, and cardiovascular diseases. *Front Pharmacol.* 3. 55
43. Zheng et al. (2013, Jan 15). Phycocyanin and phycocyanobilin from *Spirulina platensis* protect against diabetic nephropathy by inhibiting oxidative stress. *Am J Physiol Regul Integr Comp Physiol.* 304(2). R110-20.
44. Fukui M, Tanaka M, et al. (2008). Relationship between serum bilirubin and albuminuria in patients with type 2 diabetes. *Kidney Int.* 74. 1197–1201.
45. Yasuda M, Kiyohara Y, et al. (2011). High serum bilirubin levels and diabetic retinopathy: the Hisayama Study. *Ophthalmology.* 118. 1423–1428.
46. Ohnaka K, Kono S, et al. (2010). Inverse associations of serum bilirubin with high sensitivity C-reactive protein, glycated hemoglobin, and prevalence of type 2 diabetes in middle-aged and elderly Japanese men and women. *Diabetes Res Clin Pract.* 88. 103–110.
47. Han SS, Na KY, et al. (2010). High serum bilirubin is associated with the reduced risk of diabetes mellitus and diabetic nephropathy. *Tohoku J Exp Med.* 221. 133–140.
48. Choi SH, Yun KE, Choi HJ. (2011, Jun 22). Relationships between serum total bilirubin levels and metabolic syndrome in Korean adults. *Nutr Metab Cardiovasc Dis.* Epub. DOI: http://dx.doi.org/10.1016/j.numecd.2011.03.001.

49. Ikeda N, Inoguchi T, et al. (2011). Biliverdin protects against the deterioration of glucose tolerance in db/db mice. *Diabetologia.* 54. 2183–2191.
50. Fujii et al. (2010, Nov). Bilirubin and biliverdin protect rodents against diabetic nephropathy by downregulating NAD(P)H oxidase. *Kidney Int.* 78(9). 905-19.
51. Unpublished data; cited in Shuji Sasaki and Toyoshi Inoguchi. (2012 Aug). The Role of Oxidative Stress in the Pathogenesis of Diabetic Vascular Complications. *Diabetes Metab* J. 36(4). 255–261.
52. Zheng J, Inoguchi T, et al. (2013, Jan 15). Phycocyanin and phycocyanobilin from *Spirulina platensis* protect against diabetic nephropathy by inhibiting oxidative stress. *Am J Physiol Regul Integr Comp Physiol.* 304(2). R110-20.
53. Inoguchi T, Nawata H. (2005). NAD(P)H oxidase activation: a potential target mechanism for diabetic vascular complications, progressive beta-cell dysfunction and metabolic syndrome. *Curr Drug Targets.* 6. 495–501.
54. Davis SN, Granner DK. Insulin, oral hypoglycemic agents, and the pharmacology of endocrine pancreas. In: Hardman JG Limberd L, Malinoff P, Ruddon R, Gilman A, editors. (1996). *Goodman and Gilman's the pharmacological basis of therapeutics.* 9th ed. New York: McGraw Hill. 1487–518.
55. Shrivastava S, Shrivastava P, Ramasamy J. (2013). Role of self-care in management of diabetes mellitus. *J Diabetes Metab Disord.* 12. 14.
56. Szczepanska-Sadowska E, Cudnoch-Jedrzejewska A, Ufnal M, Zera T. (2010, Oct). Brain and cardiovascular diseases: common neurogenic background of cardiovascular, metabolic and inflammatory diseases. *J Physiol Pharmacol.* 61(5). 509-21.
57. Hirata T, Tanaka M, Ooike M, Tsunomura T, Sakaguchi M. (2000). Antioxidant activities of phycocyanobilin prepared from *Spirulina platensis. J. Appl. Phycol.* 12. 435-439.
58. Qureshi M, Kidd M, and Ali R. (1995). *Spirulina platensis* extract enhances chicken macrophage functions after in vitro exposure. *Journal of Nutritional Immunology.* 3(4). 35-45.
59. Iyer U, Parikh P, Mani U. (2001). Role of spirulina in control of glycemia and lipidemia in type 2 diabetes mellitus. *J Med Food.* 4. 193–199.
60. Retrieved from <http://diabetes.webmd.com/news/20081104/diabetes-patients-fish-may-help-kidneys>
61. Nkondjock A, Receveur O. (2003). Fish-seafood consumption, obesity, and risk of type 2 diabetes: an ecological study. *Diabetes and Metabolism.* 29(6). 635–42.
62. Pérez-Martínez P, et al. (2011). Mediterranean diet rich in olive oil and obesity, metabolic syndrome and diabetes mellitus. *Curr Pharm Des.* 17(8). 769-77.
63. Zhang M, Picard-Deland E, Marette A. (2013). Fish and Marine Omega-3 Polyunsatured Fatty Acid Consumption and Incidence of Type 2 Diabetes: A Systematic Review and Meta-Analysis. *International Journal of Endocrinology.* 501015. Epub 2013 Sep 8.

64. Mani U, Desai S, Iyer U. (2000). Studies on the long-term effect of spirulina supplementation on serum lipid profile and glycated proteins in NIDDM patients. *Journal of Nutraceuticals Functional and Medical Foods.* 2. 25–32.
65. Mišurcová L, Škrovánková S, Samek D, Ambrožová J, Machů L. (2012). Health Benefits of Algal Polysaccharides in Human Nutrition. *Advances in Food and Nutrition Research.* 66. 75–145.
66. Nakaya N, Homma Y, Goto Y. (1988). Cholesterol lowering effects of Spirulina. *Nutr Rep Int.* 37. 1329–37.
67. Kataoka N, Misaki A. (1983). Glycolipid isolated from *Spirulina maxima*: structure and fatty acid composition. *Agricul Biol Chem.* 47. 2349–55.
68. de Caire GZ, De Cano MS, de Mule MZ, Steyerthal N, Piantanida M. (1995). Effect of *Spirulina platensis* on glucose, uric acid and cholesterol levels in the blood of rodents. *Int J Exp Bot.* 57. 93–6.
69. Hirata T, Tanaka M, Ooike M, Tsunomura T, Sakaguchi M. (2000). Antioxidant activities of phycocyanobilin prepared from *Spirulina platensis*. *J Appl Phycol.* 12. 435–439.
70. Torres P, Miranda R, Paredes M, Mascher D. (1998). *Spirulina maxima* prevents induction of fatty liver by carbon tetrachloride in the rat. *IUBMB Life.* 44. 787–93.
71. Edwards, M. (2011, Mar 20). Algae 101, part 23. Obesity and Diabetes Moderation with Algae. Retrieved from <http://www.algaeindustrymagazine.com/algae-medical-solutions-part-5/>
72. Becker E, Jakober B, Luft D, Schmuling R. (1986). Clinical and biochemical evaluations of the algae Spirulina with regards to its application in treatment of obesity: A double blind cross over study. *Nutr Rep Int.* 4. 565–74.
73. Anwer, R., Alam, A. et al. (2012, Oct). Spirulina: possible pharmacological evaluation for insulin-like protein. *Journal of Applied Phycology.* 25. 883-889.
74. Anwer, R., Alam, A. et al. (2012, Oct). Spirulina: possible pharmacological evaluation for insulin-like protein. *Journal of Applied Phycology.* 25. 883-889.
75. Cherng J, Shih M. (2006, Feb 9). Improving glycogenesis in Streptozocin (STZ) diabetic mice after administration of green algae Chlorella. *Life Sci.* 78(11). 1181-6.
76. Hosoyamada Y, Takai T, Kato T. (1991). Effects of water-soluble and insoluble fractions of Spirulina on serum lipid components and glucose tolerance in rats. *J Jpn Soc Nutr Food Sci.* 44. 273–7.
77. Iyer UM, Parikh PM, Mani U. (2001). Role of spirulina in control of glycemia and lipidemia in type 2 diabetes mellitus. *J Med Food.* 4. 193–199.
78. Marcel, A., Ekali, L., et al. (2011, July). The effect of *Spirulina platensis* versus soybean on insulin resistance in HIV-infected patients: a randomized pilot study. *Nutrients.* 3(7). 712–724.
79. Parikh P, Mani U, Iyer U. (2001). Role of spirulina in the control of glycemia and lipidemia in type 2 diabetes mellitus. *Dig Dis Sci.* 4. 193–199.

80. Rossmeisl M, Jelenik T, et al., (2009, May). Prevention and reversal of obesity and glucose intolerance in mice by DHA derivatives. *Obesity* (Silver Spring). 17(5). 1023-31.
81. Mani S, Iyer U, Subramanian S. (1998). Studies on the effect of *Spirulina* supplementation in control of diabetes mellitus. In: Subramanian G, editor. *Cyanobacterial biotechnology*. USA: Science Publishers Inc. 301–4.
82. Layam A, Chandra L, Reddy K. (2006). Antidiabetic property of Spirulina. *Diabet Croatica*. 35. 29–33.
83. Mani U, Iyer U, Nayak U. (2002). Hypocholesterolemic effect of Spirulina in patients with hyperlipidemic nephrotic syndrome. *J Med Food*. 5. 91–6.
84. Mani U, Desai S, Iyer U. (2000). Studies on long term effect of Spirulina supplementation on serum lipid profile and glycated proteins in NIDDM patients. *J Nutraceut*. 2. 25–32.
85. Mani, U., Desai, S, and Iyer, U. (2000). Studies on the long-term effect of Spirulina supplementation on serum lipid profile and glycated proteins in NIDDM patients, *J Nutraceut*. 2. 25–32.
86. Shah, S. (1997). Role of spirulina in the management of hyperlipidemia, *MSc Dissertation Thesis*, Department of Foods & Nutrition, The M.S. University of Baroda, Vadodara India.
87. Mani, U., Iyer U. et al. (2008). Therapeutic Utility of Spirulina. In *Spirulina in Human Nutrition and Health*. Edited by M.E. Gershwin and Amha Belay. CRC Press. 73.
88. The Use of Medicines in the United States: Review of 2010. *IMS Institute for Healthcare Informatics* (2011, April).
89. Takai Y. (1991). Effect of water soluble and insoluble fractions of Spirulina over serum lipids and glucose resistance of rats. *J Jpn Soc Nutr Food Sci*. 44. 273–7.
90. Urmila Jarouliya, U., et al. (2012, March). Alleviation of metabolic abnormalities induced by excessive fructose administration in Wistar rats by Spirulina maxima. *Indian J Med Res*. 135(3). 422–428.
91. Jarouliya U, Anish Z, Kumar P, Bisen P, Prasad G. (2012, March). Alleviation of metabolic abnormalities induced by excessive fructose administration in Wistar rats by Spirulina maxima. *Indian J Med Res*. 135(3). 422–428.
92. Iyer U, Dhruv S, and Mani I. (2008). Spirulina and Its Therapeutic Implications as a Food Product. In Spirulina in Human Nutrition and Health. Edited by M.E. Gershwin and Amha Belay. CRC Press. 59-61.
93. Kim M, Kim W. (2005). The change of lipid metabolism and immune function caused by antioxidant material in the hypercholesterolemin elderly women in Korea. *The Korean Journal of Nutrition*. 38. 67–75.
94. Lee E., Park J. et al. (2008, winter). A randomized study to establish the effects of spirulina in type 2 diabetes mellitus patients. *Nutr Res Pract*. 2(4). 295–300.
95. Park J, Kim W. (2003). The effect of spirulina on lipid metabolism, antioxidant capacity and immune function in Korean elderly. *The Korean Journal of Nutrition*. 36. 287–297.

96. Park H, Lee Y, et al. (2008). A radomized double-blind, placebo-controlled study to establish the effects of spirulina in elderly Koreans. *Ann Nutr Metab.* 52. 322–328.
97. Eun Hee Lee, Ji-Eun Park, et al. (2008, winter). A randomized study to establish the effects of blue green algae in type 2 diabetes mellitus patients. *Nutr Res Pract.* 2(4). 295–300.
98. Aizzat, O. Sopiah, H. et al. (2010, Dec). Modulation of oxidative stress by Chlorella vulgaris in streptozotocin (STZ) induced diabetic Sprague-Dawley rats. *Advances in Medical Sciences.* 55(2). 281–288.
99. Dong LY, Jin J, Lu G, Kang XL. (2013 Mar 21). Astaxanthin Attenuates the Apoptosis of Retinal Ganglion Cells in db/db Mice by Inhibition of Oxidative Stress. *Mar Drugs.* 11(3). 960-74.
100. Hirata T, Tanaka M, Ooike M, Tsunomura T, Sakaguchi M. (2000). Antioxidant activities of phycocyanobilin prepared from *Spirulina platensis. J Appl Phycol.* 12. 435–439.
101. Kim M, Cheong S, Lee J, Kim M, Sok D, Kim M. (2010, Apr). Spirulina improves antioxidant status by reducing oxidative stress in rabbits fed a high-cholesterol diet. *J Med Food.* 13(2). 420-6.
102. Riss J, Décordé K, et al. (2007, Sep 19). Phycobiliprotein C-phycocyanin from *Spirulina platensis* is powerfully responsible for reducing oxidative stress and NADPH oxidase expression induced by an atherogenic diet in hamsters. *J Agric Food Chem.* 55(19). 7962-7.
103. Ku C, Pham T, et al. (2013, Apr.). Edible blue-green algae reduce the production of pro-inflammatory cytokines by inhibiting NF-κB pathway in macrophages and splenocytes. *Biochim Biophys Acta.* 1830(4). 2981-8.
104. Yu B, Wang J, et al. (2012). Spirulina is an effective dietary source of zeaxanthin to humans. *British Journal of Nutrition.* 108 (4). 611–619.
105. Paredes-Carbajal M, Torres-Duran P, et al. (1997). Effects of dietary Spirulina maxima on endothelium dependent vasomotor responses of rat aortic rings. *Life Sci.* 61. 211–9.
106. Pak W, Takayama F, et al. (2012). Anti-oxidative and anti-inflammatory effects of spirulina on rat model of non-alcoholic steatohepatitis. *Journal of Clinical Biochemistry and Nutrition.* 51 (3). 227–234.
107. Bhattacharyya S, Mehta P. (2012). The hepatoprotective potential of Spirulina and vitamin C supplemention in cisplatin toxicity. *Food & Function.* 3 (2). 164–169.
108. Hassan A, Abdel-Aziem S, Abdel-Wahhab M. (2012). Modulation of DNA damage and alteration of gene expression during aflatoxicosis via dietary supplementation of Spirulina (Arthrospira) and whey protein concentrate. *Ecotoxicology and Environmental Safety* 79. 294–300.
109. Lee S, Jeon Y. (2013, Mar 4). Anti-diabetic effects of brown algae derived phlorotannins, marine polyphenols through diverse mechanisms. *Fitoterapia.* 86C. 129-136.

110. Layam A, Reddy C. (2007). Antidiabetic property of spirulina. *Diabetol Croat.* 35. 27–48.
111. Cherng J, Shih M. (2006, Feb 9). Improving glycogenesis in Streptozocin (STZ) diabetic mice after administration of green algae Chlorella. *Life Sci.* 78(11). 1181-6.
112. Dong L, Jin J, Lu G, Kang X. (2013, Mar 21). Astaxanthin Attenuates the Apoptosis of Retinal Ganglion Cells in db/db Mice by Inhibition of Oxidative Stress. *Mar Drugs.* 11(3). 960-74.
113. Lee S, Jeon Y. (2013, Mar 4). Anti-diabetic effects of brown algae derived phlorotannins, marine polyphenols through diverse mechanisms. *Fitoterapia.* 86C. 129-136.
114. Ou Y, Lin L, Yang X, Pan Q, Cheng X. (2013, May). Antidiabetic potential of phycocyanin: Effects on KKAy mice. *Pharm Biol.* 51(5). 539-544.
115. Joventino I, Alves H, et al. (2012, Aug.). The microalga *Spirulina platensis* presents anti-inflammatory action as well as hypoglycemic and hypolipidemic properties in diabetic rats. *J Complement Integr Med.* 10(9). Article 17.
116. Mani U, Desai S, Iyer U. (2000). Studies on the long-term effect of Spirulina supplementation on serum lipid profile and glycated proteins in NIDDM patients. *J Nutraceut.* 2(3). 25-32.
117. Itawa K, Inayama T. (1990). Effect of *Spirulina platensis* on plasma lipoprotein lipase activity in fructose induced hyperlipidemic rats. *J Nutr Sci Vitaminol.* 36. 165–71.
118. Hosoyamada Y, Takai T, Kato T. (1991) Effects of water-soluble and insoluble fractions of Spirulina on serum lipid components and glucose tolerance in rats. *Journal of Japanese Society of Nutrition and Food Science.* 44. 273–277.
119. Iwata K, Inayama T, Kato T. (1990). Effects of *Spirulina platensis* on plasma lipoprotein lipase activity in fructose-induced hyperlipidemia in rats. *J Nutr Sci Vitaminol.* 36. 165–171.
120. Kushak R, Drapeau C, Van Cott E. (2000). Favorable effects of blue green algae *Aphanizomenon flos-aquae* on rat plasma lipids. *JANA.* 2. 59-65.
121. Park J, Kim W. (2003). The effect of spirulina on lipid metabolism, antioxidant capacity and immune function in Korean elderly. *The Korean Journal of Nutrition.* 36. 287–297.
122. Kato T, Takemoto K, Katayama H, Kuwabara Y. (1984). Effects of Spirulina (*Spirulina platensis*) on dietary hypercholesterolemia in rats. *J Jap Soc Nutr Food Sci.* 37. 323–332.
123. Blé-Castillo JL, Rodríguez-Hernández A, et al. (2002). *Arthrospira maxima* prevents the acute fatty liver induced by the administration of simvastatin, ethanol and a hypercholesterolemic diet to mice. *Life Sci.* 70. 2665–2673.
124. Iwata K, Inayama T, Kato T. (1990). Effects of *Spirulina platensis* on plasma lipoprotein lipase activity in fructose-induced hyperlipidemic rats. *J Nutr Sci Vitaminol.* 36. 165–71.

125. Kato T, Takemoto K, Katayama H, Kuwabara Y. (1984). Effects of Spirulina (*Spirulina platensis*) on dietary hypercholesterolemia in rats. *J Jap Soc Nutr Food Sci.* 37. 323–332.
126. Cheong SH, Kim MY, et al. (2010). *Spirulina* Prevents Atherosclerosis by Reducing Hypercholesterolemia in Rabbits Fed a High-Cholesterol Diet. *J Nutr Sci Vitaminol.* 56. 34–40.
127. Moura, L. Puga, G., et al. (2011). Exercise and spirulina control non-alcoholic hepatic steatosis and lipid profile in diabetic Wistar rats. *Lipids Health Dis.* 10. 77.
128. Guan Y, Zhao H, Ding X, Zhu Y. (2007). Analysis of the contents of elements in Spirulina from different producing areas. *Guang Pu Xue Yu Guang Pu Fen Xi.* 27. 1029–1031.
129. Gonzalez De Rivera C, Miranda-Zamora R, Pardes-Carbajal MC, Juarez Oropeza MA. (1993). Preventive effects of Spirulina maxima on the fatty liver induced by a fructose rich diet in the rat, a preliminary report. *Life Sci.* 53. 57–61.
130. Torres-Duran P, Miranda-Zamora R, et al. (1999). Studies on the preventive effect of Spirulina maxima on fatty liver development induced by carbon tetrachloride, in the rat. *J Ethanopharmacol.* 64. 141–7.
131. Moura L. Puga G., et al. (2011). Exercise and *spirulina* control non-alcoholic hepatic steatosis and lipid profile in diabetic Wistar rats. *Lipids Health Dis.* 10. 77.
132. Moura, L. Puga, G., et al. (2011). Exercise and spirulina control non-alcoholic hepatic steatosis and lipid profile in diabetic Wistar rats. *Lipids Health Dis.* 10. 77.

Chapter 11: Give Your Immune System A Fighting Chance

1. Beisel W. (1996). Nutrition in pediatric HIV infection: setting the research agenda. Nutrition and immune function: Overview. *Journal of Nutrition.* 126(10 Suppl). 2611-1 5.
2. Beisel W. (1992, March). History of nutritional immunology: introduction and overview. *Journal of Nutrition.* 122(3 Suppl). 591-6.
3. Alexander M, Newmark H, Miller R. (1985). Oral betacarotene can increase the number of OKT4 + cells in human blood. *Immunology Letter.* 9(4). 221-4.
4. Abrams, K. (1996). *Algae to the Rescue.* Logan House. 71.
5. Bendich, A. (1989). Carotenoids and the immune response. *Journal of Nutrition.* 119. 112-5.
6. Okai Y, Higashi-Okai K. (1996, Dec). Possible immunomodulating activities of carotenoids *in vitro* cell culture experiments. *International Journal of Immunopharmacology.* 18(12). 753-8.
7. Younghaung W et al. (1994, June). The study on curative effect of zinc-containing blue-green algae for zinc deficient children. Paper presented at the *Fifth International Phycological Congress*, Qingdao, China.

8. Sevilla I, Aguiree N. (1995). Study on the Effects of Super Blue-Green Algae on the Nutritional Status and School Performance of First-, Second-, and Third-Grade Children Attending the Monseñor Velez School in Nandaime, Nicaragua [*doctoral dissertation*, in Spanish]. Universidad CentroAmericana, Nicaragua.
9. Bruno J, Gittelman J, Tuchfeld B. (2001). Lowered aluminum, with better detoxification and improved cognitive, behavioral ratings by children eating Aphanizomenon flos-aquae. Unpublished manuscript.
10. Belay, A. et al. (1993) Current knowledge on potential health benefits of Spirulina. *J. Appl. Phycol.*, 5, 235-41.
11. Khan Z, Bhadouria P, Bisen, P. (2005). Nutritional and therapeutic potential of Spirulina. *Curr. Pharm. Biotechnol.*, 6(5), 373-9.
12. American Physiological Society (2010, August 27). Macrophages: The 'defense' cells that help throughout the body. *ScienceDaily*. Retrieved October 8, 2013, from http://www.sciencedaily.com/releases/2010/08/100826141232.htm
13. Vivier E, Raulet D, et al. (2011 Jan 7). Innate or adaptive immunity? The example of natural killer cells. *Science*. 331(6013). 44-9.
14. Qureshi M, Garlich J, Kidd M. (1996). Dietary *Spirulina platensis* enhances humoral and cell-mediated immune functions in chickens. *Immunopharmacology and Immunotoxicology*. 18(3). 465-76.
15. Drapeau C. (1998, Aug 8). Increased macrophage activity observed in college basketball players eating AFA using live blood analysis. *Presentation at Klamath Falls*, Oregon.
16. Hayashi O, Katoh T, Okuwaki Y. (1994, Oct). Enhancement of antibody production in mice by dietary *Spirulina platensis*. *Journal of Nutritional Science and Vitaminology* (Tokyo). 40(5). 431-41.
17. Hayashi O, Hirahashi T, Katoh T, Miyajima H, Hirano T, Okuwaki Y. (1998). Class-specific influence of dietary *Spirulina platensis* on antibody production in mice. *Journal of Nutritional Science and Vitaminology* (Tokyo). 44(6). 841-51.
18. Hart A, Zaske L, Patterson K, Drapeau C, Jensen G. (2007, Sep). Natural killer cell activation and modulation of chemokine receptor profile in vitro by an extract from the cyanophyta Aphanizomenon flos-aquae. *J Med Food*. 10(3). 435-41.
19. Manoukian R, Citton M, Huerta P, et al. (1998). Phytoceuticals: Examining the Health Benefits and Pharmaceutical Properties of Natural Antioxidants and Phytochemicals. Massachusetts: International Business Communications, Inc.
20. Jensen G, Ginsberg D, Huerta P, Drapeau C. (2000, Jan). Consumption of *Aphanizomenon flos-aquae* has rapid effects on the circulation and function of immune cells in humans. A novel approach to nutritional mobilization of the immune system. *Journal of the American Nutraceutical Association*. 2(3). 50-8.
21. Manoukian R et al. (1998). Effects of the blue-green algae *Aphanizomenon flos-aquae* (L.) Ralphs on human natural killer cells. In *Phytoceuticals*: Examining the health benefits and pharmaceutical properties of natural antioxidants and phytochemicals. IBC Library Series 1911, ch. 3.1. Boston, March 1998; 233-41.

22. Jensen G, Ginsberg D, Huerta P, Drapeau C. (2000, Jan). Consumption of *Aphanizomenon flos-aquae* has rapid effects on the circulation and function of immune cells in humans. A novel approach to nutritional mobilization of the immune system. *Journal of the American Nutraceutical Association.* 2(3). 50-8.
23. Borchers, A., et al., Spirulina and Immunity. Chpt. 8. In Spirulina in Human Nutrition and Health. Edited by M.E. Gershwin and Amha Belay. CRC Press. 180
24. Hirahashi T, et al. (2002). Actiation of the human innate immune system by Spirulina: augmentation of interferon production and NK cytotoxicity by oral administration of hot water extract of *Spirulina platensis*. *Int. Immunopharmacol.* 2 (4), 423-424).
25. Tornabene T et al. (March 1985). Lipid and lipopolysaccharide constitutents of cyanobacterium *Spirulina platensis*. *Ecol Prog Serv.* 22. 121.
26. Cheng-Wu Z, Chao-Tsi T, Zhen Z. (1994). The effects of polysaccharide and phycocyanin from *Spirulina platensis* on peripheral blood and hematopoietic system of bone marrow in mice. *Proceedings of the Second Asia-Pacific Conference on Algal Biotechnology*. National University of Singapore. 58.
27. Kovats E. (1996, March). Potentiation of HIV envelope glycoprotein and other immunogens by endotoxin and its molecular fragments. *International Journal of Immunopharmacology.* 43(3). 248-56.
28. Besednova N, Smolina T, Mikheiskaia L, Ovodova R. (1979, Dec). Immunostimulating activity of the lipopolysaccharides of blue-green algae. *Zhurnal Mikrobiiolgii, Epidemiologii I Immunobiologii* (Russian). (1 2). 75-9.
29. Abrams K. (1996.). *Algae to the Rescue.* Logan House. 75.
30. Pugh N, Ross S, ElSohly H, ElSohly M, Pasco D. (2001). Isolation of Three High Molecular Weight Polysaccharide Preparations with Potent Immunostimulatory Activity from *Spirulina platensis*, *Aphanizomenon flos-aquae*, and *Chlorella pyrenoidosa*. *Planta Med.* 67. 737-742.
31. Rivas, R. and Riquelme, C. (2012) Probiotic Biofilms. Open Access Chapter 28, InTech; http://cdn.intechopen.com/pdfs/39609/InTech-Probiotic_biofilms.pdf
32. Valencia A, Walker J. (1999, June). A multi-axial treatment paradigm for mild traumatic brain injury to achieve reparative functional metaplasticity. Paper presented at the *Third World Congress on Brain Injury*, Quebec City, Canada.
33. Qureshi M, Ali R. (1996). *Spirulina platensis* exposure enhances macrophage phagocytic function in cats. *Immunopharmacology and Immunotoxicology.* 18(3). 457-63.
34. Qureshi M et al. (1994). Immune enhancement potential of *Spirulina platensis* in chickens. *Poultry Science.* 73. 46.
35. Terziev V, Planski B, Encheva I. (1983). Use of nonspecific agents and vaccination in bronchopneumonia prevention in cattle. *Vet Med Nauki* [Bulgarian]. 20(1). 36-9.
36. UMT To Develop Protein That Can Increase Aquatic Animals And Livestock Resilience Against Diseases, *Eastern Region News*, (2012, Oct 16). Retrieved from http://www.bernama.com.my/bernama/state_news/news.php?cat=et&id=702359

37. Zhang Cheng-Wu, et al. (1994). Effects of polysaccharide and phycocyanin from spirulina on peripheral blood and hematopoietic system of bone marrow in mice. *Second Asia-Pacific Conference on Alga Biotechnology*, April 25-27
38. Hayashi O, Ono S, Ishii K, et al. (2006, Feb.). Enhancement of proliferation and differentiation in bone marrow hematopoietic cells by *Spirulina (Arthrospira) platensis* in mice. *Journal of Applied Phycology*. 18(1). 47-56
39. McCarty MF. (2007, Dec). Clinical potential of Spirulina as a source of phycocyanobilin. *J Med Food*. 10(4). 566-70.
40. Buletsa et al, (1996). Retrieved from <http://spirulinamedicalresearch.blogspot.com/>
41. Penton-Rol G, et al. (2011, Jan. 11). C-Phycocyanin ameliorates experimental autoimmune encephalomyelitis and induces regulatory T cells. *Int Immunopharmacol*. 11(1). 29-38.
42. Rocuts F, Zhang X, et al. (2010). Bilirubin promotes de novo generation of T regulatory cells. *Cell Transplant*. 19(4). 443-51.
43. Liu Y, Li P, Lu J, Xiong W, Oger J, Tetzlaff W, Cynader M. (2008, Aug 1). Bilirubin possesses powerful immunomodulatory activity and suppresses experimental autoimmune encephalomyelitis. *J Immunol*. 181(3). 1887-97.
44. Jensen, G., et al. (2001, winter). Blue-Green Algae as an Immuno-Enhancer and Biomodulator. *JANA*. 3(4).
45. Horrobin E et al. (1979). The nutritional regulation of T-lymphocyte function. *Medical Hypotheses*. 5. 969.
46. Bock K, Sabin N. (1997). *The Road to Immunity*. Pocket Books. 177.
47. Pugh N, Ross S, ElSohly H, ElSohly M, Pasco D. (2001). Isolation of Three High Molecular Weight Polysaccharide Preparations with Potent Immunostimulatory Activity from *Spirulina platensis, Aphanizomenon flos-aquae,* and *Chlorella pyrenoidosa*. *Planta Med*. 67. 737-742.
48. Borchers, A., Belay, A., Keen, L., and Gershwin, E. Spirulina and Immunity. Chpt. 8 in M.E. Gershwin and Amha Belay (2008) edition of *Spirulina in Human Nutrition and Health*.

Chapter 12: Cancer Protective for Cells Gone Wild

1. Cited by Linet M. (2000). Evolution of cancer epidemiology. *Epidemiology Review*. 22(1). 35.
2. Murakami A, Ohigashi H, Koshimizu K. (1996, Jan). Anti-tumor promotion with food phytochemicals: a strategy for cancer chemoprevention. *Bioscience, Biotechnology, and Biochemistry*. 60(1). 1-8.
3. Hocman G. (1989). Prevention of Cancer: vegetables and plants. CBP: *Comparative Biochemistry and Physiology B*. 93(2). 201-12.
4. Journal of the National Cancer Institute. (2013, Aug 15). Retrieved from <http://bit.ly/17o1kAA and http://bit.ly/1begcFn>

5. Eheman, C. et al. (2012). Annual Report to the Nation on the status of cancer, 1975-2008, feartuing cancers associated with excess weight and lack of sufficient physical activity. *Cancer.* 118. 2339 – 2366.
6. National Cancer Institute. Fact Sheet. Obesity and Cancer Risk. Key Points. Retrieved from <www.cancer.gov/cancertopics/factsheet/Risk/obestity>
7. Hart, J. (June 2013). The Link between Obesity and Cancer. *Alternative and Complementary Therapies.* 19, (3).
8. Shaharudin S, Sulaiman S, Emran N, Shahril M, Hussain S. (2011, Jan-Feb). The use of complementary and alternative medicine among Malay breast cancer survivors. *Altern Ther Health Med.* 17(1). 50-6.
9. Chen F, Zhang Q. (1995, Jan). [Inhibitive effects of spirulina on aberrant crypts in colon induced by dimethylhydrazine]. [Article in Chinese] *Zhonghua Yu Fang Yi Xue Za Zhi.* 29(1). 13-7.
10. Schwartz J, Shklar G. (1989). A cyanobacteria extract and B-carotene stimulate an antitumor host response against an oral cancer cell line. *Phytotherapy Research.* 3(6). 243-248.
11. Dasgupta T, Banejee S, Yadav P, Rao A. (2001, Oct). Chemomodulation of carcinogen metabolising enzymes, antioxidant profiles and skin and forestomach papillomagenesis by *Spirulina platensis*. *Mol Cell Biochem.* 226(1-2). 27-38.
12. Higashi-Okai K, Okai Y. (1998, Sep). Potent suppressive activity of chlorophyll a and b from green tea (Camellia sinensis) against tumor promotion in mouse skin. *Journal of UOEH* (Japanese). 20(3). 181-8.
13. Okai Y et al. (1996, Aug). Suppressive effects of chlorophyllin on mutagen-induced umu C gene expression in Salmonella typhimurium (TA 1535/pSK 1002) and tumor promoter dependent ornithine decarboxylase induction in BALB/c 3T3 fibroblast cells. *Mutation Research.* 370(1). 11-7.
14. Breinholt V et al. (1995). Dietary chlorophyllin is a potent inhibitor of aflatoxin B1 hepatocarcinogenesis in rainbow trout. *Cancer Research.* 55(1). 57-62.
15. Mansi A., Parasramka, W., et al. (2012). MicroRNA profiling of carcinogen-induced rat colon tumors and the influence of dietary spinach. *Molecular Nutrition & Food Research.* DOI: 10.1002/mnfr.201200117
16. Bailey, G. (2002). Chlorophylls and Cancer Prevention: Passing the First Hurdle, Retrieved from <http://lpi.oregonstate.edu/f-w02/chlorophylls.html>
17. Olsen J, Dragsted L, Autrup H. (1988, Sept). Cancer risk and occupational exposure to aflatoxins in Denmark. *Br J Cancer.* 58(3). 392–396.
18. Williams JH, Phillips T, et al. (2004). Human aflatoxicosis in developing countries: a review of toxicology, exposure, potential health consequences, and interventions. *The American Journal of Clinical Nutrition.* 80. 1106-1122.
19. Murugavel KG, et al. (2007 Nov). Prevalence of aflatoxin B1 in liver biopsies of proven hepatocellular carcinoma in India determined by an in-house immunoperoxidase test. *J Med Microbiol.* 56(11). 1455-9.
20. Retrieved from <http://www.cancerquest.org/cancer-causing-environmental-agents.html>

21. Bailey, G. (2002). Chlorophylls and Cancer Prevention: Passing the First Hurdle. Retrieved from <http://lpi.oregonstate.edu/f-w02/chlorophylls.html>
22. Amaro H, Barros R, Guedes A, Sousa-Pinto I, Malcata F. (2012, Dec 19). Microalgal compounds modulate carcinogenesis in the gastrointestinal tract. *Trends Biotechnol.* S0167-7799(12)00199-0.
23. Talero E, Ávila-Roman J, Motilva V. (2012). Chemoprevention with phytonutrients and microalgae products in chronic inflammation and colon cancer. *Curr Pharm Des.* 18(26). 3939-65.
24. Cui Y, Lu Z, et al. (2007). Carotene induces apoptosis and up-regulates peroxisome proliferator-activated receptor L expression and reactive oxygen species production in MCF-7 cancer cells. *Eur. J. Cancer. 43,* 2590B 2601.
25. Kotake-Nara, E, Asai A, Nagao A. (2005). Neoxanthin and fucoxanthin induce apoptosis in PC-3 human prostate cancer cells. *Cancer Lett. 220,* 75B 84.
26. Das, S.K.; Hashimoto, T.; Kanazawa, K. (2008). Growth inhibition of human hepatic carcinoma HepG2 cells by fucoxanthin is associated with down-regulation of cyclin D. *Biochim. Biophys. Acta Gen. Subj.* 1780, 743B 749.
27. Das S, Hashimoto T, et al. (2005) Fucoxanthin induces cell cycle arrest at G0/G1 phase in human colon carcinoma cells through up-regulation of p21WAF1/Cip1. *Biochim. Biophys. Acta Gen. Subj. 1726,* 328B 335.
28. Hosokawa M, Kudo M, et al. (2004). Fucoxanthin induces apoptosis and enhances the antiproliferative effect of the PPAR*y*L ligand, troglitazone, on colon cancer cells. *Biochim. Biophys. Acta Gen. Subj. 1675,* 113B 119.
29. Hosokawa M, Wanezaki S, Miyauchi K, et al. (1999). Apoptosis-inducing effect of fucoxanthin on human leukemia cell line HL-60. *Food Sci. Technol. Res. 5,* 243B 246.
30. Moreau D, Tomasoni C., et al. (2006). Cultivated microalgae and the carotenoid fucoxanthin from Odontella aurita as potent anti-proliferative agents in bronchopulmonary and epithelial cell lines. *Environ. Toxicol. Pharm.* 22, 97B 103.
31. Cha K, Koo S, Lee D. (2008). Antiproliferative effects of carotenoids extracted from Chlorella ellipsoidea and Chlorella vulgaris on human colon cancer cells. *J. Agric. Food Chem. 56.* 10521B 10526.
32. Pasquet, V., et al. (2011). Antiproliferative Activity of Violaxanthin Isolated from Bioguided Fractionation of *Dunaliella* tertiolecta Extracts. *Marine Drugs.* 9. 829.
33. Sarkar A, et al. (1995). Beta-carotene prevents lipid peroxidation and red blood cell damage in experimental hepatocarcinogenesis. *Cancer Biochemistry and Biophysics.* 15(2). 111-25.
34. Tinkler J et al. (1994). Dietary carotenoids protect human cells from damage. *Journal of Photochemistry and Photobiology.* 26(3). 282-5.
35. Garewal H et al. (1995). Emerging role of beta-carotene and antioxidant nutrients in prevention of oral cancer. *Archives of Otolaryngology.* 121(2). 141-4.
36. Le Marchand L et al. (1993). Intake of specific carotenoids and lung cancer risk. *Cancer Epidemiology.* 2. 183-7.

37. Omenn G et al. (1996, Nov). Risk factors for lung cancer and for intervention effects in CARET, the Beta-Carotene and Retinol Efficacy Trial. *Journal of the National Cancer Institute*. 88(21). 1550-9.
38. Johnson E, Schroeder W. (1996). Microbial cartenoids. *Advanced Biochemistry Engineering and Biotechnology*. 53. 119-178.
39. Mittal A, et al. (1999, Mar). Modulatory potential of Spirulina fusiformis on carcinogen metabolizing enzymes in Swiss albino mice. *Phytotherapy Research*. 13(2). 111-4.
40. Kozlenko R, Henson, R. (1996). The study of Spirulina, effects on the AIDS virus, cancer and the immune system. *Healthy & Natural*. 3(5). 66-7.
41. Fujii Y, Sakamoto S, Ben-Amotz A, Nagasawa H. (1993, Mar-Apr). Effects of beta-carotene-rich algae *Dunaliella bardawil* on the dynamic changes of normal and neoplastic mammary cells and general metabolism in mice. *Anticancer Research*. 13(2). 389-93.
42. Xue L. (1993, Nov). Experimental study on extract of *Dunaliella salina* in preventing NSAR-induced cancer of proventriculus in mice. *Chung Hua Yu Fang I Hsueh Tsa Chih* (Chinese). 27(6). 350-3.
43. Nagasawa H et al. (1991, Mar-Apr). Suppression by beta-carotene-rich algae *Dunaliella bardawil* of the progression, but not the development, of spontaneous mammary tumours in SHN virgin mice. *Anticancer Research*. 11(2). 713-7.
44. Nagasawa H et al. (1989, Jan-Feb). Inhibition by beta-carotene-rich algae *Dunaliella* of spontaneous mammary tumourigenesis in mice. *Anticancer Research*. 9(1). 71-5.
45. Combs W et al. (1989). In vivo and in vitro effects of B-carotene and algae extracts in murine tumor models. *Nutrition and Cancer*. 12(4). 371-9.
46. Nagasawa H et al. (1989). Inhibition by beta carotene rich algae *Dunaliella* of spontaneous mammary tumourigensis in mice. *Anticancer Research*. 9. 71.
47. Noda K, Tanaka K et al. (2002, Sept). Simple assay for antitumour immunoactive glycoprotein derived from Chlorella vulgaris strain CK22 using ELISA. Phytother Res. 16(6). 581-5.
48. Hasegawa T, Matsuguchi T, et al. (2002, March). Toll-like receptor 2 is at least partly involved in the antitumor activity of glycoprotein from Chlorella vulgaris. Int Immunopharmacol. 2(4). 579-89.
49. Konishi, F. Tanaka, K., et al. (1990). Enhanced resistance against Escherichia coli infection by subcutaneous administration of hot-water extract of *chlorella vulgaris* in cyclophosphamide-treated mice. *Cancer Immunology Immunotherapy*, 32(1). 1-7.
50. Noda K, Ohno N, et al. (1996). A water-soluble antitumor glycoprotein from *Chlorella vulgaris*. *Planta Med*. 62. 423-426.
51. Singh A, Singh S, Bamezai R. (1999). Inhibitory potential of *Chlorella vulgaris* (E25) on mouse skin papillomagenesis and xenobiotic detoxication system. *Anticancer Res*. 19. 1887-1892.

52. Tanaka K, Konishi F, et al. (1984) Augmentation of antitumor resistance by a strain of unicellular green algae, chlorella vulgaris. *Cancer Immunol. Immunother.* 17(2). 90-94.
53. Schwartz J, Shklar G, Suda D. (1986). Inhibition of experimental oral carcinogenesis by topical beta carotene. *Carcinogenesis.* 7(5). 711-5.
54. Schwartz J, Shklar G, Reid S, Trickler D. (1988). Prevention of experimental oral cancer by extracts of Spirulina-*Dunaliella* algae. *Nutrition and Cancer.* 11(2). 127-34.
55. Schwartz J, Sloane D, Shklar G. (1989). Prevention and inhibition of oral cancer in the hamster buccal pouch model associated with carotenoid immune enhancement. *Tumor Biology.* 10. 297-309.
56. Shklar G, Schwartz J. (1988, May). Tumor necrosis factor in experimental cancer regression with alpha-tocopherol, beta-carotene, canthaxanthin and algae extract. *European Journal of Cancer and Clinical Oncology.* 24(5). 839-50.
57. Schwartz J, Shklar G. (1987, June). Regression of experimental hamster cancer by beta-carotene and algae extracts. *Journal of Oral and Maxillofacial Surgery.* 45(6). 510-5.
58. Mathew B et al. (1995). Evaluation of chemoprevention of oral cancer with Spirulina fusiformis. *Nutrition and Cancer.* 24(2). 197-202.
59. Smith C, Zhang X. (1996). Mechanism of action of cryptophycin. *Journal of Biological Chemistry.* 271(11). 6192-8.
60. Smith C, Zhang X et al. (1994, July). Cryptophycin: a new antimicrotubule agent active against drug-resistant cells. *Cancer Research.* 54(14). 3779-84.
61. Trimurtulu G, Ohtani I, et al. (1994) Total structures of cryptophycins, potent antitumor depsipeptides from the blue-green alga Nostoc sp. strain GSV 224. *Journal of the American Chemical Society.* 116, 4729-4737.
62. Verdier-Pinard P, et al. (1998). Structure activity analysis of the interaction of curacin A, the potent colchicines site antimitotic agent, with tubulin and effects of analogs on the growth of MCF-7 breast cancer cells. *Molecular Pharmacology.* 53, 62-76.
63. Luesch H, Yoshida W, et al. (2002, Jan). Symplostatin 3, a new dolastatin 10 analogue from the marine cyanobacterium Symploca sp. VP 452. *J Nat Prod.* 65(1). 16-20.
64. Ogino J et al. (June 1996). Dendroamides, new cyclic hexapeptides from a blue-green alga. Multidrug-resistance reversing activity of dendroamide A. *Journal of Natural Products.* 59(6). 581-6.
65. Smith C et al. (1994). Reversal of multiple drug resistance by tolyporphin, a novel cyanobacterial natural product. *Oncology Research.* 6(4-5). 211-8.
66. Smith C, Carmeli S, Moore R, Patterson G. (March 1993). Scytophycins, novel microfilament-depolymerizing agents which circumvent P-glycoprotein-mediated multidrug resistance. *Cancer Research.* 53(6). 1343-7.
67. Vermeil C, Morin O. (1976, Oct). Experimental role of the unicellular algae Prototheca and Chlorella (chlorellacease) in anti-cancer immunogenesis.

Comptes Rendus Des Seances De La Societe De Biologie Et De Ses Filiales. 170(3). 646-9.

68. Mishima T et al. (1998). Inhibition of tumor invasion and metastasis by calcium spirulan (Ca-SP), a novel sulfated polysaccharide derived from blue-green algae, *Spirulina platensis*. *Clinical and Experimental Metastasis.* 16(6). 541-50.
69. Tanaka K et al. (1998, Feb). A novel glycoprotein obtained from Chlorella vulgaris strain CK22 shows antimetastatic immuno-potentiation. *Cancer Immunology and Immunotherapy.* 45(6). 313-20.
70. Hayakawa Y, Hayashi T et al. (1997, March). Calcium spirulan as an inducer of tissue-type plasminogen activator in human fetal lung fibroblasts. *Biochimica Et Biophysica Acta.* 1355(3). 241-7.
71. Noda K et al. (October 1996). A water-soluble antitumor glycoprotein from Chlorella vulgaris. *Planta Medica.* 62(5). 423-6.
72. Tokuda H et al. (1996). Inhibition of 12-O-TPA promoted mouse skin papilloma by digalactosyl diacylglycerols from the freshwater cyanobacterium Phormidium tenue. *Cancer Letters.* 104(1). 91-5.
73. Vermeil C, Morin O, Le Bodic L. (1985, Mar-Jun). The stimulation of tumoricidal peritoneal macrophages can be directly induced by peritoneal implantation of unicellar algae in humans. *Archives De L Institut Pasteur De Tunis* (French). 62(1-2). 91-4.
74. Yamamura Y. et al. (1979, Aug). Nonspecific Adjuvant Immunotherapy of Lung Cancer with Cell Wall Skeleton of Mycobacterium bovis Bacillus Calmette-Guérin. *Cancer Res.* 39; 3262.
75. Azuma I, Seya T. (2001 Jul). Development of immunoadjuvants for immunotherapy of cancer. *Int Immunopharmacol.* 1(7). 1249-59.
76. Akao Y, Ebihara T, et al. (2009, Aug). Enhancement of antitumor natural killer cell activation by orally administered Spirulina extract in mice. *Cancer Sci.* 100(8). 1494-501.
77. Carbonnelle D, et al. (1999). Antitumor and antiproliferative effects of an aqueous extract from the marine diatom Haslea ostrearia (simonsen) against solid tumors: lung carcinoma (NSCLC-N6), kidney carcinoma (E39) and melanoma (M96) cell lines. *Anticancer Research.* 19. 621-4.
78. Berge J et al. (1997). Antiproliferative effects of an organic extract from the marine diatom *Skeletonema costatum* (grev.) against a non-small-cell bronchopulmonary carcinoma line (NSCLC-N6). *Anticancer Research.* 17. 2115-20.
79. Henrikson R. (1989). *Earth Food Spirulina.* Ronore Enterprises. 69.
80. Lijjima N et al. Anti-tumor agent and method of treatment therewith. US patent pending, ref. P1150-726-A82679, Application filed 15 Sep. 1982.
81. Patent claim filed by Dainippon Ink and Chemicals and Tokyo Kenkyukai for "Anti-tumor agents containing phycobilin — also used to treat ulcers and hemorrhoid bleeding" (1983, JP58065216-A830418).
82. Cited in: Challem J. (1981). *Spirulina.* Keats Publishing. 17.

83. Singh et al. (1999). Inhibitory potential of Chlorella vulgaris (E-2S) on mouse skin papillomagenesis and xenobiotic detoxification system. *Anticancer Research.* 19. 1887-92.
84. Tanaka K et al. (1990). Oral administration of Chlorella vulgaris augments concomitant antitumor immunity. *Immunopharmacology and Immunotoxicology.* 12(2). 277-91.
85. Konishi F et al. (1985). Antitumor effect induced by a hot water extract of Chlorella vulgaris: resistance to meth-A tumor growth mediated by CE-induced polymorphonuclear leukocytes. *Cancer Immunology and Immunotherapy.* 19(2). 73-8.
86. Tanaka K et al. (1984). Augmentation of antitumor resistance by a strain of unicellular green algae, *Chlorella vulgaris. Cancer Immunology and Immunotherapy.* 17(2). 90-4.
87. Nomoto K, Yokokura T, Satoh H, Mutai M. (March 1983). Antitumor activity of Chlorella extract, PCM-4, by oral administration. *Gan To Kagaku Ryoho* (Japanese). 10(3). 781-5.
88. Morin O, Guilard R, Guihard D, Vermeil C. (1980). New approach to the study of the experimental inhibitory effect of the unicellular alga Chlorella pyrenoidosa against the murine sarcomas BP8 and L1210. *Comptes Rendus Des Seances De La Societe De Biologie Et De Ses Filiales* (French). 174(1). 74-81.
89. Neveu P et al. (1978, Dec). Modulation of antibody synthesis by an antitumor algae. *Experientia.* 34(12). 1644-5.
90. Miyazawa Y et al. (1988, Dec). Immunomodulation by a unicellular green algae (Chlorella pyrenoidosa) in tumor-bearing mice. *Journal of Ethnopharmacology.* 24(2). 135-46.
91. Tanaka K et al. (1998, Feb). A novel glycoprotein obtained from Chlorella vulgaris strain CK22 shows antimetastatic immunopotentiation. *Cancer Immunology and Immunotherapy.* 45(6). 313-20.
92. Retrieved from <http://www.oasisofhope.com/immune_stimulation.php>
93. Perumal SS, Shanthi P, Sachdanandam P. (V). Energy-modulating vitamins - a new combinatorial therapy prevents cancer cachexia in rat mammary carcinoma. *Br J Nutr.* 93(6). 901-9.
94. Perumal SS, Shanthi P, Sachdanandam P. (2005, Jun) Energy-modulating vitamins--a new combinatorial therapy prevents cancer cachexia in rat mammary carcinoma. *Br J Nutr.* 93(6). 901-9.
95. Gorban E, et al. (2000). Clinical and experimental study of Spirulina efficacy in chronic diffuse liver diseases. *Lik. Sprava.* 6. 89.
96. Ben-Amotz A et al. (1996, Nov). Natural beta-carotene and whole body irradiation in rats. *Radiation and Environmental Biophysics.* 35(4). 285-8.
97. Konishi F et al. (1996, June). Protective effect of an acidic glycoprotein obtained from culture of *Chlorella vulgaris* against myelosuppression by 5-fluorouracil. *Cancer Immunology and Immunotherapy.* 42(5). 268-74.
98. Retrieved from <http://phys.org/news/2012-12-algae-complex-anti-cancer-drug.html>

99. Federation of American Societies for Experimental Biology (2010, March 2). Researchers fishing for cancer cure discover active DHA derivatives. *ScienceDaily.* Retrieved July 20, 2013 from <http://www.sciencedaily.com/releases/2010/03/100301091554.htm>
100. Elmesery M, Algayyar M, et al. (2009). Chemopreventive and renal protective effects for docosahexaenoic acid (DHA): implications of CRP and lipid peroxides. *Cell Division.* 4:6.
101. Brasky T, Till C, et al. (2011). Serum Phospholipid Fatty Acids and Prostate Cancer Risk: Results from the Prostate Cancer Prevention Trial. *Am J. Epidemiol.* 173(12). 1429–1439.
102. Sears, B. (2013, July 11). Omega-3 fatty acids and prostate cancer? Oh, really? Retrieved from <http://zonediet.com/blog/2013/07/omega-3-fatty-acids-and-prostate-cancer-oh-really/>
103. Haas G, Delongchamps N, Brawley OW Wang C, and de la Roza G. (2008). The worldwide epidemiology of prostate cancer: perspectives from autopsy studies. *Can J Urol.* 15. 3866-3871
104. Marugame T and Mizuno S. (2005). Comparison of Prostate Cancer Mortality in Five Countries: France, Italy, Japan, UK and USA from the WHO Mortality Database (1960–2000). *Jpn J Clin Oncol* 35. 690–691.

Chapter 13: Detoxify for Life

1. Abrams K. (1996). Algae to the Rescue. Logan House. 26.
2. Liska, D. (1998, Jun).The Detoxification Enzyme Systems. Alternative Medicine Review. 3 (3). 187-198.
3. Tel-Or E, Huflejt M, Packer L. (1985, Oct). The role of glutathione and ascorbate in hydroperoxide removal in cyanobacteria. *Biochemical and Biophysical Research Communications.* 132(2). 533-9.
4. Arenesen E. (1995, Aug). Serum total homocysteine and coronary heart disease. *International Journal of Epidemiology.* 24(4). 704-9.
5. Cui J, Wakabayashi S et al. (1989, Mar). Isolation and sequence studies of cysteinyl peptides from Spirulina glutathione reductase: comparison of active site cysteine peptides with those of other flavoprotein disulfide oxidoreductases. *Journal of Biochemistry* (Tokyo). 105(3).390-4.
6. Sundquist A, Fahey R. (1989, Nov). Evolution of antioxidant mechanisms: thiol-dependent peroxidases and thioltransferase among prokaryotes. *Journal of Molecular Evolution.* 29(5). 429-35.
7. Pandit A, Sachdeva T, Bafna P. (2012). Drug-Induced Hepatotoxicity. *Jour. of Applied Pharmaceutical Science.* 02 (05). 233-243
8. Torres-Duran P et al. (1999, Feb). Studies on the preventive effect of *Spirulina maxima* on fatty liver development induced by carbon tetrachloride, in the rat. *Journal of Ethnopharmacology.* 64(2). 141-7.

9. Torres-Duran P et al. (1998, April). *Spirulina maxima* prevents induction of fatty liver by carbon tetrachloride in the rat. *Biochemistry and Molecular Biology International.* 44(4). 787-93.
10. Vadiraja B, Gaikwad N, Madyastha K. (1998). Hepatoprotective effect of C-phycocyanin: Protection for carbon tetrachloride and R(+)-pulegone-mediated hepatotoxicity in rats. *Biochemical and Biophysical Research Communications.* 249(2). 428-31.
11. Murthy K, Rajesha J, Swamy M, Ravishankar G. (2005, winter). Comparative evaluation of hepatoprotective activity of carotenoids of microalgae. J Med Food. 8(4). 523-8.
12. Vanitha A, Murthy K, et al. (2007, Mar-Apr). Effect of the carotenoid-producing alga, Dunaliella bardawil, on CCl4-induced toxicity in rats. Int J Toxicol. 26(2). 159-67.
13. Kujawska M, Jodynis-Liebert J, et al. (2007, Aug). Protective effect of Aquilegia vulgaris (L.) on carbon tetrachloride-induced oxidative stress in rats. *Indian J Exp Biol.* 45(8). 702-11.
14. Gonzalez de Rivera C et al. (1993). Preventative effect of Spirulina maxima on the fatty liver induced by a fructose-rich diet in the rat, a preliminary report. *Life Science.* 53(1). 57-61.
15. Dashwood R, Guo D. (1995). Protective properties of chlorophylls against the covalent binding of heterocyclic amines to DNA in vitro and in vivo. Paper presented at the *Princess Takamatsu Symposium.* 23. 181-9.
16. Challem J. (1981). *Spirulina.* Keats Publishing. 14.
17. Sukenik A, Takahasi H, and Modady S. (1994). Dietary lipids from marine unicellular algae enhance the amount of liver and blood omega-3 fatty acids in rats. *Annals of Nutrition and Metabolism.* 38(2). 85-96.
18. Murthy K, Rajesha J, Swamy M, Ravishankar G. (2005, winter). Comparative evaluation of hepatoprotective activity of carotenoids of microalgae. *J Med Food.* 8(4). 523-8.
19. Levin, G, Yeshurun M, Mokady S. (1997). In vivo antiperoxidative effect of 9-cis beta-carotene compared with that of the all-trans isomer. *Nutrition and Cancer.* 27(3). 293-7.
20. Gagne S. (1990). *Energetics of Food.* Spiral Science Inc. 274.
21. Challem J. (1981). *Spirulina.* Keats Publishing. 18.
22. Kutala K, Iyyapu K et al. Drug-Induced Nephrotoxicity Protection by Spirulina. In Spirulina in Human Nutrition and Health. Edited by M.E. Gershwin and Amha Belay. CRC Press (2008). Pg. 165.
23. Farooq S, Ebrahim, A, et al. (2006). Oxalate mediated nephronal impairment and its inhibition by c-phycocyanin: A study on urolithic rats. *Molecular and Cellular Biochemistry.* 284. 95–101.
24. Thamilselvan V, Menon M, Thamilselvan S. (2009, Nov). Oxalate-induced activation of PKC-α and -δ regulates NADPH oxidase-mediated oxidative injury in renal tubular epithelial cells. *Am J Physiol Renal Physiol.* 297(5). F1399–F1410.

25. Farooq S, Ebrahim, A, et al. (2006). Oxalate mediated nephronal impairment and its inhibition by c-phycocyanin: A study on urolithic rats. *Molecular and Cellular Biochemistry* 284. 95–101.
26. Yamane Y. (1988, April 15). The effect of Spirulina on nephrotoxicity in rats. Paper presented at *Annual Symposium of the Pharmaceutical Society of Japan.*
27. Lee W, Rosenbaum M. (1987). *Chlorella.* Keats Publishing. 17.
28. Pore R. (1984). Detoxification of chlordecone poisoned rats with *Chlorella* and Chlorella-derived sporopollenin. *Drug and Chemical Toxicology.* 7(1). 57-7.
29. Kutala K, Iyyapu K et al. (2008). Drug-Induced Nephrotoxicity Protection by Spirulina. In Spirulina in Human Nutrition and Health. Edited by M.E. Gershwin and Amha Belay. CRC Press. 167.
30. Kuhad A, Tirkey N, Pilkhwal S, Chopra K. (2006). Renoprotective effect of *Spirulina fusiformis* on cisplatin-induced oxidative stress and renal dysfunction in rats. *Ren Fail.* 28(3). 247-54.
31. Kuriakose GC, Kurup MG. (2008). Evaluation of renoprotective effect of *Aphanizomenon flos-aquae* on cisplatin-induced renal dysfunction in rats. *Ren Fail.* 30(7). 717-25.
32. Mohan I, Khan M, et at. (2006, Dec). Protection against cisplatin-induced nephrotoxicity by Spirulina in rats. *Cancer Chemother Pharmacol.* 58(6). 802-8.
33. Kuhad A, Tirkey N, Pilkhwal S, Chopra K. (2006 Apr). Effect of Spirulina, a blue green algae, on gentamicin-induced oxidative stress and renal dysfunction in rats. *Fundam Clin Pharmacol.* 20(2). 121-8.
34. Karadeniz A, Yildirim A, Simsek N, Kalkan Y, Celebi F. (2008) *Spirulina platensis* protects against gentamicin-induced nephrotoxicity in rats. *Phytother Res.* 22. 1506–1510.
35. Deng, R, Chow T. (2010 Aug). Hypolipidemic, Antioxidant and Antiinflammatory Activities of Microalgae Spirulina. *Cardiovasc Ther.* 28(4). e33–e45.
36. Giordano G, Costa L. (2012, Jun 24). Developmental neurotoxicity: some old and new issues. *ISRN Toxicol.* Retrieved January 6, 2014 from <http://www.ncbi.nlm.nih.gov/pmc/articles/PMC3658697/>
37. Lam H, Kwok K, et al. (2013, Apr). Long term neurocognitive impact of low dose prenatal methylmercury exposure in Hong Kong. *Environ Int.* 54. 59-64.
38. Shastri D, Kumar M, Kumar A. (1999, May). Modulation of lead toxicity by Spirulina fusiformis. *Phytother Res.* 13(3). 258-60.
39. Upasani CD, Balaraman R. (2003, Apr). Protective effect of Spirulina on lead induced deleterious changes in the lipid peroxidation and endogenous antioxidants in rats. Phytother Res. 17(4). 330-4.
40. Ponce-Canchihuamán J, Pérez-Méndez O, Hernández-Muñoz R, Torres-Durán P, Juárez-Oropeza M. (2010, Mar). Protective effects of *Spirulina maxima* on hyperlipidemia and oxidative-stress induced by lead acetate in the liver and kidney. *Lipids Health Dis.* 31(9). 35.
41. Upasani CD, Khera A, Balaraman R. (2001, Jan). Effect of lead with vitamin E, C, or Spirulina on malondialdehyde, conjugated dienes and hydroperoxides in rats. *Indian J Exp Biol.* 39(1). 70-4.

42. Upasani C, Balaraman R. (2003, Apr). Protective effect of Spirulina on lead induced deleterious changes in the lipid peroxidation and endogenous antioxidants in rats. *Phytother Res.* 17(4). 330-4.
43. Misbahuddin M, Islam A, Khandker S, Ifthaker-Al-Mahmud, Islam N, Anjumanara. (2006). Efficacy of spirulina extract plus zinc in patients of chronic arsenic poisoning: a randomized placebo-controlled study. *Clin Toxicol (Phila).* 44(2). 135-41.
44. Saha S, Misbahuddin M, Ahmed A. (2010, Jan). Comparison between the effects of alcohol and hexane extract of spirulina in arsenic removal from isolated tissues. *Mymensingh Med J.* 19(1). 27-31.
45. Bruno J, Gittelman, J, Tuchfeld B. (2000). Lowered aluminum, with better detoxification, and improved cognitive, behavioral ratings by children eating Aphanizomenon flos-aquae. Unpublished manuscript.
46. Burnatowska-Hledin M, Kaiser L, Mayor G. (1983). Aluminum, parathyroid hormone, and osteomalacia. *Special Topics in Endocrinology and Metabolism.* 5. 201-26.
47. Valencia A, Walker J. (1999, June). A multi-axial treatment paradigm for mild traumatic brain injury to achieve reparative functional metaplasticity. Paper presented at the Third World Congress on Brain Injury, Quebec City, Canada.
48. Golubkina N, et al. (2000, Jan-Feb). The selenium haemostasis during experimental anaphylaxis reaction in rats treated with reduced glutathione and selenium enriched Spirulina. *Voprosy Meditsinskoi Khimii* (Russian). 46(1). 22-7.
49. Salman B, Kerem M, et al. (2008, Jun). Effects of Cholerella sp. microalgae extract on colonic anastomosis in rats with protein-energy malnutrition. *Colorectal Dis.* 10(5). 469-78.
50. Bedirli A, Kerem M, et al. (2009, Dec). Administration of Chlorella sp. microalgae reduces endotoxemia, intestinal oxidative stress and bacterial translocation in experimental biliary obstruction. *Clin Nutr.* 28(6). 674-8.
51. Young R, Bergei J. (1980, Jan). Use of chlorophyllin in the care of geriatric patients. Journal of the American Geriatric Society. 28(1). 46-7. 47

Chapter 14: A Healthy Heart Matters

1. Bircher-Benner. (1973). Nutrition Plan for High Blood Pressure Problems, Jove Books.
2. Arenesen E. (1995, Aug). Serum total homocysteine and coronary heart disease. *International Journal of Epidemiology.* 24(4). 704-9.
3. Ho, K et al. (1993). Survival after the onset of congestive heart failure in Framingham Heart Study subjects. *Circulation.* 88. 107-115
4. Sanchez A. (2000, Mar 13). Cited in: HeartInfo.
5. Greenfield B. (1013, Oct 24). Healthy Living Does Butter Still Deserve a Bad Rap? Controversial Doc Says No. Retrieved from <http://shine.yahoo.com/healthy-living/does-butter-still-deserve-a-bad-rap--controversial-doc-says-no-002203701.html>

6. Kelly F. (1998, Mar). Use of antioxidants in the prevention and treatment of disease. *Jour of The Inter Federation of Clinical Chemistry.* 10(1). 21-3.
7. Mani, U., et al. (2008). Therapeutic Utility of Spirulina. In Spirulina in Human Nutrition and Health. Edited by M.E. Gershwin and Amha Belay. CRC Press. 75.
8. Deng R, Chow T. (2010, Aug). Hypolipidemic, Antioxidant and Antiinflammatory Activities of Microalgae *Spirulina. Cardiovasc Ther.* 28(4). e33–e45.
9. Calabrese V, Cornelius C, et al. (2010, Dec). Oxidative Stress, Redox Homeostasis and Cellular Stress Response in Ménière's Disease: Role of Vitagenes. *Neurochemical Research.* 35(12). 2208-2217.
10. Riss J, et al. (2007, Sep 19). Phycobiliprotein C-phycocyanin from *Spirulina platensis* is powerfully responsible for reducing oxidative stress and NADPH oxidase expression induced by an atherogenic diet in hamsters. *J Agric Food Chem.* 55(19);7962-7
11. Deng R, Chow T. (2010, Aug). Hypolipidemic, antioxidant, and antiinflammatory activities of microalgae Spirulina. *Cardiovasc Ther.* 28(4). e33-45.
12. Mietus-Snyder M, Malloy, M. (1998). Endothelial dysfunction occurs in children with two genetic hyperlipidemias: Improvement with antioxidant vitamin therapy. *Journal of Pediatrics.* 133(1). 35-40.
13. Paredes-Carbajal M, et al. (1997). Effects of dietary Spirulina maxima on endothelium dependent vasomotor response of rat aortic rings. *Life Science.* 61(15). 211-9.
14. Mascher D, Paredes-Carbajal M, et al. (2006, Jan). Ethanolic extract of Spirulina maxima alters the vasomotor reactivity of aortic rings from obese rats. *Arch Med Res.* 37(1). 50-7.
15. Chiu H, et al. (2006, Feb). Mechanisms involved in the antiplatelet effect of C-phycocyanin. *British Journal of Nutrition.* 95(2). 435-40.
16. Pomin V, Mourão P. (2008, Dec). Structure, biology, evolution, and medical importance of sulfated fucans and galactans. *Glycobiology.* 18(12). 1016-27.
17. Abrams K. Algae to the Rescue. Logan House. 1996. 39-40.
18. Pirt S. Biochemical Society Trans. (1980) 8. cited by Aaronson and Dubinsky. Mass production of microalgae. *Experientia.* 1982; 43(suppl). 42.
19. Shaish A, Harari A, et al. (2006, Nov). 9-cis beta-carotene-rich powder of the alga *Dunaliella bardawil* increases plasma HDL-cholesterol in fibrate-treated patients. *Atherosclerosis.* 189(1). 215-21.
20. Sano T, Tankaka Y. (1987). Effect of dried, powdered Chlorella vulgaris on experimental atherosclerosis and alimentary hypercholesterolemia in cholesterol-fed rabbits. *Artery.* 14(2). 76-84.
21. Santo T et al. (1988). Effect of lipophilic extract of Chlorella vulgaris on alimentary hyperlipidemia in cholesterol-fed rats. *Artery.* 15(4). 217-24.
22. Rolle I, Pabst W. (1980). Uber die cholesterinsenkende Wirkung der einzelligen Grunalge Scenedesmus acutus. *Nutritional Metabolism* (English abstract). 24. 291-301.
23. Conquer J, Holub B. (1995, June). Effects of supplementation with an algae source of docosahexaenoic acid on risk factors for heart disease in vegetarians.

Abstract of the 64th Congress of the European Arteriosclerosis Society. Utrecht, Netherlands. 10-13.

24. Morris D et al. (1994). Serum carotenoids and coronary heart disease. The Lipid Research Clinics Coronary Primary Prevention Trial and Follow-up Study. *Journal of the American Medical Association.* 272. 1439-41.
25. Iwata K, Inayama T, Kato T. (1990, April). Effects of *Spirulina platensis* on plasma lipoprotein lipase activity in fructose-induced hyperlipidemic rats. *Journal of Nutritional Science and Vitaminology* (Tokyo). 36(2). 165-71.
26. Han L, Li D, Xiang L, Gong X, Kondo Y, Suzuki I, Okuda H. (2006, Jan). [Isolation of pancreatic lipase activity-inhibitory component of *Spirulina platensis* and it reduce postprandial triacylglycerolemia] *Yakugaku Zasshi.* 126(1). 43-9. [Article in Japanese]
27. Martens J, Barg H, Warren M, et al. (2002). Microbial production of vitamin B_{12}. *Appl Microbiol Biotechnol.* 58. 275–285.
28. Rush University Medical Center (2011, Sept 27). Low vitamin B12 levels may lead to brain shrinkage, cognitive problems. *ScienceDaily.* Retrieved October 5, 2013, from http://www.sciencedaily.com /releases/2011/09/110926165852.htm
29. Joseph J, Loscalzo J. (2013). Methoxistasis: Integrating the Roles of Homocystein and Folic Acid in Cardiovascular Pathobiology. *Nutrients.* 5. 3235-3256.
30. Joseph J, Loscalzo J. (2013). Methoxistasis: Integrating the Roles of Homocystein and Folic Acid in Cardiovascular Pathobiology. *Nutrients.* 5. 3235-3256.
31. Joseph J, Loscalzo J. (2013). Methoxistasis: Integrating the Roles of Homocystein and Folic Acid in Cardiovascular Pathobiology. *Nutrients.* 5. 3235-3256.
32. Carmel R, Green R, Rosenblatt D, Watkins D. (2003). Update on cobalamin, folate, and homocysteine. *Hematology Am Soc Hematol Educ Program.* 62-81.
33. Baroni L, Scoglio S, Benedetti S, et al. (2009, Mar). Effect of a Klamath algae product on blood levels of vitamin B12 and homocysteine in vegan subjects: a pilot study. *Int J Vitam Nutr Res.* 79(2). 117-23.
34. Croft M, Lawrence A, et al. (2005, Nov 3). Algae acquire vitamin B12 through a symbiotic relationship with bacteria. *Nature.* 438(7064). 90-3.
35. Rodionov D, Vitreschak A, Mironov A, Gelfand, M. (2003). Comparative genomics of the vitamin B12 metabolism and regulation in prokaryotes. *J Biol Chem.* 278. 41148–41159.
36. Biotechnology and Biological Sciences Research Council (2006, Jan 14). The Secret Life of Algae. *ScienceDaily.* Retrieved October 5, 2013, from <http://www.sciencedaily.com /releases/2006/01/060114152028.htm>
37. A 'B12 shot' for marine algae. (2012, June 8). Retrieved from <http://www.terradaily.com/reports/A_B12_shot_for_marine_algae_999.html>
38. Bertrand E, Allen A. (2012). Influence of vitamin B auxotrophy on nitrogen metabolism in eukaryotic phytoplankton. *Front Microbiol.* 3. 375.

39. Yamada, et al. (1996). Content and characteristics of vitamin B12 in some seaweeds. *J. Nutr. Sci. Vitaminol.* 42. 497-505.
40. Yamada K, Yamada Y, Fukuda M, Yamada S. (1999, Nov). Bioavailability of dried asakusanori (porphyra tenera) as a source of Cobalamin (Vitamin B12). *Int J Vitam Nutr Res.* 69(6). 412-8.
41. Rauma A, Törrönen R., Hänninen O., Mykkänen H. (1995). Vitamin B-12 status of long-term adherents of a strict uncooked vegan diet ("living food diet") is compromised. *J. Nutr.* 125. 2511–2515.
42. Dagnelie P. (1997, Feb). Some algae are potentially adequate sources of vitamin B-12 for vegans, *J Nutr.* 127(2). 379; author reply 380.
43. Watanabe F, Katsura H, et al. (1999, Nov). Pseudovitamin B-12 is the predominant cobamide of an algal health food, spirulina tablets. *Journal of Agricultural and Food Chemistry.* 47 (11). 4736-4741.
44. Adarme-Vega T, Lim D, Timmins M, Vernen F, Li Y, Schenk P. (2012, Jul 25). Microalgal biofactories: a promising approach towards sustainable omega-3 fatty acid production. *Microb Cell Fact.* 11. 96.
45. Smith D et al. (1989). Eskimo plasma constituents, dihomo-linolenic acid, eicosapentaenoic acid and docosahexaenoic acid inhibit the release of atherogenic mitogens. *Lipids.* 24. 70-5.
46. Mogilanski, L. (2013, April 9). Study Shows Omega-3 Fatty Acids Extend Life. Retrieved from <http://www.thecrimson.com/article/2013/4/9/fish-omega-science-hsph/#>
47. Cohen Z, Norman H, Heimer Y. (1995). Microalgae as a source of omega three fatty acids. *Plants in Human Nutrition.* 77. 1-31.
48. Sukenik A, Takahashi H, Mokady S. (1994). Dietary lipids from marine unicellular algae enhance the amount of liver and blood omega-3 fatty acids in rats. *Annals of Nutrition and Metabolism.* 38(2). 85-96.
49. Kushak R et al. (2000, Jan). Favorable effects of blue-green algae *Aphanizomenon flos-aquae* on rat plasma lipids. *Journal of the American Nutraceutical Association.* 2(3). 59-65.
50. Krylov V et al. (2000). Retrospective epidemiological study using medical records to determine which diseases are improved by *Aphanizomenon flos-aquae* supplements. Unpublished manuscript.
51. Conquer J, Holub B. (1996). Supplementation with an algae source of docosahexaenoic acid increases (n-3) fatty acid status and alters selected risk factors for heart disease in vegetarian subjects. *Journal of Nutrition.* 126. 3032-9.
52. Reddy S, Sanders T, Obeid O. (1994). The influence of maternal vegetarian diet on essential fatty acid status of the newborn. *European Journal of Clinical Nutrition.* 48. 358-68.
53. Neuringer M, et al. (1988). The essentiality of n-3 fatty acids for the development and function of the retina and brain. *Annual Review of Nutrition.* 8. 517-41,

54. Pan W et al. (1993). Hemostatic factors and blood lipids in young Buddhist vegetarians and omnivores. *American Journal of Clinical Nutrition*. 58. 354-9.

55. Kushak R, Drapeau C, van Cott E, Winter H. (1999, Mar 21-25). Blue-green alga *Aphanizomenon flos-aquae* as a source of dietary polyunsaturated fatty acids and a hypo-cholesterolemic agent. Paper presented at the *Annual Meeting of the American Chemical Society*. Anaheim, California.

56. Rolle I, Pabst W. (1980). The cholesterol-lowering effect of the unicellular green alga - Scenedesmus acutus. *Nutrition and Metabolism*. 24(5). 291-301.

57. Hori K, Ishibashi G, Okita T. (1994). Hypo-cholesterolemic effect of blue-green alga, ishikurage (*Nostoc commune*) in rats fed atherogenic diet. *Plant Foods Human Nutrition*. 45(1). 63-70.

58. de Caire G, de Cano M, de Mule C, Steyerthal N, Piantanida M. (1995). Effect of *Spirulina platensis* on glucose, uric acid and cholesterol levels in the blood of rodents. *International Journal Experimental Botany*. 57. 93-6.

59. Delio M. (2000, Aug 30). Blue Goo and the Healthy Heart. *Wired News*.

60. Park J, et al. (2010). Astaxanthin decreased oxidative stress and inflammation and enhanced immune response in humans. *Nutrition & Metabolism*. 7. 18.

61. Khan S, et al. (2010, Oct). Novel astaxanthin prodrug (CDX-085) attenuates thrombosis in a mouse model. *Thrombosis Research*. 126(4). 299-305.

62. Preuss H, Echard B, Bagchi D, et al. (2009). Astaxanthin lowers blood pressure and lessens activity of the renin-angiotensin system in Zucker Fatty Rats. *J Functional Foods*. 1. 13-22

63. Nishikawa Y, Minenaka Y, Ichimura M, et al. (2005). Effects of astaxanthin and vitamin C on the prevention of gastric ulceration in stressed rats. *J Nutr Sci Vitaminology*. (Tokyo) 51. 135-41.

64. It S, Ogata E, Yamada M, Showa Denko. Anti-stress agent for animals and a method of reducing stress in animals. *United States patent. 5937790*.

65. Asami S, Yang Z, Yamashita E, Otoze H. Anti-stress composition. *United States patent*. 6265450.

66. Retrieved from <http://www.algaeindustrymagazine.com/fuji-building-30million-astaxanthin-facility-in-washington-state/>

67. Stavin, M. (2005, Jan 4). Algae derivative stimulates heart regeneration. Retrieved from <http://www.haaretz.com/print-edition/news/algae-derivative-stimulates-heart-regeneration-1.146052>

68. Stavin, M. (2005, Jan 4). Algae derivative stimulates heart regeneration. Retrieved from <http://www.haaretz.com/print-edition/news/algae-derivative-stimulates-heart-regeneration-1.146052>

69. Stavin, M. (2005, Jan 4). Algae derivative stimulates heart regeneration. Retrieved from <http://www.haaretz.com/print-edition/news/algae-derivative-stimulates-heart-regeneration-1.146052>

70. Leor J, et al. (2009). Intracoronary Injection of In Situ Forming Alginate Hydrogel Reverses Left Ventricular Remodeling After Myocardial Infarction in Swine. *J. Am. Coll. Cardiol.* 54(11). 1014-23

Chapter 15: Antiviral, Antibacterial, and Antifungal Effects

1. Rivas R, Riquelme C. (2012) Probiotic Biofilms. Open Access Chapter 28, InTech. Retrieved from <http://cdn.intechopen.com/pdfs/39609/InTech-Probiotic_biofilms.pdf>
2. Borowitzka M. (1995). Microalgae as sources of pharmaceuticals and other biologically active compounds. *J. Appl. Phycol.* 7. 3–15.
3. Woodruff H. (1980, June). Natural products from microorganisms. *Science.* 208(449). 1225-9.
4. Witvrouw M, De Clercq E. (1997, Oct). Sulfated polysaccharides extracted from sea algae as potential antiviral drugs. *General Pharmacology.* 29(4). 497-511.
5. Smelcerovic A, Knezevic-Jugovic Z, Petronijevic Z. (2008). Microbial polysaccharides and their derivatives as current and prospective pharmaceuticals. *Curr Pharm Des.* 14(29). 3168-95.
6. Bewick P et al. (1984). *Chlorella: The Emerald Food.* Ronin Publishing.
7. Schwimmer, M. Schwimmer, D. (1955). *The Role of Algae and Plankton in Medicine.* New York, Grune & Stratton. 42
8. Schmidt R. (1997). Treatment for Leprosy. *Price-Pottenger Nutrition Foundation Health Journal.* 21(3).
9. Martinez-Nadal F. (1970, Aug). Antimicrobial activity of *Spirulina maxima.* Paper presented at the *Tenth International Congress of Microbiology,* Mexico City.
10. Ostensvik O, Skulberg O, Underdal B, Hormazabal V. (1998). Antibacterial properties of extracts from selected planktonic freshwater cyanobacteria – a comparative study of bacterial bioassays. *Journal of Applied Microbiology.* 84(6). 1117-24.
11. Kreitlow S, Mundt S, Lindequist U. (1999, April). Cyanobacteria — a potential source of new biologically active substances. *Journal of Biotechnology.* 70(1-3). 61-3.
12. Frankmolle W, et al. (1992, Sept). Antifungal cyclic peptides from the terrestrial blue-green alga *Anabaena laxa. I.* Isolation and biological properties. *Journal of Antibiotics* (Tokyo). 45(9). 1451-7.
13. Aaronson A, Dubinsky Z. (1982). Mass production of microalgae. *Experientia.* 43(Suppl.). 44.
14. Shalaby E. (2011, Sep). Algae as promising organisms for environment and health. *Plant Signal Behav.* 6(9). 1338–1350.
15. Gustafson K, et al. (1989). AIDS antiviral sulfolipids from cyanobacteria. *Journal of the National Cancer Institute.* 81. 1254-8.
16. Cardellina J et al. (1993, July). A chemical screening strategy for the dereplication and prioritization of HIV-inhibitory aqueous natural products extracts. *Journal of Natural Products.* 56(7). 1123-9.
17. Lau A, et al. (1993, April). Inhibition of reverse transcriptase activity by extracts of cultured blue-green algae (cyanophyta). *Planta Medica.* 59(2). 148-51.
18. Ayehunie S, Belay A, Baba T, Ruprecht R. (1998, May). Inhibition of HIV-1 replication by an aqueous extract of *Spirulina platensis* (Arthrospira platensis).

Journal of Acquired Immune Deficiency Syndrome and Human Retrovirology. 18(1). 7-12.

19. Hayashi et al. (1996). Calcium spirulan, an inhibitor of enveloped virus replication, from a blue-green algae *Spirulina platensis*. *Journal of Natural Products.* 59. 83-7.
20. Ayehunie S, Belay A, Baba T, Ruprecht R. (1998, May). Inhibition of HIV-1 replication by an aqueous extract of *Spirulina platensis* (*Arthrospira platensis*). *Journal of Acquired Immune Deficiency Syndrome and Human Retrovirology.* 18(1). 7-12.
21. Loya S, Reshef V, Mizrachi E et al. (1998). The inhibition of the reverse transcriptase of HIV-1 by the natural sulfoglycolipids from cyanobacteria: contribution of different modalities to their high potency. *Journal of Natural Products.* 61(7). 891-5.
22. Schaeffer D, Krylov V. (2000). Anti-AIDS activity of extracts and compounds from algae and cyanobacteria. *Ecotoxicology and Environmental Safety.* 45(3). 208-27.
23. Gandhi M, Boyd M et al. (2000). Properties of cyanovirin-N (CV-N): inactivation of HIV-1 by sessile cyanovirin-N. *Developments in Biological Standardization.* 102. 141-8.
24. Hayashi T, Hayashi K, Maeda M, Kojima I. (Jan 1996). Calcium spirulan, an inhibitor of enveloped virus replication, from a blue-green alga *Spirulina platensis*. *Journal of Natural Products.* 59(1). 83-7.
25. Hayashi K, Hayashi T, Kojima I. (1996). A natural sulfated polysaccharide, calcium spirulan, isolated from *Spirulina platensis*: in vitro and ex vivo evaluation of anti-herpes simplex virus and anti-human immunodeficiency virus activities. *AIDS Research and Human Retroviruses.* 12(15). 1463-71.
26. Larsen L, Moore R, Patterson G. (1994, Mar). Beta-carotenes from the blue-green alga Dichothrix baueriana. *Journal of Natural Products.* 57(3). 419-21.
27. Hayashi K et al. (1993). An extract from *Spirulina platensis* is a selective inhibitor of *Herpes simplex* virus type 1 penetration into HeLa cells. *Phytotherapy Research.* 7. 76-80.
28. Guzman-Murillo M, Ascencio F. (2000, June). Anti-adhesive activity of sulphated exopolysaccharides of microalgae on attachment of red sore disease-associated bacteria and *Helicobacter pylori* to tissue culture cells. *Letters of Applied Microbiology.* 30(6). 473-8.
29. NIH/National Cancer Institute (2003, Mar 5). Anti-HIV Protein from Blue-green Algae also Inhibits Ebola Infection. *ScienceDaily.* Retrieved September 8, 2013, from http://www.sciencedaily.com /releases/2003/03/030305081547.htm
30. Shih, S., Tsai, K. et al. (2003). Inhibition of Enterovirus 71-Induced Apoptosis by Allophycocyanin Isolated from a Blue Green Algae *Spirulina platensis*. *Journal of Medical Virology* 70. 119-125.
31. Lahitova N, Doupovcova M, Zvonar J, Chandoga J, Hocman G. (1994). Antimutagenic properties of fresh-water blue-green algae. *Folia Microbiologica.* 39(4). 301-3.

32. Hasegawa T et al. (1994). Augmentation of the resistance against *Listeria monocytogenes* by oral administration of a hot water extract of Chlorella vulgaris in mice. *Immunopharmacology and Immunotoxicology.* 16(2). 191-202.
33. Haegawa T et al. (1995). Hot water extracts of Chlorella vulgaris reduce opportunistic infection with Listeria monocytogenes in C57BL/6 mice infected with LP-BM5 murine leukemia viruses. *International Journal of Immunopharmacology.* 17(6). 505-12.
34. Tanaka K et al. (1986). Augmentation of host-defense by a unicellular green algae, Chlorella vulgaris, to Escherichia coli infection. *Infection and Immunity.* 53. 267.
35. Ibusuki K Minamishima Y. (1990). Effect of Chlorella vulgaris extracts on murine cytomegalovirus infections. *Natural Immunity and Cell Growth Regulation.* 9. 121.
36. Konishi F et al. (1990). Enhanced resistance against Escherichia coli infection by subcutaneous administration of the hot-water extract of Chlorella vulgaris in cyclophosphamide-treated mice. *Cancer Immunology and Immunotherapy.* 32(1). 1.
37. Hasegawa T et al. (1990). Accelerated restoration of the leukocyte number and augmented resistance against Escherichia coli in cyclophosphamide-treated rats orally administered with a hot water extract of Chlorella vulgaris. *International Journal of Immunopharmacology.* 12. 883.
38. Fabregas J et al. (1999, Nov). In vitro inhibition of the replication of Haemorrhagic septicaemia virus and African swine fever virus by extracts from marine microalgae. *Antiviral Research.* 44(1). 67-73.
39. Krylov V et al. (2000). Retrospective epidemiological study using medical records to determine which diseases are improved by *Aphanizomenon flos-aquae* supplements. Unpublished manuscript.
40. Drapeau C. (1998, Aug 8). Results of Survey of AFA Consumers. *Presentation at Klamath Falls*, Oregon.
41. Chiueh et al. (2009, Feb 12) Method for inhibiting infection and reproduction of influenza type A WSN virus. *United States Patent* Application. Pub. No.: 2009/0042801 A1.
42. McCarty MF, Barroso-Aranda J, Contreras F. (2010, Jan). Practical strategies for targeting NF-kappaB and NADPH oxidase may improve survival during lethal influenza epidemics. *Med Hypotheses.* 74(1). 18-20.
43. Swine Flu Impact On Anti-Viral Sales – Use of Tamiflu And Relenza Will Come From Stockpiles. (2009, April 27). *Zacks Investment Research.* Retrieved from <http://www.dailymarkets.com/stock/2009/04/27/swine-flu-impact-on-anti-viral-sales-use-of-tamiflu-and-relenza-will-come-from-stockpiles/>
44. Chiueh et al. (Feb. 12, 2009) Method for inhibiting infection and reproduction of influenza typa A wsn virus. *United States Patent* Application. Pub. No.: 2009/0042801 A1.

45. Uglitskikh A, Tsokova N, et al. (2006, Jan-Feb). [Experience with a selenium-containing biological active supplement used in children with pneumonias in an intensive care unit] *Anesteziol Reanimatol.* (1). 45-8. Article in Russian.
46. Sommer A, et al. (1984). Increased risks of respiratory disease and diarrhea in children with pre-existing mild vitamin A deficiency. *American Journal of Clinical Nutrition.* 40. 1090-95.
47. Pinnock C et al. (1986). Vitamin A status in children who are prone to respiratory tract infections. *Australian Pediatric Journal.* 22(2). 95-9.
48. Nikitin A, Simvolokov S. (1989). The efficacy of the natural antibacterial preparation chlorophyll in the combined treatment of acute lung abscesses. *Ter Arkh* (Russian). 61(8). 113-6.
49. Bowers W. (1947). Chlorophyll in wound healing and suppurative disease. *American Journal of Surgery.* 73. 37.
50. American Museum of Natural History (2013, May 23). Scientists offer first definitive proof of bacteria-feeding behavior in green algae. *ScienceDaily.* Retrieved September 8, 2013, from <http://www.sciencedaily.com /releases/2013/05/130523143741.htm>
51. Kay RO. (1991). Microalgae as food and supplement. Crit Rev Food Sci Nutr. 30. 555–573

Chapter 16: Allergy and Asthma Relief

1. Schmidt M, Smith L, Sehnert K. (1993). *Beyond Antibiotics: Healthier Options for Families.* North Atlantic Books. 74.
2. Cited in: Challem J. (1981). *Spirulina.* Keats Publishing. 17-8.
3. Kim H, Lee E, Cho H, Moon Y. (1998). Inhibitory effect of mast cell-mediated immediate-type allergic reactions in rats by Spirulina. *Biochemistry and Pharmacology.* 55(7). 1071-6.
4. Yang H, Lee E, Kim H. (1997). *Spirulina platensis* inhibits anaphylactic reaction. *Life Sciences.* 61(13). 1237-44.
5. Hayashi O et al. (1998, Dec). Class specific influence of dietary *Spirulina platensis* on antibody production in mice. *Journal of Nutritional Science and Vitaminology* (Tokyo). 44(6). 841-51.
6. Drapeau C. (Aug 8, 1998). *Results of Survey of AFA Consumers.* Presentation at Klamath Falls, Oregon.
7. Bruno J, Gittelman J, Tuchfeld B. (2001). Lowered aluminum, with better detoxification and improved cognitive, behavioral ratings by children eating Aphanizomenon flos-aquae. Unpublished manuscript.
8. Mao T, Water, V, Gershwin, M. (2005). Effects of a Spirulina-Based Dietary Supplement on Cytokine Production From Allergic Rhinitis Patients. *Jour of Medicinal Food.* 8(1). 27-30.
9. Cingi C, Conk-Dalay M, Cakli H, Bal C. (2008, Oct). The effects of spirulina on allergic rhinitis. *European Archives of Oto-Rhino-Laryngology.* 265(10). 1219-23.

10. Chen L, Zhang S, Huang D, Tan J, He S. (2005). Experimental study of *Spirulina platensis* in treating allergic rhinitis in rats, *Journal of Central South University*. 30(1). 96–98.
11. Kushak R, et al. (Jan 2000). Ravorable effects of blue-green alge *Aphanizomenon flos-aquae* on rat plasma lipids. *Journal of the American Nutraceutical Association*. 2(3). 59-65.
12. Bisgaard H. (1984). Leukotrienes and prostaglandins in asthma. *Allergy*. 69. 413-20.
13. Neuman I, Nahum H, Ben-Amotz A. (1999, June). Prevention of exercise-induced asthma by a natural isomer mixture of beta-carotene. *Annals of Allergy and Asthma Immunology*. 82(6). 549-53.
14. Senevirathne M, Kim SK. (2011). Marine macro- and microalgae as potential agents for the prevention of asthma: hyperresponsiveness and inflammatory subjects. *Adv Food Nutr Res*. 64. 277-86.
15. M. Kalafati, et al. (Jan 2010). Ergogenic and Antioxidant Effects of Spirulina Supplementation in Humans. *Medicine and Science. Sports and Exercise*. 42(1); 142-51.
16. Samb A, et al. (2002). Heme oxygenase modulates oxidant-signaled airway smooth muscle contractility: role of bilirubin. *Am J Physiol Lung Cell Mol Physiol*. 283. L596 – L603.
17. Wilks A. (2002, Aug). Heme oxygenase: evolution, structure, and mechanism. *Antioxid Redox Signal*. 4(4). 603-14.
18. Taille C. et al. (2003). Heme oxygenase inhibits human airway smooth muscle proliferation via a bilirubin-dependent modulation of ERK1/2 phosphorylation. *J Bio Chem*. 278. 27160-27168.
19. McCarty M. (2007, Dec). Clinical potential of Spirulina as a source of phycocyanobilin. *J Med Food*. 10(4). 566-70.
20. Cited in: Challem J. (1981). *Spirulina*. Keats Publishing. 17.
21. Abrams K. (1998). *Attention Deficit Hyperactivity Disorder: A Nutritional Approach*. Book Crafter. 48-9.
22. Morse, D., and Choi A. (2005). Heme Oxygenase-1: From Bench to Bedside. *Am J Respir Crit Care Med*. 172. 660-670.
23. Morse, D., and Choi A. (2005). Heme Oxygenase-1: From Bench to Bedside. *Am J Respir Crit Care Med*. 172. 660-670.
24. Basu S, Salley S, Whittlesey, G, Klein M. (1994, Jul-Sep). Feasibility studies for a photosynthetic artificial lung. Optimization of parameters affecting photosynthesis. *ASAIO Journal*. 40(3). M743-6.

Chapter 17: Safely Using Edible Microalgae

1. Hunter B. (1971). Consumer Beware. Simon and Schuster.
2. Spolaore P, Joannis-Cassan C, Isambert A. (2006, Feb). Commercial applications of microalgae. *Journal of Bioscience and Bioengineering*. 101(2). 87–96

3. Center for Food Safety and Applied Nutrition. (1993). MMWR. Morbidity and Mortality Weekly Report. 42(41).
4. Carmichael W, An J. (1999, Nov). Using an enzyme linked immunosorbent assay (ELISA) and a protein phosphatase inhibition assay (PPIA) for the detection of microcystins and nodularins. *Natural Toxins.* 7(6). 377-385.
5. Iverson F, Truelove J. (1994). Toxicology and seafood toxins: domoic acid. *Natural Toxins.* 2(5). 334-9.
6. Drapeau C. (1999). *Aphanizomenon flos-aquae:* blue green-algae. Cell Tech. 8.
7. Henrickson R. (1989). *Earth Food: Spirulina.* Ronore Enterprises. 90.
8. Nakashima M et al. (1989, May-Jun). Extraction of light filth from Spirulina powders and tablets: collaborative study. *Journal - Association of Official Analytical Chemists.* 72(3). 451-3.
9. Gilroy D, et al. (2000). Assessing potential health risks from microcystin toxins in blue-green algae dietary supplements. *Environmental Health Perspectives.* 108(5). 437.
10. Miller T, Beversdorf L, et al. (2013). Spatiotemporal Molecular Analysis of Cyanobacteria Blooms Reveals Microcystis-Aphanizomenon Interactions. *PLoS One.* 8(9). e74933.
11. Phinney J, Peek C. (1961). Klamath Lake, an instance of natural enrichment. In Transactions of the Seminar on Algae and Metropolitan Wastes. *U.S. Public Health Service.* 22-27.
12. *Aphanizomenon flos-aquae* Review Of The Literature Regarding Neurotoxicity; http://www.stemtechbiz.co.nz/media/pdf/AFA-Review-of-Literature.pdf
13. Sawyer, P.J., Gentile J.H., and J.J. Sasner. (1968). Demonstration of a toxin from *Aphanizomenon flos-aquae* (L.) Ralfs, *Can. J. Microbiol.*, vol. 14, pp. 1199-1204.
14. Li R, Carmichael W, Liu Y, Watanabe M. (2000). Taxonomic re-evaluation of *Aphanizomenon flos-aquae* NH-5 based on morphology and 16S rRNA gene sequences. *Hydrobiologia.* 438. 99-105.
15. Li R, Carmichael W, Liu Y, Watanabe M. (2000). Taxonomic re-evaluation of *Aphanizomenon flos-aquae* NH-5 based on morphology and 16S rRNA gene sequences. *Hydrobiologia.* 438. 99-105.
16. *Aphanizomenon flos-aquae* Review Of The Literature Regarding Neurotoxicity; <http://www.stemtechbiz.co.nz/media/pdf/AFA-Review-of-Literature.pdf>
17. Hitzfeld et al. (2000, March). Cyanobacterial toxins in drinking water treatment. *Environmental Health Perspectives.* 108(1). 119.
18. Schaeffer D, Malpas P, Barton L. (1999, Sept). Risk assessment of microcystin in dietary Aphanizomenon flos-aquae. *Ecotoxicology and Environmental Safety.* 44(1). 73-80.
19. Barton et al. (1997, June). Determination of a safe level of microcystins as a contaminant of Aphanizomenon flos-aquae: a new algal dietary supplement for humans. Cited by Bergwall R. The whole truth about some wonderful foods. New Life Options.
20. Repavich W, Sonzogni W, et al. (1990). Cyanobacteria (blue-green algae) in Wisconsin waters: Acute and chronic toxicity. *Water Research.* 24. 222-31.

21. Falconer I, et al. (1994). Toxicity of Microcystis aeruginosa in drinking water to growing pigs, as an animal model for human injury and risk assessment. *Environmental Toxicology and Water Quality*. 9. 131.
22. Cited in: Dietrich D, Hoeger S. (2005). Guidance values for microcystins in water and cyanobacterial supplement products (blue-green algal supplements): a reasonable or misguided approach? *Toxicology and Applied Pharmacology*. 203. 273-289.
23. Jensen G. (1999). Personal communication.
24. Jensen G. (1999). Personal communication.
25. Cox P, Sacks O. (2002). Cycad neurotoxins, consumption of flying foxes, and ALS-PDC disease in Guam. *Neurology*. 58(6). 956–959.
26. Dietrich D, et al. (2008). Toxin mixture in cyanobacterial blooms—a critical comparison of reality with current procedures employed in human health risk assessment. *Adv Exp Med Biol*. 619. 885–912.
27. Personal Communication with Dr. Jerry Anderson. (2014, Jan 27). New Earth Company.
28. Personal Communication with Dr. Wayne Carmichael. (2012, Aug 12).
29. Press Release by Cyanotech Corporation, (2005, June 8). Retrieved from <http://www.cyanotech.com/news/news_060805.html>
30. Tiberg E et al. (1995). Allergy to green algae (Chlorella among children). *Journal of Allergy and Clinical Immunology*. 96. 257-9.
31. Tiberg E. (1987). Microalgae as aeroplankton and allergens. *Advances in Aerobiology*. 51. 171-3.
32. Jitsukawa K, Suizu R, Hidana A. (1984, May). Chlorella photosensitization. New phytophotodermatosis. *International Journal of Dermatology*. 23(4). 263-8.
33. Becker, E. (1995). Microalgae, Biotechnology and Microbiology, Cambridge University Press.
34. Becker E. (1980). Comparative toxicological studies with algae in India, Thailand, and Peru. In *Algae biomass*, Shelef, G. and Soeder, C.J., Eds., Elsevier/North-Holland Biomedical Press, Tubingen, West Germany. 767.

Part III: Algae to the Rescue: Planetary Medicine

Chapter 18: Global Food Supply and Ecological Restoration

1. Pisciotta J, Zou Y, Baskakov I. (2010). Light-Dependent Electrogenic Activity of Cyanobacteria. *PLoS ONE* 5(5). e10821. doi:10.1371/journal.pone.0010821
2. Gomez-Alvarez V, King GM, Nu¨sslein K (2007). Comparative bacterial diversity in recent Hawaiian volcanic deposits of different ages. *FEMS Microbiol Ecol* 60. 60 70
3. Wilcox H. (1982). The ocean as a supplier of food and energy. *Experientia*. 43(suppl). 37-41.

4. Iyengar and Nair. (2000, April 17) Global outlook on nutrition and the environment: meeting the challenges of the next millennium. *Science of the Total Environment*. 249. 331-316.
5. Wilcox H. (1982). The ocean as a supplier of food and energy. *Experientia*. 43(suppl). 37-41.
6. Cited by: Postel S. (1989, May-June). Land's End. *Worldwatch Magazine*. 13.
7. Wenzlau, S. (2013, April 11). Global Food Prices Continue to Rise. Retrieved from <http://www.worldwatch.org/global-food-prices-continue-rise-0>
8. Wenzlau, S. (2013, April 11). Global Food Prices Continue to Rise. Retrieved from <http://www.worldwatch.org/global-food-prices-continue-rise-0>
9. Aldridge T, Schulbach H. (1979, fall). Water requirements for food production. *Soil and Water*. (38) University of California Cooperative Extension.
10. Bewick E. et al. (1984). *Chlorella: The Emerald Food*. Ronin Publishing. 16.
11. Henrikson R. (1989). *Earth Food Spirulina*. Ronore Enterprises. 148.
12. Mark Edwards, (2012). Why Imagine Our Algae Future? Chapter In: Henrikson, R. and Edwards, M. *Imagine Our Algae Future: Visionary Algae Architecture and Landscape Designs*. Ronore Enterprises, Inc. 11.
13. Simpore J, Kabore F, et al. (2006, Jan). Nutrition rehabilitation of undernourished children utilizing Spiruline and Misola. *Nutr J*. 23(5). 3.
14. Branger B, et al. (2003, May). Spiruline as a food supplement in case of infant malnutrition in Burkina-Faso. [Article in French] *Arch Pediatr*. 10(5). 424-31.
15. Sevilla I, Aguiree N. (1995). The Nicaragua Report Summary. Recieved 11/13/13 from <http://www.newearth.com/science/studies/>
16. Sevilla I, Aguiree N. (1995). Study on the Effects of Super Blue-Green Algae on the Nutritional Status and School Performance of First-, Second-, and Third-Grade Children Attending the Monsenor Velez School in Nandaime, Nicaragua [dissertation, in Spanish]. Universidad CentroAmericana, Nicaragua.

Chapter 19: Microalgae as Fertilizer and Animal Feed

1. 1981 UN FAO [Food and Agriculture Organization] report, "Blue-Green Algae in Rice Production." Cited by Henderson: Blue-Green Algae for Rice Production. *FAO Soils Bulletin*. Rome.
2. Olness A, Archer D. (2005). Effect of organic carbon on available water in soil. *Soil Science*, 170. 90–101.
3. Daniel Marano. (2014, Jan/Feb). Rich Dirt, Poor Dirt. *Psychology Today*. 40.
4. Martin J. (1971). Decomposition and binding action of polysaccharides in soil. *Soil Biology & Biochemistry* 3, 33-41.
5. The Effects of Microalgae on Soils: A Literature Review. K. Boym, Ed. Soil Technology Corp. pg. 3; Retrieved November 2013 from <http://www.soiltechcorp.com/DEV/images/uploads/product_PDFs/Literature_Review_-_Effects_of_Microalgae_on_Soils.pdf>

6. Zhang Y, Wang H, et al. (2006, June). The microstructure of microbiotic crust and its influence on wind erosion for a sandy soil surface in the Gurbantunggut Desert of Northwestern China. *Geoderma.* 132(3–4). 441–449
7. The Effects of Microalgae on Soils: A Literature Review. K. Boym, Ed. Soil Technology Corp. pg. 3. Retreived from <http://www.soiltechcorp.com/DEV/images/uploads/product_PDFs/Literature_Review_-_Effects_of_Microalgae_on_Soils.pdf>
8. Henrikson, R. and Edwards, M. (2012) Imagine Our Algae Future: Visionary Algae Architecture and Landscape Designs. Ronore Enterprises, Inc.
9. Berry J. (2010). Marine and Freshwater Microalgae as a Potential Source of Novel Herbicides, Herbicides and Environment, A. Kortekamp (Ed.), InTech. Retrieved from <http://www.intechopen.com/books/herbicides-and-environment/marine-and-freshwater-microalgae-as-apotential-source-of-novel-herbicides>
10. Pisciotta J, Zou Y, Baskakov I. (2010). Yang, Ching-Hong. Ed. Light-Dependent Electrogenic Activity of Cyanobacteria. *PLoS ONE* 5 (5). e10821.
11. Dr. Ghazi Khan, Electro-culture. Retrieved November 2013 from <http://ecosustainablevillage.com.ip01-web23.net/new_page_11.htm>
12. Barclay W et al. (1998). Production of docosahexaenoic acid from microalgae and its benefits for use in animal feeds. *World Review of Nutrition and Diet.* 83. 61-76.
13. Venkataraman L, Somasekaran T, Becker E. (1994, July). Replacement value of blue-green alga (*Spirulina platensis*) for fishmeal and a vitamin-mineral premix for broiler chicks. *British Poultry Science.* 35(3). 373-81.
14. Kapoor R, Mehta U. (1993, Jan.). Effect of supplementation of blue-green alga (Spirulina) on outcome of pregnancy in rats. *Plant Foods in Human Nutrition.* 43(1). 29-35.
15. Ross E, Dominy W. (1990, May). The nutritional value of dehydrated, blue-green algae for poultry. *Poultry Science.* 69(5). 794-800.
16. Schubert, L. (1988). The use of Spirulina and Chlorella as food sources for animals and humans. *Progressive Phycological Research.* 6. 237-54.
17. Planski B, Poshchakov E. (1982). Dynamics of weight gain in calves fed with microalgae. *Veterinarnomeditsinski Nauki* (Bulgarian). 19(3). 52-5.
18. Meske C, Pruss H. (1977). Microalgae as a component of fish meal-free fish food. Fortschritte in der *Tierphysiologie und Tierernahrung* (German). 8. 71-81.
19. Fevrier C, Seve B. (1975). Incorporation of a Spiruline (*Spirulina maxima*) in swine food. *Ann Nutr Aliement* (French). 29(6). 625-50.
20. Arasaki S, Arasaki T. (1983). *Low Calorie, High Nutrition Vegetables From the Sea: To Help You Look and Feel Better.* Japan Publications. 17.
21. Yamamoto, T. & Ishibashi M. (1971). The content of trace elements in seaweeds. *Proc. int. Seaweed Symp.* 7. 511–514.
22. Kapoor R, Mehta U. (1993, Jan). Effect of supplementation of blue green alga (Spirulina) on outcome of pregnancy in rats. *Plant Foods Hum Nutr.* 43(1). 29-35.

23. Fevrier C. and Seve B. (1975). Incorporation of a spiruline (*Spirulina maxima*) in swine food. [Article in French.] *Ann Nutr Aliment*. 29(6). 625-50.
24. Pabst W, Payer H, Rolle I, Soeder C. (1978). Multigeneration feeding studies in mice for safety evaluation of the microalga, Scenedesmus acutus. Biological and hematological data, *Food Cosmet. Toxicol.* 16. 249-254
25. Harrison M. (1986). Doctoral dissertation. The value of waste-grown microalgae as a protein supplement for starting, growing, and finishing swine. Retrieved December 2013 from http://ufdc.ufl.edu/AA00003807/00001/1j.
26. Bewick E et al. (1984). Chlorella: The Emerald Food. Ronin Publishing. 77.
27. Bewick E et al. (1984). Chlorella: The Emerald Food. Ronin Publishing. 78.
28. Oswald W, Golueke. (1973). An algal regenerative system for single-family farms and villages. *Compost Science Journal of Waste Recycling*.
29. Henrikson R, Edwards M. (2012). *Imagine Our Algae Future*. Ronore Enterprises, Inc.
30. AlgaeIndustryMagazine.com. Retrieved from <http://www.algaeindustrymagazine.com/special-report-spirulina-part-6/>
31. Super Blue Green Algae Report (2004). Rocky Mountain Wildlife Conservation Center publisher.

Chapter 20: Microalgae Water Purification and Conservation

1. 3 Things that most of the world can't do. Retrieved Nov. 13, 2013 from <http://water.org/water-crisis/water-facts/sanitation/>
2. World Health Organization. Safe Water and Global Health". Who. int. 2008-06-25. Retrieved 2010-07-25 from <www.whoint>
3. United Nations Development Programme. (2006). Human Development Report 2006, Beyond Scarcity: Power, poverty and the global water crisis. Retrieved from <http://hdr.undp.org/en/reports/global/hdr2006>
4. Harvard School of Public Health News. (2013, winter). Distilling the truth about water. Retrieved from<http://www.hsph.harvard.edu/news/magazine/distilling-the-truth-about-water/>
5. World Water Day: The Surprising Amount of Water Required to Produce 13 Items from Chocolate to Wine. (2011, May 25). The Huffington Post. Retrieved from <http://www.huffingtonpost.com/green/>
6. Torcellini, P. Long N. Judkoff R. (2003, Dec). Consumptive Water Use for U.S. Power Production. NREL/TP-550-33905. Retrieved from <http://www.nrel.gov/docs/fy04osti/33905.pdf>
7. Robbins J. (1987). Diet for a New America. Stillpoint. 367.
8. Retrieved from <http://en.wikipedia.org/wiki/Ogallala_Aquifer>
9. Chow D. (2013, Aug 26). Water Woes: Vast US Aquifer Is Being Tapped Out. Retrieved from <http://www.livescience.com/39186-kansas-aquifer-water-depletion.html>

10. Chow D. (2013, Aug 26). Water Woes: Vast US Aquifer Is Being Tapped Out. Retrieved from <http://www.livescience.com/39186-kansas-aquifer-water-depletion.html>
11. Fields D, Robin H. (1985, Jan). America's appetite for meat is ruining our water. *Vegetarian Times*.
12. Robbins J. (1987). Diet for a New America. Stillpoint. 367.
13. Gardella F. (1999). The Meat Industry . . . Is it worth it? Seminar at University of California, Irvine. Retrieved from http://darwin.bio.uci.edu/sustain/global/sensem/MeatIndustry.html
14. Pimentel, D. (2001). Ecological Integrity: Integrating Environment, Conservation and Health, Island Press, Washington DC.
15. Retrieved from <http://fourthfloorwalkup.com/2010/04/23/what-did-you-do-for-mother-earth-today/#sthash.Ohdf01nI.dpuf>
16. Source of life running out: water scientists. (2013, May 24). Retrieved from <http://phys.org/news/2013-05-source-life-scientists.html#jCp.>
17. Matusiak K. (1976). Studies on the purification of wastewater from the nitrogen fertiler industry by intensive algal cultures. I. Growth of Chlorella vulgaris in wastes. *Acta Microbiologica Polonica*. 25(3). 233-42.
18. Matusiak K et al. (1976). Studies on the purification of wastewater from the nitrogen fertilizer industry by intensive algal cultures. II. Removal of nitrogen from the wastewater. *Acta Microbiologica Polonica*. 25(4). 361-73.
19. Hammouda O, Gaber A, Abdel-Raouf N. (1995, Aug). Microalgae and wastewater treatment. *Exotoxicological Environmental Safety*. 31(3). 205-10.

Chapter 21: Microalgae Chemical Know-how and Toxic Clean-Up

1. Stauber J, Rampton S. (1995). *Toxic Sludge is Good for You: Lies, Damn Lies and the Public Relations Industry*. Common Courage Press.
2. Nealon, C. (2013). UB studies the slimy organism's commercial potential and how capturing it can clean our rivers and lakes. Retrieved from <www.buffalo.edu/news/releases/2013/07/027.html>
3. Lee J et al. (2000, spring). Methods to enhance tolerances of Chlorella KR-1 to toxic compounds in flue gas. *Applied Biochemistry and Biotechnology*. (84-86). 329-42.
4. Hammouda O, Gaber A, Abdel-Raouf N. (1995, Aug). Microalgae and wastewater treatment. *Exotoxicological Environmental Safety*. 31(3). 205-10.
5. Lambolez L et al. (1994, Aug). The environmental risks of industrial waste disposal: an experimental approach including acute and chronic toxicity studies. *Exotoxicology and Environmental Safety*. 28(3). 3317-28.
6. Chan S, Chow H, Wong M. (1991, Sept). Microalgae as bioabsorbents for treating mixture of electroplating and sewage effluent. *Biomedical Environmental Science*. 4(3):250-61.
7. Kaplan D. (2013). Absorption and Adsorption of Heavy Metals by Microalgae. Chapter in Handbook of Microalgal Culture: Applied Phycology and

Biotechnology, Second Edition. Edited by Amos Richmond and Qiang Hu. John Wiley & Sons, Ltd. Published.

8. Kaplan D. (2013). Absorption and Adsorption of Heavy Metals by Microalgae. Chapter in Handbook of Microalgal Culture: Applied Phycology and Biotechnology, Second Edition. Edited by Amos Richmond and Qiang Hu. John Wiley & Sons, Ltd. Published.
9. Chong A, Wong Y, Tam N. (2000, July). Performance of different microalgal species in removing nickel and zinc from industrial wastewater. *Chemosphere.* 41(1-2). 251-7.
10. Darnall D. (1991). Removal and recovery of heavy metal ions from wastewaters using a new biosorbent; AlgaSORB. *Innovative Hazardous Waste Treatment Technology.* 3. 65-72.
11. Savvaidis I. (1998, April). Recovery of gold from thiourea solutions using microorganisms. *Biometals.* 11(2). 145-51.
12. Vymazal J. (1984). Short-term uptake of heavy metals by periphyton algae. *Hydrobiolgia.* 119. 171-9.
13. Holan Z, Volesky B, Prasetyo I. (1993). Biosorption of cadmium by biomass of marine algae. *Biotechnology and Bioengineering.* 41. 819-25.
14. Wong M, Pak D. (1992, June). Removal of Cu and Ni by free and immobilized microalgae. *Biomedical and Environmental Science.* 5(2). 99-108.
15. Horikoshi T et al. (1979). Uptake of uranium by various cell fractions of Chlorella regularis. *Radioisotopes.* 28(8). 485-7.
16. Matsumoto et al. (2000, spring). Floating cultivation of marine cyanobacteria using coal fly ash. *Applied Biochemistry and Biotechnology.* (84-86). 51-7.
17. Zhang L, Huang G, Yu Y. (1998, July). Immobilization of microalgae for biosorption and degradation of butyltin chlorides. *Artificial Cells, Blood Substitutes, and Immobilization Biotechnology.* 26(4). 399-410.
18. Negoro M et al. (1991, spring). Growth of microalgae in high CO_2 gas and effects of SOX and NOX. *Applied Biochemistry and Biotechnology.* (28-29). 877-86.
19. Converti A, Scapazzoni S, Lodi A, Carvalho J. (2006, Jan). Ammonium and urea removal by *Spirulina platensis* . *J Ind Microbiol Biotechnol.* 33(1). 8-16.
20. Arsenic. Retrieved from http://kineticonaz.com/Arsenic.html
21. Jiang Y, Purchase D, Jones H, Garelick H. (2011, Sept). Effects of arsenate (AS5+) on growth and production of glutathione (gsh) and phytochelatins (PCS) in Chlorella vulgaris. *Int J Phytoremediation.* 13(8). 834-44.
22. Bonaventure C, Johnson F. (1997, Feb). Healthy environments for healthy people: bioremediation today and tomorrow. *Environmental Health Perspective.* 105(suppl)1. 5-20.
23. Retrieved from <http://www.cbsnews.com/8301-505123_162-42849644/7-scary-facts-about-pharmaceutical-pollution-of-drinking-water/?pageNum=2>
24. Retrieved from <http://www.cbsnews.com/8301-505123_162-42849644/7-scary-facts-about-pharmaceutical-pollution-of-drinking-water/?pageNum=2>

25. Retrieved from <http://www.algaeindustrymagazine.com/synechocystis-6803-just-might-algal-strain/>
26. Fast Forward: A Taste of the Future. (2013, June). *Smithsonian.* 108.

Chapter 22: Future of Microalgae a Sustainable Energy Source

1. Arasaki S, Arasaki T. (1983). *Vegetables from the Sea.* Japan Publ. Inc.
2. Increasing Atmospheric Carbon Dioxide. University of Delaware's College of Marine and Earth Studies. Retrieved December 2013 from <http://co2.cms.udel.edu/Increasing_Atmospheric_CO2.htm>
3. Schenk PM, Thomas-Hall SR, Stephens E, Marx UC, Mussgnug JH, Posten C, Kruse O, Hankamer B. (2008). Second generation biofuels: high-efficiency microalgae for biodiesel production. *Bioenergy Res.* 1(1). 20–43.
4. Henrikson, R. and Edwards, M. (2012) Imagine Our Algae Future: Visionary Algae Architecture and Landscape Designs. Ronore Enterprises, Inc. 9.
5. Svoboda S. (2007, July 1). The Greenest Green Fuel. Retrieved Dec 2013 from <http://www.popsci.com/scitech/article/2007-07/greenest-green-fuel>
6. Flammini, A. (2008, Oct). Biofuels and the underlying causes of high food prices. Received from <http://www.globalbioenergy.org/fileadmin/user_upload/gbep/docs/BIOENERGY_INFO/0810_Flammini_-_Biofuels_and_the_underlying_causes_of_high_food_prices_GBEP-FAO.pdf>
7. Retrieved from <http://www.petrosuninc.com/algae-biofuels.html>
8. Peak Energy Blog. (2008, Jan 17). Algae Biofuel Plant in Arizona. Retrieved Dec. 17, 2013 from <http://peakenergy.blogspot.com/2008/01/algae-biofuel-plant-in-arizona.html>
9. Henrikson R. (2012). *Imagine Our Algae Future: Visionary Algae Architecture and Landscape Designs.* Ronore Enterprises, Inc.
10. Henrikson R. (2013). *Algae Microfarms.* Ronore Enterprises, Inc.
11. Bachofen R. (1982). The production of hydrocarbons by Botryococcus braunii. *Experientia.* 43(suppl). 53.
12. Rezanka T, Azhradnik J, Podojil M. (1982). Hydrocarbons in green and blue-green algae. *Folia Microbiologica.* 27(6). 450-4.
13. Tornabene T. (1982). Microorganisms as hydrocarbon producers. *Experientia.* 43(suppl). 52.
14. Parmar A, Singh NK, Pandey A, Gnansounou E, Madamwar D. (2011 Nov) Cyanobacteria and microalgae: a positive prospect for biofuels. *Bioresour Technol.* 102(22). 10163-72
15. Bewicke D. (1982). *The Chlorella Story.* Planet Publications. 28.
16. Rifkin J. (2003). *The Hydrogen Economy.* Tarcher.
17. Tasios Melis. (2000, Jan). It was enormously exciting, it was unbelievable. Plant Physiology. 22. 127-136. Retrieved from <http://www.fuelcellsworks.com/Supppage4611.html>
18. Retrieved from <http://www.shamengo.com/en/pionnier/82-pierre-calleja>

19. Fermentalg. CO_2-absorbing algae street lamp. (2012, May 30). Retrieved from <http://wattnow.org/2205/fermentalg-co2-absorbing-algae-street-lamp>
20. Pisciotta J, Zou Y, Baskakov I. (2010). Yang, Ching-Hong. ed. Light-Dependent Electrogenic Activity of Cyanobacteria. *PLoS ONE*. 5(5). e10821.
21. Dickey G. (2010, April 13). Stanford researchers find electrical current stemming from plants. Retrieved <http://news.stanford.edu/news/2010/april/electric-current-plants-041310.html.>
22. Holmes J. (2012, July 3). America's 'Great Green Fleet.' Retrieved from <http://thediplomat.com/flashpoints-blog/2012/07/03/americas-great-green-fleet/>
23. McMahon J. (2012, Feb. 26). Obama's Favorite Algae Company. Retrieved from <http://www.forbes.com/sites/jeffmcmahon/2012/02/26/obamas-favorite-algae-company/2/>
24. Coren, M. (2012, April 27). NASA's Got A Brand New Bag, And It's Full Of Algae. <http://www.fastcoexist.com/1679739/nasas-got-a-brand-new-bag-and-its-full-of-algae>
25. Goldenberg S. (2010, Feb). Algae to solve the Pentagon's jet fuel problem. US scientists believe they will soon be able to use algae to produce biofuel for the same cost as fossil fuels. Retrieved Dec. 08, 2013 from <http://www.theguardian.com/environment/2010/feb/13/algae-solve-pentagon-fuel-problem>
26. Hernandez P. (2011, Nov. 8). United Airlines Flight Powered by Algae Jet Fuel. Retrieved Dec. 12, 2013 from <http://app1.kuhf.org/articles/1320714247-United-Airlines-Flight-Powered-By-Algae-Jet-Fuel.html>
27. Gunther M. (2010, Sept 21). Richard Branson: Algae is Tomorrow's Jet Fuel. Retrieved Dec. 12, 2013 from <http://www.greenbiz.com/blog/2010/09/21/richard-branson-algae-jet-fuel>
28. Thomas J. (2006, June 26). New Company to Produce Biodiesel from Algae. Retrieved January 5, 2014 from <http://www.treehugger.com/renewable-energy/new-company-to-produce-biodiesel-from-algae.html>
29. Venteris E, Skaggs R, et al. (2013). A GIS Cost Model to Assess the Availability of Fresh water, Seawater, and Saline Groundwater for Algal Biofuel Production in the United States. *Environmental Science & Technology*. 47(9). 4840-4849
30. Going Green: Equipped to Grow Serious Amounts of Pond Scum for Fuel. *ScienceDaily*. May 21, 2013; Retrieved from <http: //www.sciencedaily.com /release s/2013/05/130521140916.htm>
31. New study: Algae biofuel cuts CO_2 emissions by up to 68 percent; Retrieved from <http://www.biodieselmagazine.com/articles/9336/new-study-algae-biofuel-cuts-co2-emissions-by-up-to-68-percent>

Chapter 23: Microalgae, Carbon Fixation and Climate Change

1. Henrickson R. (1989). *Earth Food Spirulina*. Ronore Enterprises.
2. Biotechnology and Biological Sciences Research Council (2006, January 14). The Secret Life of Algae. *ScienceDaily*. Retrieved January 20, 2014, from <http://www.sciencedaily.com /releases/2006/01/060114152028.htm>
3. Dismukes G, Klimov V, Baranov S, Kozlov Y, DasGupta, Tyryshkin A. (2001, Feb. 27). The origin of atmospheric oxygen on Earth: the innovation of oxygenic photosynthesis. *Proc Natl Acad Sci USA*. 98(5). 2170-5.
4. Paul J. (1996, Nov). Carbon cycling: molecular regulation of photosynthetic carbon fixation. *Microbial Ecology*. 32(3). 231-45
5. Pisciotta J, Zou Y, Baskakov I. (2010). Light-Dependent Electrogenic Activity of Cyanobacteria. *PLoS ONE* 5(5). e10821. doi:10.1371/journal.pone.0010821
6. Energy – Consumption'!A1 "Consumption by fuel, 1965–2008" (XLS). Statistical Review of World Energy 2009, BP. 31 July 2006; Retrieved from <http://www.bp.com/statisticalreview>
7. Pisciotta J, Zou Y, Baskakov I (2010). Yang, Ching-Hong. ed. Light-Dependent Electrogenic Activity of Cyanobacteria. *PLoS ONE* 5 (5). e10821. doi:10.1371/journal.pone.0010821
8. Elderfield H, Rickaby R. (2000, May 18). Oceanic Cd/P ratio and nutrient utilization in the glacial southern ocean. *Nature*. 405(6784). 305-10.
9. Grantham R. Seeking a biological solution for the greenhouse dilemma. Institut d'Evolution Moleculaire, Universite Claude Bernard, Lyons, France. 1988.
10. Watson A et al. (2000, Oct). Effect of iron supply on southern ocean CO 2 uptake and implications for glacial atmospheric CO2. *Nature*. 407(6805). 730-3.
11. Russell, K. (2013, July 23). The CIA Wants To Control The Weather Through 'Geoengineering' Business insider, July 23, 2013. Retrieved from <http://finance.yahoo.com/news/cia-wants-control-weather-geoengineering-143020268.html>
12. Russell, K. (2013, July 23). The CIA Wants To Control The Weather Through 'Geoengineering' Business insider. Retrieved from <http://finance.yahoo.com/news/cia-wants-control-weather-geoengineering-143020268.html>
13. Biello D. (2012, Oct 24). Pacific Ocean Hacker Speaks Out. Retrieved from <http://www.scientificamerican.com/article.cfm?id=questions-and-answers-with-rogue-geoengineer-carbon-entrepreneur-russ-george>
14. Bertrand E, Allen A. (2012, Oct). Influence of vitamin B auxotrophy on nitrogen metabolism in eukaryotic phytoplankton. *Frontiers in Microbiology*. 3. 375.
15. Field C., et al. (1998). Primary production of the biosphere: integrating terrestrial and oceanic components. *Science*. 281. 237-240.
16. Church M, Hutchins D, Ducklow H. (2000). Limitation of bacterial growth by dissolved organic matter and iron in the Southern ocean. *Applied Environmental Microbiology*. 66(2). 455-66.